Excursions in Biblical Chronology

Lee W. Casperson

Copyright © 2018 Lee W. Casperson

All rights reserved. No part of this book may be used or reproduced by any means, graphic, electronic, or mechanical, including photocopying, recording, taping or by any information storage retrieval system without the written permission of the publisher except in the case of brief quotations embodied in critical articles and reviews.

Because of the dynamic nature of the Internet, any web addresses or links contained in this book may have changed since publication and may no longer be valid.

Published by Lee Casperson
Ewing, NJ 08638

Cover image from "A Map of the Terrestrial Paradise" by Emanuel Bowen
Used with permission of Armenica.org

ISBN: 978-1-7323189-0-8 (sc)
ISBN: 978-1-7323189-1-5 (hc)

Library of Congress Control Number: 2018905954

Printed in the United States of America

Table Of Contents

LIST OF ILLUSTRATIONS..vii

PREFACE...xi

PART 1. THE GARDEN AND THE FLOOD

1. IN THE BEGINNING.. 2

2. ADAM AND EVE...43

3. FROM ADAM TO NOAH..82

4. RELATIVE TIMES AND PLACES...................................103

5. URARTU AND LAKE VAN...133

6. LAND OF EDEN...164

PART 2. CAUSES AND EFFECTS OF THE FLOOD

7. LAKE VAN AND THE FLOOD....................................... 200

8. VOLCANIC ERUPTIONS AND EGYPT.......................244

9. EVIDENCE OF THE BIBLICAL FLOOD.....................299

10. THE GILGAMESH EPOCH..331

11. UTNAPISHTIM'S FLOOD...368

12. FROM ADAM TO ABRAM...411

PART 3. ADVENTURES IN EGYPT

13. ARRIVAL IN EGYPT..444

14. CONQUEST OF EGYPT.. 474

15. ARGONAUTS IN EGYPT......................................504

16. MOSES AND HERACLES.....................................534

17. JOSHUA AND JASON..595

PART 4. CHRONOLOGY IN CONTEXT

18. BEFORE THE BEGINNING.................................. 650

19. THE DISCIPLE AND THE PROPHET......................695

20. SCIENCE AND CHRONOLOGY............................ 726

LIST OF ILLUSTRATIONS

Tables

Table 3.1. Biblical chronology of the antediluvian patriarchs from Adam to the beginning of the flood. The numbers shown are the indicated ages of the patriarchs at the births of their sons.. 85

Table 3.2. Comparison of the Jubilees chronology and the Samaritan chronology for the patriarchs from Adam to the flood... 88

Table 3.3. Chronological patterns in the creation stories and in the commandments of Exodus and Leviticus, starting at the top from a possible form ascribed there to Moses........... 92

Table 3.4. Approximate chronology of the antediluvian patriarchs from Adam to the beginning of the flood according to *Jasher*. The numbers shown are the ages of the patriarchs at the births of their sons... 96

Table 3.5 Approximate chronology of the antediluvian patriarchs from Adam to the beginning of the flood according to Book I of Josephus's *Antiquities of the Jews*. The numbers shown are the ages of the patriarchs at the births of their sons... 99

Table 4.1. Chronology of the flood in the lifetime of Noah.... 108

Table 6.1. Known eruptions of the Nemrut volcano more recent than 5000 B.C... 193

Table 7.1. Known eruptions of the Ararat volcano more recent than 5000 B.C... 207

Table 7.2. Greatest recorded rainfall for various time intervals... 209

Table 8.1. Anatolian dendrochronology in relative tree-ring numbers and proposed calendar years B.C. for the time period around a major growth anomaly..........................264

Table 8.2. Volcanic eruptions in the northern hemisphere between 1500 B.C. and 1600 B.C. that have an explosivity index of four or greater..267

Table 8.3. Anatolian dendrochronology in relative tree-ring numbers and proposed calendar years B.C. for the time period around two growth anomalies............................271

Table 9.1. Recent asteroid impacts................................306

Table 11.1. Comparison of the biblical and Gilgamesh flood stories...379

Table 12.1. Biblical chronology of the patriarchs from Noah to Abram. The numbers shown are the ages of these patriarchs at the births of some of their biblically most important sons..414

Table 12.2 Approximate biblical chronology for the period from Noah to Abram. The Age column corresponds to the Masoretic ages shown previously in Table 12.1. The column Adjusted Age indicates the approximate ages of the patriarchs at the births of their sons, and the column Date indicates (except for the flood line) the dates of birth of those sons... 424

Table 12.3. Reign lengths of the antediluvian rulers of the Sumerian King List. The columns labeled actual reign are calculated from the stated reigns as described in the text below..431

Table 12.4. Approximate biblical chronology of the antediluvian patriarchs from Adam to Shem. The column Adjusted Age indicates the estimated approximate ages of the patriarchs at the births of their listed sons...................... 434

Table 12.5. Approximate chronology of the rulers of Ebla during the archive period..437

Table 13.1. Approximate chronology for the period from the entry of Abram and the Hyksos into Egypt to the beginning of the Eighteenth Dynasty of Egypt and the expulsion of the Hyksos..447

Table 17.1. Genealogies from Jacob to Moses (Exodus 6:16-20) and Joshua at the time of the exodus. This doubtful genealogy from Ephraim to Joshua is a tentative interpretation of 1 Chronicles 7:20-27............................605

Table 17.2. Statements by Joshua concerning the relationship of God to Joshua and to the Hebrews............................642

Figures

Figure 2.1. Impression of a cylinder seal believed to be from the general region of Mesopotamia. (Reproduced with permission).. 56

Figure 4.1. A Map of the Terrestrial Paradise by Emanuel Bowen, 1694?-1767. (Reproduced with permission)........... 119

Figure 4.2. Expanded version of the Terrestrial Paradise and neighboring region, from the map by Bowen shown in Figure 4.1... 124

Figure 5.1. Garden of Eden friezes on the north wall of the church at Akhtamar, including Adam and Eve at the tree of life, and the serpent tempting Eve at the tree of the knowledge of good and evil.......................................139

Figure 5.2. Bathymetric map of modern Lake Van, including also the neighboring cities of Ahlat, Ercis, Tatvan, and Van, and the sites of the two volcanoes Nemruth (Nimrod, etc.) and Syphan (Sipan, etc.). Current water depth in meters is indicated. (Reproduced with permission.)....................... 141

Figure 7.1. Highest official cumulative rainfall for various periods of time, represented by the circles.......................... 210

Figure 8.1. Relative tree-ring thickness for a growth anomaly in Anatolia that may be associated with the tempest during the reign of Ahmose I... 265

Figure 8.2. Relative tree-ring thickness for two growth anomalies in Anatolia that may be associated with events near the time of the exodus.. 272

Figure 8.3. Electrostatic fields and one of their non-hazardous consequences.. 282

Figure 10.1. A Babylonian boundary stone of the twelfth century B.C., showing a scorpion man in Register 5, counting scenes downward from the top. (Reproduced with permission).. 348

Figure 10.2. Right-hand side of Relief Panel 10 of the Bronze door reliefs of Shalmaneser III at Tell Balawat (Imgur-Enlil). This relief includes scenes representing the visit of Shalmaneser to the Tigris Tunnel area in the king's seventh year 852 B.C. (Reproduced with permission)..................... 354

Figure 10.3. A cave with stalactites and stalagmites in the Tigris Tunnel region.. 356

Figure 15.1. Map of the Nile Delta and North Sinai with possible exodus routes indicated. (Reproduced with permission).. 517

Figure 19.1. Painting of Pliny the Younger working on a writing project while seated next to his mother in their home at Misenum. They are being scolded by a friend of his uncle Pliny the Elder for their lack of concern about the imminent eruption of Mt. Vesuvius seen in the background. (Reproduced with permission)...................................... 713

PREFACE

The Bible is a collection of ancient writings that is read and studied for many purposes. These writings provide a framework for stories of people, places, and events ranging in time from, in broadest terms, the creation of the universe in the distant past and on into the indefinite future. The texts included in the biblical collection vary among religious groups, and the interpretations and perceived relevance of these books for modern society also vary widely. In a previous study the author investigated the chronology of the Bible for the period of time from Abram's visit to Egypt until the end of the monarchy (L. W. Casperson, *Patterns of Biblical Chronology*). The emphasis was on the chronological patterns that can be employed in the establishment of a plausible chronology for that period. Several of the specific historical events that those patterns help to elucidate were also considered. A reader could be puzzled at the outset by the occurrence of the word "Excursions" in the title of this study. As used here, this term denotes a "deviation from a direct, definite, or proper course" (Merriam-Webster). Thus, the resources employed are not limited to conventional Bible-related literature and methods. For illustration, a few of these excursions are mentioned briefly in this preface. Most of these are chronological in nature, while others will emphasize events and interpretations that have not been widely studied in the past.

One purpose of the present study is to extend the period of interest to the entire biblical era. Some Bible readers might imagine that Genesis provides all of the information that is available concerning the beginning of the universe, and similarly Revelation might tell us all we can know about the end of the earth and of humanity. However, further biblical texts, together with ancient and modern resources from outside of the Bible, can provide much additional information on these and other subjects. Numerous quotations are included to indicate the nature of the sources from which this information may be extracted.

Other readers might take the contrary point of view that only the methods of science can provide any useful information about the early history of the universe or the likely future of the earth and its inhabitants. It is suggested here that there are limitations to each of these approaches when applied in isolation to Bible-related topics. On the other hand, extra-biblical information and methods can sometimes help to confirm and lend geographical and chronological precision to many biblical stories. Among these stories are, for example, the garden of Eden, the flood, the migration to Egypt, and the exodus. Comparing the biblical texts with other information, it is possible to infer relatively precise dates for events such as those just mentioned.

The purpose of these studies is limited to the interpretation of chronological information associated with the Bible. It is not intended here to suggest any modifications to the already very clear biblical teachings concerning moral and religious obligations. It has become increasingly popular in recent years for supposedly religious individuals and organizations to tolerate or encourage activities and behaviors that are forbidden by any honest reading of the Bible. An accompanying trend toward self-centeredness can also undermine the educational and charitable purposes of churches and other religion-based institutions, and many of these are experiencing declining membership.

The initial focus here is on the time period from Adam to Abram. As a first example, it will be noted that there are several other creation stories in ancient literature that have connections to the Bible-related texts. The biblical story of creation is compared especially to the creation stories of ancient cultures of the Middle East including Babylon, Ebla, and Greece. One of the earliest of these appears in the literature of Ebla, and the Eblaite language is closely related to Hebrew. The Bible isn't explicit concerning the place where Adam and Eve were created (or born), but it was not in the garden of Eden. Rather, they are said to have been placed in the garden of Eden after their creation elsewhere. The garden itself is said to have been at some location to the east of the narrator of the story, and other evidence suggests that the garden of Eden may have been near the Ercis gulf of Lake Van.

The biblical chronologies of the creation story aren't always consistent with each other. Thus, in the Masoretic version used in most western Bibles Genesis 2.2 states the following: "And on the seventh day God finished his work which he had done." Other Bible-related sources state that God's work was finished on the sixth day rather than the seventh. Also, genealogies with the ages of the patriarchs from Adam to the flood at the births of their sons differ between the Masoretic, Samaritan, and Septuagint versions of the Bible. Other chronologies are provided by Jubilees (book from the Ethiopian Orthodox Bible), Josephus's *Antiquities of the Jews*, and *The Book of Jasher*.

Based in part on tree-ring evidence, the biblical flood seems likely to have been caused by heavy rainfall and a rapid increase in the water level of Lake Van in about the year 2035 B.C. There is documentary evidence of the flooding of rivers in China and Sumer at about the same time as this proposed date for the biblical flood. The Babylonian equivalent of the story of Noah's flood is included in the *Gilgamesh Epic*, and information about that epic is also consistent with the previously mentioned flood date. Assuming that the patriarchs between Noah and Abraham lived more typical modern life spans, an approximate chronology is developed for this period. Starting from the flood date, it is also possible to obtain an approximate chronology for the dates of biblical events back to Adam.

The period of the Hebrew presence in Egypt was considered in our previous study, but additional information concerning this period is introduced here. The recorded effects in Egypt of volcanic eruptions in the Mediterranean area lead to fixed dates for several significant events including the entry into Egypt by Abram in about 1727 B.C., the journey of Jacob and his family to Egypt in about 1625 B.C., the tempest in Egypt in about 1554 B.C., and the exodus of the Hebrews from Egypt in about 1440 B.C. Based on textual records, other intermediate dates and identifications for the Egyptians and the Hebrews can also be inferred.

Many scientists, especially ones of an atheistic persuasion, have very different ideas from those emphasized in this book concerning the beginning and ending of the universe (and other events in between). Some of their ideas will be indicated in the last chapters here. In the presumed absence of a creator, such scientists have developed chronological conjectures for the evolution of the universe and its inhabitants, but for example their postulated creation of the universe by itself may seem less than compelling. On the other hand, the inference by scientists that the earth will one day be burned up may be compatible with several biblical texts.

Acknowledgments

The author is pleased to acknowledge the guidance and encouragement provided many years ago by his parents and others on topics relating to the Bible and science. He is also especially appreciative of the assistance and patience of his wife Susan throughout this project.

PART I

THE GARDEN AND THE FLOOD

1. IN THE BEGINNING

"In the beginning God created the heavens and the earth. The earth was without form and void, and darkness was upon the face of the deep; and the spirit of God was moving over the face of the waters" (Genesis 1:1–2 RSV).

1.1 Introduction

The first chapter of Genesis is a brief description of God's creation of the universe, from the "heavens and the earth" to the first human beings. This creation story serves as the reference point for all of the discussions in the present chapter. It is probably evident that stories from the earliest times would likely involve some of the largest uncertainties, and readers should not expect many simple and unambiguous results.

The biblical creation story is not the only such story known to exist. Explanations for the creation of the universe in general and the world as we know it have been developed at various times and by many different cultures. These stories typically describe the origins of such things as the sun, moon, stars, plants, animals, and human beings, as well as the relationship of humans to their gods. A few examples of creation stories are considered in this chapter, beginning with that of the Bible. Most non-biblical creation accounts are very different from the biblical version, but some have significant similarities.

The creation story given in Genesis is reviewed in Section 1.2. In broadest outline this creation story indicates that God created the universe in a period of seven days. Thus this story is used in part to emphasize the importance of remembering the seven-day week and especially the holiness of the seventh day. The period of seven days may not originally have been associated with the creation story, and this possibility is discussed in Section 1.3. A Babylonian creation story is summarized in Section 1.4, and a brief Eblaite version is considered in Section 1.5. A later Greek creation account is discussed in Section 1.6. In addition to such ancient stories of how the universe was created, some scientists have developed explanations that they claim do not require the existence of a creator. A brief summary of such a description of the origin of the universe is included in Section 20.2.

1.2 The biblical account

As known to even casual readers of the Bible, the opening verses of the book of Genesis include an outline of the history of the universe, from the beginning of time to the existence of the physical world as we know it today. The interpretation of that history will be of recurring interest throughout this chapter, so we begin with a quotation of the familiar biblical creation story:

> In the beginning God created the heavens and the earth. The earth was without form and void, and darkness was upon the face of the deep; and the Spirit of God was moving over the face of the waters.

And God said, "Let there be light"; and there was light. And God saw that the light was good; and God separated the light from the darkness. God called the light Day, and the darkness he called Night. And there was evening and there was morning, one day.

And God said, "Let there be a firmament in the midst of the waters, and let it separate the waters from the waters." And God made the firmament and separated the waters which were under the firmament from the waters which were above the firmament. And it was so. And God called the firmament Heaven. And there was evening and there was morning, a second day.

And God said, "Let the waters under the heavens be gathered together into one place, and let the dry land appear." And it was so. God called the dry land Earth, and the waters that were gathered together he called Seas. And God saw that it was good.

And God said, "Let the earth put forth vegetation, plants yielding seed, and fruit trees bearing fruit in which is their seed, each according to its kind, upon the earth." And it was so. The earth brought forth vegetation, plants yielding seed according to their own kinds, and trees bearing fruit in which is their seed, each according to its kind. And God saw that it was good. And there was evening and there was morning, a third day.

And God said, "Let there be lights in the firmament of the heavens to separate the day from the night; and let them be for signs and for seasons and for days and years, and let them be lights in the firmament of the heavens to give light upon the earth." And it was so. And God made the two great lights, the greater light to rule the day, and the lesser light to rule the night; he made the stars also. And God set them in the firmament of the heavens to give light upon the earth, to rule over the day and over the night, and to separate the light from the darkness. And

God saw that it was good. And there was evening and there was morning, a fourth day.

And God said, "Let the waters bring forth swarms of living creatures, and let birds fly above the earth across the firmament of the heavens." So God created the great sea monsters and every living creature that moves, with which the waters swarm, according to their kinds, and every winged bird according to its kind. And God saw that it was good. And God blessed them, saying, "Be fruitful and multiply and fill the waters in the seas, and let birds multiply on the earth." And there was evening and there was morning, a fifth day.

And God said, "Let the earth bring forth living creatures according to their kinds: cattle and creeping things and beasts of the earth according to their kinds." And it was so. And God made the beasts of the earth according to their kinds and the cattle according to their kinds, and everything that creeps upon the ground according to its kind. And God saw that it was good.

Then God said, "Let us make man in our image, after our likeness; and let them have dominion over the fish of the sea, and over the birds of the air, and over the cattle, and over all the earth, and over every creeping thing that creeps upon the earth." So God created man in his own image, in the image of God he created him; male and female he created them. And God blessed them, and God said to them, "Be fruitful and multiply, and fill the earth and subdue it; and have dominion over the fish of the sea and over the birds of the air and over every living thing that moves upon the earth." And God said, "Behold, I have given you every plant yielding seed which is upon the face of all the earth, and every tree with seed in its fruit; you shall have them for food. And to every beast of the earth, and to every bird of the air, and to everything that creeps on the earth, everything that has the breath of life, I have given every green plant for food." And it was so. And God saw everything that he had made, and behold, it was very good. And there was evening and there was morning, a sixth day.

Thus the heavens and the earth were finished, and all the host of them. And on the seventh day God finished his work which he had done, and he rested on the seventh day from all his work which he had done. So God blessed the seventh day and hallowed it, because on it God rested from all his work which he had done in creation (Genesis 1:1–2:3).[1]

One of the best-known features of this account is its conspicuous association with the seven-day week, and this aspect will be considered first. It should be noted that except for the creation story, the week may have been introduced many centuries later. The first appearance of the word "week" in the Bible seems to be in the well-known story of Jacob's marrying Leah and Rachel, the two daughters of Laban. Jacob served Laban seven years for his first wife and another seven years for his second:

So Jacob served seven years for Rachel, and they seemed to him but a few days because of the love he had for her.

Then Jacob said to Laban, "Give me my wife that I may go in to her, for my time is completed." So Laban gathered together all the men of the place, and made a feast. But in the evening he took his daughter Leah and brought her to Jacob; and he went in to her. (Laban gave his maid Zilpah to his daughter Leah to be her maid.) And in the morning, behold, it was Leah; and Jacob said to Laban, "What is this you have done to me? Did I not serve with you for Rachel? Why then have you deceived me?" Laban said, "It is not so done in our country, to give the younger before the first-born. Complete the week of this one, and we will give you the other also in return for serving me another seven years." Jacob did so, and completed her week; then Laban gave him his daughter Rachel to wife. (Laban gave his maid Bilhah to his daughter Rachel to be her maid.) (Genesis 29:20-29).

The word week in this context can apparently be interpreted to mean a period of seven years rather than seven days. The same story is recorded in Jubilees:[2]

> And he [Jacob] went on his journey, and came to the land of the east, to Laban, the brother of Rebecca, and he was with him, and served him for Rachel his daughter one week.[3] And in the first year of the third week [of the forty-fourth jubilee] he [Jacob] said unto him [Laban]: "Give me my wife, for whom I have served thee seven years;" and Laban said unto Jacob. "I will give thee thy wife" (Jubilees 28:1-2).

The idea of a week of years also appears at later places in the Bible:

> And you shall count seven weeks of years, seven times seven years, so that the time of the seven weeks of years shall be to you forty-nine years (Leviticus 25:8).

> Seventy weeks of years are decreed concerning your people and your holy city, to finish the transgression, to put an end to sin, and to atone for iniquity, to bring in everlasting righteousness, to seal both vision and prophet, and to anoint a most holy place. Know therefore and understand that from the going forth of the word to restore and build Jerusalem to the coming of an anointed one, a prince, there shall be seven weeks. Then for sixty-two weeks it shall be built again with squares and moat, but in a troubled time. And after the sixty-two weeks, an

anointed one shall be cut off, and shall have nothing; and the people of the prince who is to come shall destroy the city and the sanctuary. Its end shall come with a flood, and to the end there shall be war; desolations are decreed. And he shall make a strong covenant with many for one week; and for half of the week he shall cause sacrifice and offering to cease; and upon the wing of abominations shall come one who makes desolate, until the decreed end is poured out on the desolator (Daniel 9:24-27).

The clear references to the traditional seven-day week rather than the seven-year week seem to commence in Exodus, and some of these references are indicated here:

On the sixth day they gathered twice as much bread, two omers apiece; and when all the leaders of the congregation came and told Moses, he said to them, "This is what the Lord has commanded: 'Tomorrow is a day of solemn rest, a holy sabbath to the Lord; bake what you will bake and boil what you will boil, and all that is left over lay by to be kept till the morning' " (Exodus 16:22-23).

Moses said, "Eat it today, for today is a sabbath to the Lord; today you will not find it in the field. Six days you shall gather it; but on the seventh day, which is a sabbath, there will be none (Exodus 16:25-26).

See! The Lord has given you the sabbath, therefore on the sixth day he gives you bread for two days; remain every man of you in his place, let no man go out of his place on the seventh day." So the people rested on the seventh day (Exodus 16:29-30).

Remember the sabbath day, to keep it holy. Six days you shall labor, and do all your work; but the seventh day is a sabbath to the Lord your God; in it you shall not do any work, you, or your son, or your daughter, your manservant, or your maidservant, or your cattle, or the sojourner who is within your gates (Exodus 20:8-10).

Six days shall work be done, but the seventh day is a sabbath of solemn rest, holy to the Lord; whoever does any work on the sabbath day shall be put to death (Exodus 31:15).

Six days shall work be done, but on the seventh day you shall have a holy sabbath of solemn rest to the Lord; whoever does any work on it shall be put to death (Exodus 35:2).

Given the emphasis on the seven-day week in Exodus, it is of interest to consider the possibility that the implication in Genesis of a seven-day creation period might not be authentic. In other words, an original creation story lacking any association with the week might have been modified to support this much later concept. To consider such a possibility, one can look for any irregularities in the creation formulas that would suggest that the days of the week were later additions.

First of all, it may be noted that most of the paragraphs in the creation story quoted above present God as speaking his creation into existence using opening and closing words such as the following: "And God said, 'Let there be. . . .' And there was. . . . And God saw that it was good. . . . And there was evening and there was morning, a . . . day." It should be noted however that not all of the creation segments as quoted above include the closing words "And there was evening and there was morning, a . . . day." In short, there appear to be more creation events in the story than there are days of what we call a week to which they could easily have been assigned by the writer. To be specific, the creation activities of Genesis 1.1 – 2.3 as quoted earlier in this section are not all explicitly assigned to numbered days of the week. This suggests the possibility that in an earlier creation account the days of the week may not have been mentioned.

1.3 A week-free version

It may be noted that the period of seven days occurs at several places in the earliest biblical stories, but it is usually not explicitly identified with a seven-day week. Thus, in Genesis, the seven-day time interval is also associated with events of the flood (Genesis 7:4, 7:10, 8:10, 8:12), the pursuit of Jacob by his father-in-law Laban (Genesis 31:23), and the period of mourning at the death of Jacob (Genesis 50:10). On the other hand, the seven-day period is mentioned frequently beginning in Exodus, including especially its application as the time period of the week and sometimes claiming its origin in the seven days of creation (Exodus 16:22-30; 20:8-11; 23:12; 31:12-17; 34:21; 35:2-3). The origin and use of the seven-day week has been commented on as follows:

> The seven-day week is probably of Israelite origin. At least no such week is attested prior to Israel's. . . . From Israel's earliest days the seven-day cycle culminating in a seventh-day sabbath ran uninterruptedly through the months and years (Exodus 34:21, 23:12). Its recurrence was independent of lunar and solar cycles, and it has remained constant in all Israelite and Jewish calendars.[4]

1. In the Beginning

It is possible that the seven-day week became part of the official calendar for Israel with the writing of Exodus.

Traditionally, Moses is considered to be the author or editor of the first five books of the Bible including of course Genesis and Exodus. Adapting the creation story to the perhaps later-introduced seven-day week would add impact to any sabbath-day requirements. Thus, as God rested on the seventh day, the Hebrews were required to rest on the sabbath day, and in principle the violation of this requirement carried the death penalty (Exodus 31:14-15, 35:2). Such severe punishment was revoked by Jesus, as he reminded his followers that "The sabbath was made for man, not man for the sabbath" (Mark 2:27). On a different topic, it should be noted that in the Masoretic text it is stated that "on the seventh day God finished his work which he had done" (Genesis 2:2). This statement could be interpreted to imply that God himself worked on the seventh day, since finishing one's work would involve work. This seeming contradiction doesn't occur in other versions of the creation story, as discussed in more detail in Section 3.4.

It may also be recalled that the idea of God ever ceasing from work seems to be contrary to Jesus's own statements after he healed a lame man on the sabbath day. There we are reminded that God is always working:

> Jesus said to him, "Rise, take up your pallet, and walk." And at once the man was healed, and he took up his pallet and walked.
> Now that day was the sabbath. So the Jews said to the man who was cured, "It is the sabbath, it is not lawful for you to carry your pallet." But he answered them, "The man who healed me said to me, 'Take up your pallet, and

walk.' " They asked him, "Who is the man who said to you, 'Take up your pallet, and walk'?" Now the man who had been healed did not know who it was, for Jesus had withdrawn, as there was a crowd in the place. Afterward, Jesus found him in the temple, and said to him, "See, you are well! Sin no more, that nothing worse befall you." The man went away and told the Jews that it was Jesus who had healed him. And this was why the Jews persecuted Jesus, because he did this on the sabbath. But Jesus answered them, "My Father is working still, and I am working." This was why the Jews sought all the more to kill him, because he not only broke the sabbath but also called God his Father, making himself equal with God (John 5:8-18).

The above discussion has included the observation that references to the seven-day week in the Genesis creation story might appear to predate the actual introduction of the seven-day week as described in Exodus. This concept has been suggested before, and previous authors have argued "that there was an earlier Hebrew version of the cosmogony in which that [seven days] scheme did not exist."[5] Also, various attempts have been made to reconcile the number and order of the creation events:

> But on the whole the evidence seems to warrant the conclusions: that the series of works and the series of days are fundamentally incongruous, that the latter has been superimposed on the former during the Hebrew development of the cosmogony, that this change is responsible for some of the irregularities of the disposition, and that it was introduced certainly not later than P [the priestly source in the documentary hypothesis], and in all probability long before his time."[6]

1. In the Beginning

Assuming that the chronology of the seven-day week was a later addition to an already existing creation story, it would be of interest to consider how the creation story of Genesis quoted above might have appeared if the references to the days of the week were removed. The story is reproduced here with those references deleted and other comments also removed. For a later purpose in this chapter the creation phases are slightly re-ordered. Thus, the activities of the fourth day are moved earlier, and the sentences indicating the reaction of the creator to his creation are also omitted:

> In the beginning God created the heavens and the earth. The earth was without form and void, and darkness was upon the face of the deep; and the Spirit of God was moving over the face of the waters.
> And God said, "Let there be light"; and there was light. [A]nd God separated the light from the darkness. God called the light Day, and the darkness he called Night.
> And God said, "Let there be a firmament in the midst of the waters, and let it separate the waters from the waters." And God made the firmament and separated the waters which were under the firmament from the waters which were above the firmament. And it was so. And God called the firmament Heaven.
> And God said, "Let there be lights in the firmament of the heavens to separate the day from the night; and let them be for signs and for seasons and for days and years, and let them be lights in the firmament of the heavens to give light upon the earth" And it was so. And God made the two great lights, the greater light to rule the day, and the lesser light to rule the night; he made the stars also. And God set them in the firmament of the heavens to give light upon the earth, to rule over the day and over the night, and to separate the light from the darkness.

The Garden And The Flood

And God said, "Let the waters under the heavens be gathered together into one place, and let the dry land appear." And it was so. God called the dry land Earth, and the waters that were gathered together he called Seas.

And God said, "Let the earth put forth vegetation, plants yielding seed, and fruit trees bearing fruit in which is their seed, each according to its kind, upon the earth." And it was so. The earth brought forth vegetation, plants yielding seed according to their own kinds, and trees bearing fruit in which is their seed, each according to its kind.

And God said, "Let the waters bring forth swarms of living creatures, and let birds fly above the earth across the firmament of the heavens." So God created the great sea monsters and every living creature that moves, with which the waters swarm, according to their kinds, and every winged bird according to its kind.

And God said, "Let the earth bring forth living creatures according to their kinds: cattle and creeping things and beasts of the earth according to their kinds." And it was so. And God made the beasts of the earth according to their kinds and the cattle according to their kinds, and everything that creeps upon the ground according to its kind.

Then God said, "Let us make man in our image, after our likeness; and let them have dominion over the fish of the sea, and over the birds of the air, and over the cattle, and over all the earth, and over every creeping thing that creeps upon the earth." So God created man in his own image, in the image of God he created him; male and female he created them.

Thus the heavens and the earth were finished, and all the host of them.

The above shortened account will be recalled again in the following sections of this chapter. First, we will consider for

1. In the Beginning

comparison the often mentioned but very different account recorded by the Babylonians and Assyrians.

1.4 The Babylonian account

According to the biblical records, some of the most ancient ancestors of the Hebrews lived in or near Mesopotamia. Also, many of the world's earliest written records had their origin in that area, including various stories relating to the creation of the universe. Perhaps not surprisingly, it is sometimes assumed that the creation account in the book of Genesis must be an adaptation of earlier Babylonian versions: "The discovery of the Babylonian versions of the Creation- and Deluge-traditions has put it beyond reasonable doubt that these are the originals from which the biblical accounts have been derived."[7] Contrary to this statement, findings in Ebla, to be discussed in Section 1.5 below, have yielded a creation story that is similar to the biblical creation account and was written several centuries before the Babylonian versions.

Among the Babylonian creation stories, *Enûma elish* has attracted the most attention. This epic was recorded on seven clay tablets and includes over one thousand lines of text.[8] The date of its writing can be estimated: [9]

> [I]f we consider that the two main objects of the epic are to justify Marduk's ascendancy to supreme rulership over all the Babylonian divinities and to support Babylon's claim to pre-eminence above all the other cities in the country, as we have seen, and that Babylon rose to political supremacy during the First Babylonian Dynasty (1894-1595),[10] particularly under the energetic king Hammurabi (1792-1750), and that during this dynasty

Marduk became the national god,[11] it would seem that the poem, in approximately its present form, was composed some time during the First Babylonian dynasty.

According to Heidel and others, the writing of the epic most likely occurred during the reign of Hammurabi. In the popular "short chronology" Hammurabi reigned during about 1728-1686 B.C.[12] Thus, *Enûma elish* might be considered to date from roughly 1700 B.C.

A major purpose of *Enûma elish* seems to be the glorification of Marduk, one of the most powerful of the numerous gods of the Babylonian pantheon, and Marduk's creation of the universe is largely incidental to that purpose. The first gods were Apsu and Tiamat, and they had numerous descendant gods. Major conflicts arose, and eventually the god Ea killed Apsu, his own great grandfather and the father of all of the gods. Tiamat then became a tyrant and had many followers among the gods, but she was killed by Ea's son Marduk, the hero of the epic. The gruesome lines of *Enûma elish* relevant to creation are not particularly coherent in the surviving text, and only a paraphrase is needed here. To give some idea of the style and content of the text we quote Heidel's summary of Marduk's creation activities after the death of Tiamat and the capture of her followers:

> After having strengthened his hold upon the captive gods, he [Marduk] returned to Tiamat, split her skull with his unsparing club, cut her arteries, and caused the north wind to carry her blood southward to out-of-the-way places. Finally, he divided the colossal body of Tiamat into two parts to create the universe. With one half of her corpse he formed the sky, with the other he fashioned the earth, and then established Anu, Enlil, and Ea in their respective domains.

1. In the Beginning

Next, he created stations in the sky for the great gods; he organized the calendar, by setting up stellar constellations to determine, by their rising and setting, the year, the months, and the days; he built gates in the east and in the west for the sun to enter and to depart; in the very center of the sky he fixed the zenith; he caused the moon to shine forth and intrusted the night to her. . . .

The imprisoned gods, who had joined the ranks of Tiamat, were made the servants of the victors, for whose sustenance they had to provide. However, their menial task proved so burdensome that they asked Marduk for relief. As Marduk listened to the words of the captive gods, he resolved to create man and to impose on him the service which the defeated deities had to render. In consultation with Ea, it was then decided to kill the ringleader [Kingu] of the rebels, to create mankind with his blood, and to set the captive gods free. In a solemn court Kingu was indicted. He it was who "created the strife," who "caused Tiamat to revolt and prepare for battle." Accordingly, Kingu was bound and brought before Ea. With the aid of certain gods, Ea severed his arteries and created mankind with his blood, acting on the ingenious plans of Marduk. Man now had to take over the work of the defeated army of gods and feed the host of Babylonian divinities.[13]

While the Babylonian epic mentions in a qualitative way a few acts that might be associated with creation, it is clear that creation is not the main emphasis of the epic:

Yet, *Enûma elish* is not primarily a creation story at all. . . . The brief and meager account of Marduk's acts of creation is in sharp contrast to the circumstantial description of his birth and growth, his preparations for battle, his conquest of Tiamat and her host, and the elaborate and pompous proclamation and explanation of his fifty names. If the creation of the universe were the prime purpose of the epic, much more emphasis should have been placed on this point.[14]

Besides its lack of focus on the creation period, the creation details tend to be very different from those of the biblical account:

> Our examination of the various points of comparison between *Enûma elish* and Genesis 1:1–2:3 shows quite plainly that the similarities are really not so striking as we might expect, considering how closely the Hebrews and the Babylonians were related. In fact, the divergences are much more far-reaching and significant than are the resemblances, most of which are not any closer than what we should expect to find in any two more or less complete creation versions (since both would have to account for the same phenomena and since human minds think along much the same lines) which might come from entirely different parts of the world and which might be utterly unrelated to each other.[15]

A particular difference between the Babylonian and biblical creation accounts is the dominance in the Babylonian stories of polytheism as opposed to its nonexistence or faint reflection in the Genesis version (discussed in Section 2.2). In order to assume that the mostly monotheistic biblical creation story is based on *Enûma elish*, one would have to imagine more generally that the many polytheistic indications of the Babylonian creation story and other myths were expunged in the assumed later Hebrew reworkings of their texts. A few examples of such dubious assumptions follow:

> The story of Paradise . . . (as well as the story of the Tower of Babel) were originally genuine myths – stories

of the gods; and if they no longer deserve that appellation, it is because the spirit of Hebrew monotheism has exorcised the polytheistic notions of deity, apart from which true mythology cannot survive. The few passages where the old heathen conception of godhead still appears (Genesis 1:26, 3:22, 24, 6:1ff., 11:1ff.), only serve to show how completely the religious beliefs of Israel have transformed and purified the crude speculations of pagan theology, and adapted them to the ideas of an ethical and monotheistic faith.[16]

A similar sentiment is quoted in the following:

It is true, at all events, that the Babylonian parallel [to the biblical flood story] serves as a "measure of the unique grandeur of the idea of God in Israel, which was powerful enough to purify and transform in such a manner the most uncongenial and repugnant features" of the pagan myth.[17]

The reasons for making such assumptions are not compelling, as will be suggested in the following section.

There are, nevertheless, some minor similarities between *Enûma elish* and the Genesis creation story, and various explanations for these similarities have been proposed:[18]

[One] theory, designed to account for the similarities between *Enûma elish* and Genesis 1:1–2:3, was developed by Clay.[19] He contended that *Enûma elish* was an amalgamation of a Semitic myth coming from a region called Amurru (i.e., northwestern Mesopotamia, Syria, and Palestine) and a Sumerian myth presumably from the city of Eridu and that the elements which *Enûma elish* and Genesis 1:1–2:3 have in common were importations from Amurru, which, he concluded, were carried to Babylonia by Western Semites emigrating to that land.[20]

In other words, the minimal commonality between the biblical and Babylonian creation stories may have been a consequence of importations from Syria to Babylon. This possibility may be supported in part by the much earlier Eblaite creation account (undiscovered until long after the time of Clay) discussed in the following section, and also by indications considered in Section 6.5 that early occupants of a place called Eden had migrated to northwestern Mesopotamia.

Besides the many differences in detail between the biblical creation story and the fragments in *Enûma elish*, one can also perceive a fundamental qualitative difference in the style and nature of these two works. This point has been emphasized by Heidel:[21]

> [A] comparison of the Babylonian creation story with the first chapter of Genesis makes the sublime character of the latter stand out in even bolder relief. *Enûma elish* refers to a multitude of divinities emanating from the elementary world-matter; the universe has its origin in the generation of numerous gods and goddesses personifying cosmic spaces or forces in nature, and in the orderly and purposeful arrangement of pre-existent matter; the world is not *created* in the biblical sense of the term but *fashioned* after the manner of human craftsmen; as for man, he is created with the blood of a deity that might well be called a devil among the gods, and the sphere of activity assigned to man is the service of the gods. In Genesis 1:1–2:3, on the other hand, there stands at the very beginning *one* God, who is not co-united and coexistent with an eternal world-matter and who does not first develop Himself into a series of separate deities but who creates matter out of nothing and exists independently of all cosmic matter and remains one God to the end. Here

the world is created by the sovereign word of God, without recourse to all sorts of external means. God speaks, and it is done; he commands, and it stands fast. Add to this the doctrine that man was created in the image of a holy and righteous God, to be the lord of the earth, the air, and the sea, and we have a number of differences between *Enûma elish* and Genesis 1:1–2:3 that make all similarities shrink into utter insignificance.[22] These exalted conceptions in the biblical account of creation give it a depth and dignity unparalleled in any cosmogony known to us from Babylonia or Assyria.

In spite of such major differences, it is true that the Bablyonian and biblical texts involved, at least in part, the same geographical region. Thus, it should not be surprising to find other more minor associations not related to creation. An example of such an association might be the breastplate worn by a god or priest of the Babylonians or a leader of the Hebrews as a symbol of their authority and as a means for them to render just judgments and to divine the future. In *Enûma elish* Tiamat gave authority to Kingu: "She exalted Kingu; in their midst she made [him gr]eat. . . . She gave him the tablet of destinies, she fastened (it) upon his breast, (saying:) 'As for thee, thy command shall not be changed, [the word of thy mouth] shall be dependable!' "[23]

On the other hand, Hebrew priests and leaders, beginning with Aaron, sometimes also wore a breastplate including the Urim and Thummim which could be employed to determine the answers to questions:

And Aaron shall bear the names of the children of Israel in the breastplate of judgment upon his heart, when he goeth in unto the holy place, for a memorial before the

Lord continually. And thou shalt put in the breastplate of judgment the Urim and the Thummim; and they shall be upon Aaron's heart, when he goeth in before the Lord: and Aaron shall bear the judgment of the children of Israel upon his heart before the Lord continually (Exodus 28:29-30 KJV).

That there is likely a relationship between the usage and purpose of the breastplates of these two peoples has long been recognized: "[I]t is quite possible that the mythological account of the Tablets of Destiny and the Old Testament Urim and Thummim, both shaping the destiny of king and nation, revert to the same fountainhead and origin."[24] If it is true that these two breastplate stories have a common origin, it doesn't mean that either of them originated directly from the other. Although suggested a century ago by Clay, more recent discoveries related to Ebla have strengthened the notion that some features of the biblical creation account could have originated in the Syria region prior to the writing of *Enûma elish*.

Summarizing more explicitly here, it may be worth considering the possibility that the "crude speculations" and "repugnant features" of the Babylonian creation and other myths mentioned above were in some cases later corruptions of the more dignified Bible-related stories rather than imagining that the Bible stories might be purified versions of earlier Babylonian myths. This possibility is supported by the more recently discovered Ebla documents discussed in Section 1.5, which are understood to be several centuries older than the Babyonian texts. A self-consistent interpretation of Bible-related texts concerning other early events suggests that those texts were also written long before the very different versions from Babylon. An estimate for the date of the garden of Eden stories will be developed in Section 12.5.

1.5 The Eblaite account

One of the most significant recent developments in the understanding of the early Middle East was the discovery, beginning in the 1960s, of the long lost city and empire of Ebla in what is present day Syria. "It has been said quite justifiably that the Ebla discoveries have revealed a new language, a new history and a new culture."[25] At least eleven thousand cuneiform tablets were represented by the findings at Tell Mardikh (Ebla) as of 1980, and these are the largest third-millennium archives ever uncovered.[26] In spite of initial enthusiasm, significant difficulties have arisen in the detailed archeological investigations at Ebla, as well as in the interpretation of texts that have been discovered and in the publication of results. One concern is that there may initially have been an excessive emphasis in relating the discoveries on the ground to Bible-associated stories and peoples. This could in principle have led to claims that would be dramatic and appealing to popular audiences but not necessarily well supported by an objective consideration of the data.

On the other hand, there is also said to have been significant interference by the Syrian government in an effort to minimize the appearance of results that might support biblical texts in general and the antiquity in Syria of ancestors of the Hebrews in particular:

> In the March/April 1979 issue (p. 37), *BAR* [*Biblical Archaeology Review*] charged that the Syrian government was attempting to "affect the scholarly interpretation of

the Ebla tablets." Many independent sources have also reported that the Syrians were angered at the emphasis placed in the West on the tablets' alleged Biblical connections. For the Syrian authorities, the importance of the tablets lies in the light they shed on the great Syrian past, what they call "proto-Syrian" history. The Syrians attribute the Western interest in the Biblical connections to a Zionist plot. Moreover, they make no secret about how they feel. . . .

Even an unsubtle scholar will get the point: If you want to continue working on the Ebla tablets, you better give a "balanced" picture. You are certainly not free to direct your attention exclusively to the tablets' possible Biblical connections. As the Syrian ambassador stated, "We are able to close the whole thing down [the Ebla mission], but we don't want this." *BAR* continues to find this Syrian interference in the scholarly enterprise highly objectionable. . . .

It is a bit frightening that Matthiae [director of the Italian mission to Ebla] is quoted in a Syrian publication as stating that, "These allegations [linking the Ebla tablets with the Bible] were propagated by Zionist-American centres to be exploited for atrocious purposes aimed at proving the expansionist and colonialist views of the Zionist leaders." How far will Matthiae and other scholars go to please the Syrians? Will they delay publication of texts which have particular significance for Biblical studies? Will they stress pure Near Eastern history? Or avoid discussion of whatever is potentially politically objectionable? . . .

Paolo Matthiae . . . has taken the unusual step of writing to a number of scholarly journals advising them not to publish any tablets without written authorization from him. . . . Editors who have received the letter say they cannot recall ever having received a similar letter from any other archaeologist or scholar.[27]

This modern conflict is reminiscent of the ancient efforts of the Greeks in Egypt to minimize the historical significance and antiquity of the Hebrews in comparison to the Greeks themselves. On the other hand, the Hebrews sometimes sought to emphasize or exaggerate their own antiquity.[28]

In spite of such difficulties, it seems appropriate to recall here at least a few of the surprising results from the discoveries at Ebla, especially if those results might be related to the chronological questions being considered. As indicated in the previous section, one of the most striking differences between the biblical and the Babylonian creation stories is in their representations of God. The Genesis account suggests the early monotheistic beliefs of the ancestors of the Hebrews and the survival of their beliefs even after those ancestors had lived for many years among the polytheistic Babylonians. As inferred in the studies by Clay, noted in Section 1.4, such beliefs could have been maintained in part by Northwest Semites, perhaps during early times when the ancestors of the Hebrews dwelt with or were otherwise associated with their relatives in northwestern Mesopotamia. A few thoughts on the possible migratory paths of those ancestors are included in Sections 5.5 and 6.5. Such ideas also suggest the possibility that monotheism, or at least henotheism (the worship of one god while accepting that there might be others), was perhaps a common practice in the lands that came to be associated with the Amorites. Certain discoveries from the archives of Ebla may support these ideas, and the brief Eblaite story of creation is possibly significant. The introductory part of this creation story or hymn, dated to about 2500 B.C., was translated and commented on in a report by Pettinato:

> Lord of heaven and earth:
> the earth was not, you created it,
> the light of the day was not, you created it,
> the morning light you had not [yet] made exist.[29]

According to Pettinato, the above comments should be understood to represent the "official" religion of Ebla, whereas the "popular" religion is more complicated and information is incomplete: "From pure henotheism we pass to rank polytheism when we look at the religion of the masses. The available data come from the state archives and refer to official worship and reflection of some thinkers but scarcely reflect the popular beliefs that in every culture present some peculiarities."[30] "The available sources put membership in the Ebla pantheon at about 500 divinities, but they do not spell out its structure or the relationships between the individual gods."[31] This ambiguity of the Eblaite religion is reminiscent of the long-running tension between the official Hebrew monotheism and the polytheistic influences that the Hebrews found so hard to resist. Numerous examples of the Hebrews worshiping other gods may readily be identified in the Bible. In fact, such polytheistic influences may perhaps be discerned already in the Hebrew creation and tower of Babel stories, and this situation will be remarked upon in Section 2.2. In a not-so-amicable process, Pettinato resigned his position as chief epigrapher (interpreter of ancient inscriptions) of the archeological excavations at Ebla and was replaced by Archi. Following that transition, there were published disagreements concerning translations of the Ebla archives.[32]

The hymn quoted above has been considered by many authors, and Merrill had the following comments on the text as translated by Pettinato:

The full implications of this poem would require a separate lengthy study but at least a few observations can be made. First, the tenor of the hymn is almost monotheistic in spirit. Creation is attributed to only one god. Second, the order (heaven, earth, light, and morning) is identical to that in Genesis 1:1–5. Third, the inference is that creation is *ex nihilo* [out of nothing], not the manufacture of things from an original and eternal primordial substance. Fourth, the epithet of the god as the "effective word" following the statement of his creative work is identical to the biblical concept of God who created by the spoken word (Genesis 1:3) and who, in fact, is that Word Himself (John 1:1–3).[33]

There could be other aspects of Hebrew culture that one might imagine to have originated following the exodus but that have been found to have existed already in parallel form among the Eblaites a thousand years earlier. According to Exodus 23:10–11 and other texts, the Hebrews were to adopt a sabbatical year calendar system involving recurring seven-year periods.[34] Based on the Ebla archives Pettinato has suggested that kings there may have been elected for a fixed term of seven years with the possibility of re-election.[35] This would be a similar practice to the U.S. system in which presidents are elected for four-year terms. A seven-year cycle in Ebla would resemble chronologically the much later seven-year sabbatical cycle of the post-exodus Hebrews that is considered in Jubilees.

The archives of Ebla also have a bearing on the age of the Bible and the writings it contains. Thus, there has been uncertainty in the past concerning even the ability of early Hebrews to write, contrary to claims made in their own writings.[36] The discoveries at Ebla reinforce the idea that writing was already a long established skill at the time of the early Bible authors in at least the most important locations of Syria-Canaan and Babylon:

> A century ago writing was considered a relatively late cultural arrival in Israel, so that there was a marked tendency to date most of the biblical books to a late period; preferably to the sixth-century Exile and to the immediately following period. The discovery in 1929 of the Ras Shamra tablets with their long poems composed with exquisite skill and refinement led to a reevaluation of the age and level of writing skills in Canaan. As a result, critics are raising the dates of many of the poetic compositions in the Old Testament. The history and practice of writing in Canaan are now seen to be even older, going back to the beginning of the third millennium; this means that the sacred poets had fallen heir to a venerable literary tradition, and the modern critic would be well advised not to underestimate their technical capacity.[37]

It is not just the ability of the Hebrews to write that has been supported by the Ebla discoveries. Bilingual vocabularies found at Ebla have already clarified the meanings of many words and phrases of the Bible. These new resources will permit the writing of more comprehensive grammars and dictionaries of Hebrew and other Canaanite languages.[38] One would like to believe that there will be continuing discoveries to add to the wealth of insights that already have been gleaned from the relatively recent discoveries at Ebla. Unfortunately the reality might not be so promising. Organized research ended when the Syrian civil war began in 2011.[39] The excavations currently underway at Ebla are being carried out by looters.

1.6 A Greek account

As noted in Section 1.4 above, the biblical version of the creation story is relatively simple and direct in comparison with the accounts recorded by the Babylonians, which are blended with other topics and permeated with mythological allusions. It is not clear that the biblical story is dependent on the Babylonian accounts in any significant way. On the other hand, the creation story from Ebla discussed in Section 1.5 is much simpler and clearer; but its brevity, together with translation controversies, currently makes association with the biblical version a somewhat difficult and subjective process.

Another creation story of possible interest is included in the *Argonautica* by Apollonius Rhodius (Apollonius of Rhodes). Apollonius was probably born in Alexandria, Egypt; and he served as scholar and for a time as head of the Library at Alexandria. His *Argonautica* was written in his youth in about the middle of the third century B.C. When he recited his epic poem in public, it was condemned; and as a result he moved from Egypt to the island of Rhodes. While there, he rewrote the *Argonautica* and recited it to great acclaim. He then returned to Alexandria, where his revised epic was also a success.[40]

As a starting point for this discussion (and to save some page turning), it may be helpful to provide a further abridged version of the first part of the Genesis creation account as given at the end of Section 1.3. Omitted here are mainly the creation of light, plants, and man:

> In the beginning God created the heavens and the earth. The earth was without form and void, and darkness

The Garden And The Flood

was upon the face of the deep; and the Spirit of God was moving over the face of the waters.

And God said, "Let there be a firmament in the midst of the waters, and let it separate the waters from the waters." And God made the firmament and separated the waters which were under the firmament from the waters which were above the firmament. And it was so. And God called the firmament Heaven.

And God said, "Let there be lights in the firmament of the heavens to separate the day from the night; and let them be for signs and for seasons and for days and years, and let them be lights in the firmament of the heavens to give light upon the earth" And it was so. And God made the two great lights, the greater light to rule the day, and the lesser light to rule the night; he made the stars also. And God set them in the firmament of the heavens to give light upon the earth, to rule over the day and over the night, and to separate the light from the darkness.

And God said, "Let the waters under the heavens be gathered together into one place, and let the dry land appear." And it was so. God called the dry land Earth, and the waters that were gathered together he called Seas.

And God said, "Let the waters bring forth swarms of living creatures, and let birds fly above the earth across the firmament of the heavens." So God created the great sea monsters and every living creature that moves, with which the waters swarm, according to their kinds, and every winged bird according to its kind.

And God said, "Let the earth bring forth living creatures according to their kinds: cattle and creeping things and beasts of the earth according to their kinds." And it was so. And God made the beasts of the earth according to their kinds and the cattle according to their kinds, and everything that creeps upon the ground according to its kind.

1. In the Beginning

The location of the creation story in Apollonius's epic is at a time just before the Argonauts started out on their long and difficult voyage. On the evening before their departure some of the Argonauts were quarrelsome, so Orpheus played his harp and sang a creation song to calm them.[41] "Men say that he by the music of his songs charmed the stubborn rocks upon the mountains and the course of rivers,"[42] and his song had the desired effect. The Argonauts worshiped Zeus, and then they went to sleep to be ready for their morning departure.

The creation account given us by Apollonius is somewhat more comprehensive than that of the Ebla records that have been reported to present. The first part of this version can almost be considered as a short and non-theistic adaptation of the Genesis story. Thus, references to God are not included, but otherwise the topics are similar to the first part of the Genesis version as quoted in brief above. On the other hand, the latter part resembles conventional Greek mythology:

> Orpheus lifted his lyre in his left hand and made essay to sing.
> He sang how the earth, the heaven and the sea, once mingled together in one form, after deadly strife were separated each from other; and how the stars and the moon and the paths of the sun ever keep their fixed place in the sky; and how the mountains rose, and how the resounding rivers with their nymphs came into being and all creeping things.
> And he sang how first of all Ophion and Eurynome, daughter of Ocean, held the sway of snowy Olympus, and how through strength of arm one yielded his prerogative to Cronos and the other to Rhea, and how they fell into the waves of Ocean; but the other two meanwhile ruled

31

over the blessed Titan-gods, while Zeus, still a child and with the thoughts of a child, dwelt in the Dictaean cave; and the earth-born Cyclopes had not yet armed him with the bolt, with thunder and lightning; for these things give renown to Zeus.

He [Orpheus] ended, and stayed his lyre and divine voice.[43]

It is, of course, the first part of the creation story quoted above that is of particular interest. To emphasize the relationship of this story segment to the Genesis creation account, the two versions are superposed below. Written first are short segments of Apollonius's story shown in italics. These are followed immediately by the corresponding but longer segments from the abridged Genesis account given at the beginning of this section and shown in plain text:

He sang how the earth, the heaven and the sea, once mingled together in one form,
    In the beginning God created the heavens and the earth. The earth was without form and void, and darkness was upon the face of the deep; and the Spirit of God was moving over the face of the waters.

after deadly strife were separated each from other;
    And God said, "Let there be a firmament in the midst of the waters, and let it separate the waters from the waters." And God made the firmament and separated the waters which were under the firmament from the waters which were above the firmament. And it was so. And God called the firmament Heaven.

and how the stars and the moon and the paths of the sun ever keep their fixed place in the sky;

And God said, "Let there be lights in the firmament of the heavens to separate the day from the night; and let them be for signs and for seasons and for days and years, and let them be lights in the firmament of the heavens to give light upon the earth." And it was so. And God made the two great lights, the greater light to rule the day, and the lesser light to rule the night; he made the stars also. And God set them in the firmament of the heavens to give light upon the earth, to rule over the day and over the night, and to separate the light from the darkness.

and how the mountains rose,
And God said, "Let the waters under the heavens be gathered together into one place, and let the dry land appear." And it was so. God called the dry land Earth, and the waters that were gathered together he called Seas.

and how the resounding rivers with their nymphs came into being
And God said, "Let the waters bring forth swarms of living creatures, and let birds fly above the earth across the firmament of the heavens." So God created the great sea monsters and every living creature that moves, with which the waters swarm, according to their kinds, and every winged bird according to its kind.

and all creeping things.
And God said, "Let the earth bring forth living creatures according to their kinds: cattle and creeping things and beasts of the earth according to their kinds." And it was so. And God made the beasts of the earth according to their kinds and the cattle according to their kinds, and everything that creeps upon the ground according to its kind.

There are clearly significant similarities between the biblical creation story and the short creation account sung by Orpheus in the *Argonautica*. There could be various explanations for these similarities. One possibility is that Apollonius had actually read Genesis. He had a close association with the library at Alexandria when he was in Egypt before and after his visit to Rhodes. As it happens, the Pentateuch, or first five books of the Bible, were being translated from Hebrew into Greek for the library at approximately the time when Apollonius was there. The translation project eventually led to the Septuagint version of the Old Testament. It is not unreasonable to imagine that Apollonius would have been aware of this project and may have had an early look at the Greek version of Genesis.

Unfortunately, as mentioned above in Section 1.5, the Egyptians were often in disagreement with the Hebrews, and an example of this hostility has been noted previously in connection with corruptions of *The History of Egypt* written by the Egyptian historian Manetho,[44] who lived at about the same time as Apollonius. Thus, one could speculate that the inclusion of too favorable a representation of Hebrew history in the first version of the *Argonautica* may have led to the public disapproval of that work and the departure of Apollonius for Rhodes. The brief creation story of interest here may be a remnant from the first edition, with its following curiously-placed Greek mythology serving perhaps to defuse concerns of Greek readers.

There is a second possibility for the seeming similarity between the Genesis story of creation (which traditionally may have been edited by Moses in about 1450 B.C.[45]) and the story written by Apollonius twelve hundred years later in about 250 B.C. As a starting point here, a few more words need to be said about the origin of the *Argonautica*. The Argonaut story actually

1. In the Beginning

originated long before Apollonius, and his version of the creation account may have come from an earlier edition. Most records are now lost, but there remains an Argonaut story written by the Greek author Pindar in about 466 B.C.[46] Pindar also mentions Orpheus's participation in the Argonaut expedition. While Pindar's account lacks the creation story, other versions including that story may have been available in the library at Alexandria.

If the brief record of the creation of the universe ascribed to a song of the poet Orpheus was actually written by him, then one could enquire further whether Orpheus might somehow have obtained this account from Moses. Curiously, this question could have the simplest possible answer, as reported by Artapanus:

> He [Palmanothes, Amenophis I] begat a daughter Merris [Tumerisi], whom he betrothed to a certain Chenephres [Chebron, Akheperenre, Thutmose II], king of the regions above Memphis (for there were at that time many kings in Egypt); and she being barren took a supposititious child from one of the Jews, and called him Moüsos (Moses): but by the Greeks he was called, when grown to manhood, Musaeus. And this Moses, they said, was the teacher of Orpheus.[47]

In this quotation the inserted names in brackets represent royal identifications inferred previously.[48] If Artapanus was correct, then it is natural to consider the otherwise dubious-sounding possibility that Orpheus could have learned his creation story, as quoted by Apollonius, directly from Moses. This curious concept isn't quite as unlikely as it might seem. It will be suggested in Chapter 16 that the names Moses and Heracles may sometimes refer to the same person, and Orpheus and Heracles were shipmates on the expedition of the Argonauts.

The Garden And The Flood

In summary, the creation story included in the *Argonautica* has significant similarities to the creation story in Genesis. Assuming that these similarities are not merely a coincidence, several possible explanations have been proposed. Though the general form of the creation account in the *Argonautica* is similar to the more lengthy version in Genesis, there are of course differences in detail. For example, the Genesis story reports that dry land appeared in the midst of the seas (Genesis 1:9-10), while the *Argonautica* recalls "how the mountains rose." This difference seems less significant if one refers also to the creation statements in the Prayer of Moses in addition to the stories in Genesis 1: "Before the mountains were brought forth, or ever thou hadst formed the earth and the world, from everlasting to everlasting thou art God" (Psalm 90:2). Thus, Orpheus's singing of how "the mountains rose" is not so different from Moses's singing of how "the mountains were brought forth."

That poets such as Orpheus or Moses might think to compose songs about creation has more modern parallels, and following is a familiar creation hymn by Isaac Watts published in 1715:

I Sing the Mighty Power of God

Isaac Watts

   I sing the mighty pow'r of God, that made the mountains rise,
   That spread the flowing seas abroad, and built the lofty skies.
   I sing the wisdom that ordained the sun to rule the day;
   The moon shines full at His command, and all the stars obey.

1. In the Beginning

> I sing the goodness of the Lord, who filled the earth with food,
> Who formed the creatures through the Word, and then pronounced them good.
> Lord, how Thy wonders are displayed, where'er I turn my eye,
> If I survey the ground I tread, or gaze upon the sky.
>
> There's not a plant or flower below, but makes Thy glories known,
> And clouds arise, and tempests blow, by order from Thy throne;
> While all that borrows life from Thee is ever in Thy care;
> And everywhere that we can be, Thou, God art present there.

In this hymn as in the *Argonautica* the mountains are said to have risen.

A much later (and more doubtful) version of the *Argonautica* is actually attributed to Orpheus and is usually referred to as the *Orphic Argonautica*. In this work, in contrast to the statement of Artapanus, Musaeus was taught by Orpheus instead of the other way around. A few lines in the text that imply this relationship are quoted below:

> And I say to you, beloved Musaeus, son of Antiophemus, he [Jason] ordered me [Orpheus] to prepare quickly for an appropriate sacrifice.

> But, in truth, what is the reason for making a long tale of this, Musaeus, . . . They [the Argonauts] would have forgotten about the expedition had I [Orpheus] not called them back to the dark ship with my restraining words and soothing song, making them long for their oars and demand earnestly for resumption of their task.

The Garden And The Flood

> He took up the rudders in his hand, and steered the ship by the streams of Parthenius, which they call Callichorus. I [Orpheus] made mention of this to you, Musaeus, in a lofty conversation. Sailing by the outer headland, we came near the land of the Paphlagonians.
>
> Now I [Orpheus] will describe to you, O Musaeus, everything the Minyans [Argonauts] did and suffered: Argus, one of the bellicose sons of Phrixus whom Chaliope bore. . . .
>
> Of this story, intelligent Musaeus, you have heard: how once, Persephone's sisters led her through a great and wide wood, . . .[49]

1.7 Conclusion

One of the best-known stories in the Bible is the creation account at the beginning of Genesis, but other creation stories were also known in antiquity. Genesis is often considered to have been written or edited by Moses, but whether or not this is true some of the component stories could have dated from much earlier times. In this chapter the biblical creation stories have been compared to those from Babylon and Ebla, as well as a Greek record. The Babylonians left us a bizarre and fragmented creation account in their epic *Enûma elish*. The creation fragments in that polytheistic story seem to be secondary to other purposes. An earlier (c. 2500 B.C.) and more dignified creation account appears at the beginning of a record found at Ebla. That brief text has several features in common with the Hebrew story. The much later Greek account in the *Argonautica* could be based in part on the biblical version.

1. In the Beginning

1. In the absence of other indications, all biblical citations are to the Revised Standard Version (RSV) of the Bible.
2. *The Book of Jubilees or The Little Genesis*, Translated from the Ethiopic Text by R. H. Charles, with an Introduction by G. H. Box, (The Macmillan Company, New York, 1917).
3. "i.e. seven years," *The Book of Jubilees*, op. cit., p. 147, footnote 1.
4. B. E. Shafer, "Week," *The Interpreter's Dictionary of the Bible, An Illustrated Encyclopedia* (Abingdon Press, Nashville, 1976), Supplementary Volume, p. 946.
5. J. Skinner, *A Critical and Exegetical Commentary on Genesis* (Charles Scribner's Sons, New York, 1910), p. 10.
6. Ibid.
7. Ibid., p. ix.
8. A. Heidel, *The Babylonian Genesis, The Story of Creation*, Second Edition (The University of Chicago Press, Chicago, 1951), p. 1.
9. Ibid., p. 14.
10. Ibid., p. 14. At this point Heidel acknowledges provisionally accepting a chronology suggested by Sidney Smith, *Alalakh and Chronology* (London, 1940), p. 29. A lower chronology is often preferred today.
11. Ibid., p. 14. "The real beginning of Marduk's advancement dates to the reign of Hammurabi. . . ."
12. L. W. Casperson, *Patterns of Biblical Chronology* (Westbow Press, Bloomington, IN, 2012), p. 618, and references.
13. A. Heidel, op. cit., p. 9.
14. Ibid., pp. 10, 11.
15. Ibid., p. 130.
16. J. Skinner, op. cit., p. ix.
17. Ibid., pp. 178-179.
18. A. Heidel, op. cit., pp. 130-131.

19. Ibid., p. 131. "For a detailed presentation of his arguments see his works: A. T. Clay, *Amurru, the Home of the Northern Semites, A Study Showing that the Religion and Culture of Israel are Not of Babylonian Origin,* (The Sunday School Times Company, Philadelphia, 1909); A. T. Clay, *The Empire of the Amorites* (Yale University Press, New Haven, 1919); A. T. Clay, *A Hebrew Deluge Story in Cuneiform, and Other Epic Fragments in the Pierpont Morgan Library* (New Haven, 1922); A. T. Clay, *The Origin of Biblical Traditions: Hebrew Legends in Babylonia and Israel* (Yale University Press, New Haven, 1923)."
20. Ibid., "See especially, A. T. Clay, *Amurru, the Home of the Northern Semites,* op. cit., pp. 53 f."
21. A. Heidel, op. cit., pp. 139–140.
22. Ibid., p. 140. "Cf. J. Skinner, *A Critical and Exegetical Commentary on Genesis* (Charles Scribner's Sons, New York, 1910), pp. 6 f.; A. Dillmann, *Genesis, Critically and Exegetically Expounded*, Translated by W. B. Stevenson, In Two Volumes (Edinburgh, 1897), Volume I, p. 43."
23. Ibid., p. 24 (*Enûma elish*, Tablet 1, lines 147, 156-157) and p. 32 (Tablet 3, lines 38, 47-48).
24. "Urim and Thummim," *The Jewish Encyclopedia, A Descriptive Record of the History, Religion, Literature, and Customs of the Jewish People from the Earliest Times*, Isidore Singer, Managing Editor (KTAV Publishing House, Inc., New York, 1906), Volume 12, pp. 384-386; p. 385.
25. P. Matthiae, *Ebla, An Empire Rediscovered* (Doubleday and Company, Inc., Garden City, New York, 1981), p. 11.
26. "Ebla update, The known, the unknown and the debatable," *Biblical Archaeology Review*, Volume 6, Number 3, pp. 48-50 (May/June 1980).
27. Ibid.

28. L. W. Casperson, op. cit., Section 7.2, "Relative chronology," pp. 137-139; Section 9.4, "Manetho, etc.," pp. 194-210.
29. G. Pettinato, *The Archives of Ebla, An Empire Inscribed in Clay* (Doubleday and Company, Inc., Garden City, New York, 1981), p. 259.
30. Ibid., p. 260.
31. Ibid., p. 245.
32. See for example, A. Archi, "The epigraphic evidence from Ebla and the Old Testament," *Biblica*, Volume 60, Number 4, pp. 556-566 (1979); G. Pettinato, "Ebla and the Bible," *Biblical Archeologist*, Volume 43, Number 4, pp. 203-216 (Autumn, 1980).
33. E. H. Merrill, "Ebla and biblical historical inerrancy," in *Vital Apologetic Issues, Examining Reason and Revelation in Biblical Perspective*, R. B. Zuck, General Editor (Kregel Resources, Grand Rapids, MI, 1995), Chapter 13, pp. 177-193; pp. 189-190.
34. L. W. Casperson, op. cit., Section 2.2, "The Sabbatical Sequence," pp. 28-29; Section 8.5, "Similarities and a difference," pp. 175-180.
35. G. Pettinato, *The Archives of Ebla*, op. cit., p. 72.
36. L. W. Casperson, op. cit., Section 1.5, "The original authors," pp. 9-18.
37. M. Dahood, "Ebla, Ugarit, and the Bible," Afterword to G. Pettinato, *The Archives of Ebla, An Empire Inscribed in Clay* (Doubleday and Company, Inc., Garden City, New York, 1981), pp. 271-321; p. 316.
38. Ibid., pp. 317-318.
39. C. J. Chivers, "Grave robbers and war steal Syria's history," *New York Times*, p. A1, 7 April 2013.
40. *Apollonius Rhodius, The Argonautica*, with an English Translation by R. C. Seaton (Harvard University Press, Cambridge, Massachusetts, 1980), pp. vii-xiv.
41. Ibid., pp. 35, 37, 39; Book 1, lines 460-511.

42. Ibid., p. 5; Book 1, lines 26-27.
43. Ibid., pp. 37, 39; Book 1, lines 494-511.
44. L. W. Casperson, op. cit., Section 7.2, "Relative chronology," pp. 137-139; Section 9.4, "Manetho, etc.," pp. 194-210.
45. This date represents an approximate time between Moses's meeting the priest of Midian (Exodus 2:15-22) and the exodus from Egypt: L. W. Casperson, op. cit., Table 16.1, p. 408.
46. *The Odes Of Pindar Including The Principal Fragments*, With an Introduction and an English Translation by Sir John Sandys, The Loeb Classical Library, Edited by T. E. Page and W. H. D. Rouse (The Macmillan Co., New York, 1915), "Pythian IV, For Arcesilas Of Cyrene," pp. 196-231.
47. Eusebius, *Preparation for the Gospel*, Translated from a revised text by E. H. Gifford, Part 1 (Oxford at the Clarendon Press, 1903), Book 9, Chapter 27, p. 462; L. W. Casperson, op. cit., Appendix A: Artapanus, pp. 622-629; p. 624, paragraphs 7 and 8.
48. L. W. Casperson, op. cit., pp. 309-310 (Palmanothes = Amenophis I); pp. 312-317 (Merris = Tumerisi); pp. 318-321 (Chenephres = Thutmose II).
49. Pseudo-Orpheus, *The Orphic Argonautica*, English Translation by Jason Colavito (2011), five brief excerpts quoting Orpheus teaching Musaeus.

2. ADAM AND EVE

"So God created man in his own image, in the image of God he created him; male and female he created them" (Genesis 1:27).

2.1 Introduction

The previous chapter has considered some of the implications of the biblical creation story and it's possible relationship to other creation accounts from the Middle East. The purpose of this chapter is to consider more closely some of the people and events mentioned in the Bible near the time of creation. The first people were, of course, Adam and Eve and their immediate descendants. Several stories about these people will be recalled here, including information from a few less conventional sources.

The creation of Adam and Eve is discussed in Section 2.2, together with mention of their expulsion from the garden of Eden. Some curiosities in the text about a seeming plurality associated with references to God are also noted. For comparison the creation story from Jubilees is considered in Section 2.3. In contrast to the Genesis account, Jubilees mentions explicitly that angels were created on the first day to serve God. Their existence may help to clarify other indications of multiple heavenly beings.

The location of the creation of Adam and Eve is considered in Section 2.4.

The familiar story of the temptation of Adam and Eve in the garden of Eden is considered briefly in Section 2.5, and some of the possible reasons for the murder of Abel by his brother Cain are considered in Section 2.6. As discussed in Section 2.7, the angels seem sometimes to be referred to as sons of God. Like human beings they were capable of serious sins. Such fallen angels experienced God's wrath, and the Bible indicates that human beings who engage in similar sins will also suffer punishment. Early examples of such fallen humans occurred already among Adam and Eve and their first descendants. Polytheism presented a different challenge to the ancestors of the Hebrews, and some considerations of this topic are included in Section 2.8.

2.2 Creation of Adam and Eve

The question of where human beings came from has intrigued people of many cultures since the earliest recorded times. As indicated already in Section 1.1 and summarized in Chapters 18 and following, it is the view of many scientists and others today that the universe, including all of its animate and inanimate components, somehow created itself. This view is, of course, required if the existence of a creator is disallowed. Thus, for atheists (those who don't believe in the existence of God), the idea of the universe with its inhabitants creating itself would seem to be a necessary placeholder for a more detailed but as-yet-unknown explanation.

2. Adam and Eve

On the other hand, the Bible tells us that after his creation of the universe with its varied plants and animals on earth God created human beings:

> Then God said, "Let **us** make man in **our** image, after **our** likeness; and let them have dominion over the fish of the sea, and over the birds of the air, and over the cattle, and over all the earth, and over every creeping thing that creeps upon the earth." So God created man in his own image, in the image of God he created him; male and female he created them (Genesis 1:26-27).

These brief statements include the essential biblical answer to the question of the origin of human beings.

Besides providing a brief explanation of human origins, however, the above quotation also suggests a mystery concerning the nature of God. Thus, in the biblical creation account, as represented by Genesis 1:26 above, the use of the plural words **us** and **our** (set in boldface for emphasis in the above and following quotations) might seem to imply a multiplicity of gods. Polytheism was prevalent at the times of the writings of many early cultures such as those of Babylon, Ebla, Egypt, and Greece mentioned in Chapter 1. Thus, one could consider the possibility that a source document or tradition used during the original development and writing of the Genesis creation story included more than one supernatural being. However, the idea of a multiplicity of gods seems not to be the best interpretation of the available texts: "The plural **us, our** probably refers to the divine beings who surround God in his heavenly court and in whose image man was made."[1]

A similar suggestion of the plurality of God occurs in the story of the expulsion of Adam and Eve from the garden:

> Then the Lord God said, "Behold, the man has become like one of **us**, knowing good and evil; and now, lest he put forth his hand and take also of the tree of life, and eat, and live for ever" – therefore the Lord God sent him forth from the garden of Eden, to till the ground from which he was taken. He drove out the man; and at the east of the garden of Eden he placed the cherubim, and a flaming sword which turned every way, to guard the way to the tree of life (Genesis 3:22-24).

Yet another such instance occurs in the story of the tower of Babel:

> And the Lord came down to see the city and the tower, which the sons of men had built. And the Lord said, "Behold, they are one people, and they have all one language; and this is only the beginning of what they will do; and nothing that they propose to do will now be impossible for them. Come, let **us** go down, and there confuse their language, that they may not understand one another's speech." So the Lord scattered them abroad from there over the face of all the earth, and they left off building the city (Genesis 11:5-8).

A much later parallel may be found in the story of the commissioning of the prophet Isaiah:

> And I heard the voice of the Lord saying, "Whom shall I send, and who will go for **us**?" (Isaiah 6:8).

2.3 Jubilees account of creation

It was suggested in the previous section that the words **us** and **our** in Genesis 1:26 and elsewhere in the Bible could sometimes be understood to imply that in some ancient source documents there may have been a pantheon consisting perhaps of a multiplicity of supernatural beings. This same ambiguity is indicated in Jubilees,[2] except that in Jubilees there is clearly only one God, and the other subordinate supernatural beings are identified explicitly as spirits or angels. Interestingly, the responsibilities of many of the angels in Jubilees (the angels of fire, wind, clouds, light, darkness, snow, thunder and lightning, animals, etc.) are not so different from those of gods in some polytheistic religions. Here, however, it would seem that words like **us**, **we**, and **our** in Genesis or Jubilees may be intended to reflect either God's previously created angels or the association of God with those angels. Thus, an earlier version of Genesis may, like Jubilees, have included a statement about the creation of angels and their assigned responsibilities.

For completeness, a Jubilees reference to the origin of angels, who were created on the first day to serve the Lord, is included below. The frequent plurality of God with his angels is also indicated in the succeeding paragraphs:

> For on the first day He created the heavens which are above and the earth and the waters and all the spirits which serve before Him – the angels of the presence, and the angels of sanctification, and the angels [of the spirit of fire and the angels] of the spirit of the winds, and the angels of the spirit of the clouds, and of darkness, and of snow and of hail and of hoar frost, and the angels of the voices and of the thunder and of the lightning, and the

angels of the spirits of cold and of heat, and of winter and of spring and of autumn and of summer, and of all the spirits of His creatures which are in the heavens and on the earth, (He created) the abysses and the darkness, eventide (and night), and the light, dawn and day, which He hath prepared in the knowledge of His heart. And thereupon **we** saw His works, and praised Him, and lauded before Him on account of all His works; for seven great works did He create on the first day (Jubilees 2:2-3).

And He gave **us** a great sign, the Sabbath day, that **we** should work six days, but keep Sabbath on the seventh day from all work. And all the angels of the presence, and all the angels of sanctification, these two great classes — He hath bidden **us** to keep the sabbath with Him in heaven and on earth. And He said unto **us**: "Behold, I will separate unto Myself a people from among all the peoples, and these will keep the Sabbath day, and I will sanctify them unto Myself as my people, and will bless them; as I have sanctified the Sabbath day and do sanctify (it) unto Myself, even so shall I bless them, and they will be My people and I shall be their God (Jubilees 2:17-19).

And thus He created therein a sign in accordance with which they should keep Sabbath with **us** on the seventh day, to eat and to drink, and to bless Him who hath created all things as He hath blessed and sanctified unto Himself a peculiar people above all peoples, and that they should keep Sabbath together with **us** (Jubilees 2:21).

And every one who observeth it and keepeth Sabbath thereon from all his work, will be holy and blessed throughout all days like unto **us**. Declare and say to the children of Israel the law of this day both that they should keep Sabbath thereon, and that they should not forsake it in the error of their hearts; (and) that it is not lawful to do any work thereon which is unseemly, to do thereon their own pleasure, and that they should not prepare thereon anything to be eaten or drunk, and (that it is not lawful) to draw water, or bring in or take out thereon through their gates any burden, which they had not prepared for themselves on the sixth day in their dwellings. And they shall not bring in nor take out from house to house on that day; for that day is more holy and blessed than any jubilee day of the jubilees: on this **we** kept Sabbath in the heavens before it was made known to any flesh to keep Sabbath thereon on the earth (Jubilees 2:28-30).

And on the six days of the second week **we** brought, according to the word of God, unto Adam all the beasts, and all the cattle, and all the birds, and everything that moveth on the earth, and everything that moveth in the water, according to their kinds, and according to their types: the beasts on the first day; the cattle on the second day; the birds on the third day; and all that which moveth on the earth on the fourth day; and that which moveth in the water on the fifth day (Jubilees 3:1).

As quoted above, Jubilees 2:18 suggests that the two greatest classes of angels may include "all the angels of the presence, and all the angels of sanctification." Among the angels of the presence may be Gabriel: "I am Gabriel, who stand in the presence of God" (Luke 1:19). Concerning the creation of Eve, the Genesis and Jubilees accounts are very similar:

Genesis
And the Lord God said, It is not good that the man should be alone; I will make him an help meet for him (Genesis 2:18, KJV).

Jubilees

And the Lord said unto **us**: "It is not good that the man should be alone: let **us** make a helpmeet for him" (Jubilees 3:4).

In the Genesis version the Lord's audience is not identified, while in Jubilees the first **us** would seem to refer to the angels while the second **us** might include the Lord together with his angels. The similarity of these Genesis and Jubilees statements suggests that they were not written entirely independently of each other.

Comparable interpretations are implied in the Jubilees version of the tower of Babel story:

And the Lord our God said unto **us**: "Behold, they are one people, and (this) they begin to do, and now nothing will be withholden from them. Go to, let **us** go down and confound their language, that they may not understand one another's speech, and they may be dispersed into cities and nations, and one purpose will no longer abide with them till the day of judgment." And the Lord descended, and **we** descended with Him to see the city and the tower which the children of men had built. And He confounded their language, and they no longer understood one another's speech, and they ceased then to build the city and the tower (Jubilees 10:22-24).

Again, the first **us** and the **we** seem to refer to the angels, while the second **us** seems to refer to God together with his angels.

2.4 Where in creation

One of the primary subjects of Sections 2.2 and 2.3 has been the creation of Adam and Eve. Before concluding that topic, it may be appropriate to say what little can be said about where that creation is considered to have taken place. It seems often to be understood that Adam and Eve were created in the garden of Eden. Perhaps surprisingly, the early biblical texts do not support that understanding. The first Genesis statement, quoted previously, doesn't suggest any location for this creation:

> Then God said, "Let **us** make man in **our** image, after **our** likeness; and let them have dominion over the fish of the sea, and over the birds of the air, and over the cattle, and over all the earth, and over every creeping thing that creeps upon the earth." So God created man in his own image, in the image of God he created him; male and female he created them (Genesis 1:26-27).

Subsequent statements on the creation of Adam indicate explicitly that after he was created he was brought to the garden from somewhere else:

> [T]hen the Lord God formed man of dust from the ground, and breathed into his nostrils the breath of life; and man became a living being. And the Lord God planted a garden in Eden, in the east; and there he put the man whom he had formed. . . . The Lord God took the man and put him in the garden of Eden to till it and keep it (Genesis 2:7-8, 15).

Perhaps the most that could be said based on these words about the location of the garden is either that it was east of the author of the text and his expected readership or possibly that it was east of where Adam was created.

The location of the creation of Eve is also not specific:

> Then the Lord God said, "It is not good that the man should be alone; I will make him a helper fit for him.". . .

The Garden And The Flood

> So the Lord God caused a deep sleep to fall upon the man, and while he slept took one of his ribs and closed up its place with flesh; and the rib which the Lord God had taken from the man he made into a woman and brought her to the man (Genesis 2:18, 21-22).

Thus Eve is said to have been created somewhere from one of Adams ribs, and then she was brought to him. In the Jubilees version of the creation story, it is very clear that neither Adam nor Eve were created in the garden of Eden:

> And after Adam had completed forty days in the land where he had been created, **we** brought him into the Garden of Eden to till and keep it, but his wife **they** brought in on the eightieth day, and after this she entered into the Garden of Eden (Jubilees 3:9).

An explicit statement on the placement of Adam and Eve in the garden is given in *Jasher*: "And the Lord God took Adam and his wife, and he placed them in the garden of Eden to dress it and to keep it; . . ."[3]

In short, it isn't obvious from Bible-related references where Adam and Eve were actually considered to have been created. The implication would seem to be that they originated at some distance from the garden. Either they were created somewhere else as stated in the Bible (with Jubilees and *Jasher*), or in a non-biblical speculation they could have been relocated to the garden area as natural descendants of other human beings who lived elsewhere. For the present the emphasis here is, of course, on the biblical story. The creation of the universe and its inhabitants is the event most remote in time from our everyday experience, and its detailed interpretation does not necessarily affect the meaning or importance of later biblical messages. More substantial information is available on the location of the garden of Eden, and that subject will be considered in Section 4.4.

2.5 Temptation of Adam and Eve

The first story in the Bible after the creation accounts that relates to the early human beings is the story of the temptation and disobedience of Adam and Eve. This subject will also be recalled in later discussions, and for reference several verses are quoted here:

> And the Lord God planted a garden in Eden, in the east; and there he put the man whom he had formed. And out of the ground the Lord God made to grow every tree that is pleasant to the sight and good for food, the tree of life also in the midst of the garden, and the tree of the knowledge of good and evil (Genesis 2:8-9).
>
> The Lord God took the man and put him in the garden of Eden to till it and keep it. And the Lord God commanded the man, saying, "You may freely eat of every tree of the garden; but of the tree of the knowledge of good and evil you shall not eat, for in the day that you eat of it you shall die (Genesis 2:15-17).
>
> Now the serpent was more subtle than any other wild creature that the Lord God had made. He said to the woman, "Did God say, 'You shall not eat of any tree of the garden'?" And the woman said to the serpent, "We may eat of the fruit of the trees of the garden; but God said, 'You shall not eat of the fruit of the tree which is in the midst of the garden, neither shall you touch it, lest you die.' " But the serpent said to the woman, "You will not die. For God knows that when you eat of it your eyes will be opened, and you will be like God, knowing good and evil." So when the woman saw that the tree was good for

The Garden And The Flood

food, and that it was a delight to the eyes, and that the tree was to be desired to make one wise, she took of its fruit and ate; and she also gave some to her husband, and he ate (Genesis 3:1-7).

The story goes on to describe God's reaction to the disobedience of Adam and Eve. Severe punishments were assigned to the serpent, to Eve, and to Adam (Genesis 3:14-19). Then they were expelled from the garden:

> Then the Lord God said, "Behold, the man has become like one of us, knowing good and evil; and now lest he put forth his hand and take also of the tree of life, and eat, and live for ever" – therefore the Lord God sent him forth from the garden of Eden, to till the ground from which he was taken. He drove out the man; and at the east of the garden of Eden he placed the cherubim, and a flaming sword which turned every way, to guard the way to the tree of life (Genesis 3:22-24).

The natures and identities of the supernatural trees in the midst of the garden have long been debated. It is possible, of course, that the supernatural aspects were gradually introduced by creative writers over many successive transcriptions of the stories. For example, it is said of the tree of the knowledge of good and evil that "in the day that you eat of it you shall die" (Genesis 2:17). The Genesis story provides no example of that actually happening. On the contrary, Adam and Eve both ate of the tree of the knowledge of good and evil, and they both seem to have lived normal (or longer than normal) lifetimes. However, there may well have been toxic plants within the garden of Eden. Thus, poison hemlock is widespread over the world and was used to poison Socrates in Greece. Poison hemlock and yew, for example, are both found in the United States and also near the likely site of the garden of Eden. It would seem sensible that Adam and Eve would have been warned of poisonous trees, and in any case they seem not to have ingested a lethal dose.

On the other hand, the tree of life is said to have had the potential effect of allowing Adam and Eve to live forever. However, the Genesis stories provide no indication of that happening. Thus, one might be tempted to assume that there is no evidence for any plant that could be considered a tree of life. On the contrary however, it was discovered in more modern times (several centuries ago) that certain plants have the ability to compensate for dietary deficiencies and restore otherwise terminally ill patients to full health. From their discovery, these plants were named tree of life, and they also grow near the likely location of the garden of Eden. This subject is considered further in Section 11.7.

If the garden of Eden were a real place, then one might hope to find additional tangible indications of its existence. Other information on the temptation story is available from outside of the Bible and related texts. A possible illustration is provided by an early cylinder seal, and a picture of the seal's impression is shown in Fig. 2.1.[4] This seal has sometimes been interpreted to show a representation of Adam and Eve facing a fruit-bearing tree in the garden of Eden. In the background is a large serpent that could be tempting Eve to eat from the tree of the knowledge of good and evil (Genesis 3:1-6).

The Garden And The Flood

Figure 2.1 Impression of a cylinder seal believed to be from the general region of Mesopotamia. (Reproduced with permission)

The cylinder seal has been said to date from about 2200 to 2100 B.C. As will be suggested later, this date could be before the flood of Noah's time, but after the time that seems to be implied by the texts for the occupation of what came to be called the garden of Eden. Other images that seem to represent Adam and Eve and the serpent will be considered in Section 5.2.

What Adam and Eve ate in the garden of Eden and after their departure from that garden may be of some interest. The Genesis account informs us that the garden included many fruit trees, all but one of which (the tree of the knowledge of good and evil) they were welcome to use as a food source. The hanging fruit in Fig. 2.1 could perhaps be interpreted as figs. In any case, the Bible informs us that fig trees were present in the garden. After Adam and Eve had sinned they are said to have sewed fig leaves together to make clothing for themselves (Genesis 3:7). The use of figs for food in the vicinity of the garden of Eden is also mentioned many times in the non-biblical writing sometimes

2. Adam and Eve

called *The First Book of Adam and Eve*. Verses in this book, listed by chapter and verse, that mention figs one or more times include the following:[5] 36:1,2,4,6; 37:1,3; 38:4; 39:1,2,3,4; 41:1,2; 43:11; 61:18; 62:2,3,4; 63:1,3,4,5,6,7,8; 64:1,2,4,7,8; 66:5,8. The location of the garden will be of recurring interest in these investigations. It would seem that, at least during the time of Adam and Eve, fig trees grew abundantly in the vicinity of the garden of Eden. Given that background, it should also be noted for future consideration that the site of the ancient garden of Eden seems to be located in what is now the country of Turkey: "Turkey is the number one producer of both fresh and dried figs, holding more than twenty percent of the world's total output."[6]

2.6 Cain and Abel
While life in the garden of Eden sounds appealing, one might wonder what the first inhabitants did to keep themselves occupied. Perhaps not surprisingly, they were required to care for the garden:

> The Lord God took the man and put him in the garden of Eden to till it and keep it (Genesis 2:15).

> And in the first week of the first jubilee, Adam and his wife were in the Garden of Eden for seven years tilling and keeping it, and **we** gave him work and **we** instructed him to do everything that is suitable for tillage. And he tilled (the garden), and was naked and knew it not, and was not ashamed, and he protected the garden from the birds and beasts and cattle, and gathered its fruit, and ate, and put aside the residue for himself and for his wife [and put aside that which was being kept] (Jubilees 3:15-16).

The Garden And The Flood

As in earlier sections of this chapter, the plural word **we** as associated with God and his angels is written in boldface.

Several chronologies have been given for the time period shortly after creation, but Jubilees, as quoted above, suggests perhaps that Adam and Eve may have lived in the garden for about seven years. After their transgression, they were expelled from the garden. Then they continued in similar farming activities, but life was not as easy as it had been:

> And to Adam he said, "Because you have listened to the voice of your wife, and have eaten of the tree of which I commanded you, 'You shall not eat of it,' cursed is the ground because of you; in toil you shall eat of it all the days of your life; thorns and thistles it shall bring forth to you; and you shall eat the plants of the field. In the sweat of your face you shall eat bread till you return to the ground, for out of it you were taken; you are dust, and to dust you shall return." (Genesis 3:17-19).

Another early Bible story involves Cain and Abel, the first children of Adam and Eve. In this version Cain continued in the occupation of his father:

> Now Abel was a keeper of sheep, and Cain a tiller of the ground. In the course of time Cain brought to the Lord an offering of the fruit of the ground, and Abel brought of the firstlings of his flock and of their fat portions. And the Lord had regard for Abel and his offering, but for Cain and his offering he had no regard. So Cain was very angry, and his countenance fell. The Lord said to Cain, "Why are you angry, and why has your countenance fallen? If you do well, will you not be accepted? and if you do not do well, sin is couching at the door; its desire is for you, but you must master it."

> Cain said to Abel his brother. "Let us go out to the field." And when they were in the field, Cain rose up against his brother Abel, and killed him (Genesis 4:2-8).

Thus, Cain murdered Abel because of his anger that God didn't approve of his offering. However, there seems not to be a clear indication here of what was unsatisfactory about the offering.

In an enhanced version of this story, *Jasher* suggests an additional aspect of the conflict between Cain and Abel:

> And in some time after, Cain and Abel his brother went one day into the field to do their work; and they were both in the field, Cain tilling and ploughing his ground, and Abel feeding his flock; and the flock passed that part which Cain had ploughed in the ground, and it sorely grieved Cain on this account. And Cain approached his brother Abel in anger, and he said unto him, what is there between me and thee that thou comest to dwell and bring thy flock to feed in my land. And Abel answered his brother Cain and said unto him, what is there between me and thee, that thou shalt eat the flesh of my flock and clothe thyself with their wool? And now therefore, put off the wool of my sheep with which thou hast clothed thyself, and recompense me for their fruit and flesh which thou has eaten, and when thou shalt have done this, I will then go from thy land as thou hast said.[7]

The blame for the conflict here is shared by both parties. The idea seems to be that Abel's sheep had been wandering onto Cain's land, where they presumably trampled the ground and ate the crops. On the other hand, Cain had been killing Abel's sheep, eating their flesh and making clothing of their wool. It may be recalled that part of Adam's responsibility, as quoted earlier in this section, was also to protect his crops: "[A]nd he protected the garden from the birds and beasts and cattle. . ." (Jubilees 3:16).

It is not clear how the *Jasher* version of this story should actually be associated with the biblical version, but the concepts described are not unrealistic. Conflicts between farmers (of crops) and ranchers (of sheep or cattle) have probably occurred since the very introduction of these occupations. The old nursery rhyme *Little Boy Blue* may inform us of this problem: "The sheep's in the meadow, the cow's in the corn." A light-hearted song intended to represent the early days of the American West also addresses the same conflict:

> The farmer and the cowman should be friends,
> Oh, the farmer and the cowman should be friends.
> One man likes to push a plough, the other likes to chase a cow,
> But that's no reason why they cain't be friends.[8]

The context of this song appears to be as in *Jasher* that early farmers wanted to protect their crops, while ranchers preferred to be able to move their livestock over an open range.

In yet another version of the Cain and Abel story, one reason for the conflict was the age-old concept of jealousy over a beautiful woman. The story is given in two slightly different forms and the main parts of both are included here:[9]

> Then Adam said to Eve, "Behold the children are grown up;[10] we must think of finding wives for them." Then Eve answered, "How can we do it?" Then Adam said to her, "We will join Abel's [twin] sister in marriage to Cain, and Cain's [twin] sister to Abel." Then said Eve to Adam, "I do not like Cain because he is hardhearted;

but let them bide until we offer up unto the Lord in their behalf." And Adam said no more.

Meanwhile Satan came to Cain in the figure of a man of the field, and said to him, "Behold Adam and Eve have taken counsel together about the marriage of you two; and they have agreed to marry Abel's sister [Aklia][11] to thee, and thy sister [Luluwa][12] to him. . . . At these words of Satan Cain opened his ears, and leant towards his speech. And he did not remain in the field, but he went to Eve, his mother, and beat her, and cursed her, and said to her, "Why are ye about taking my sister to wed her to my brother? Am I dead?" His mother, however, quieted him, and sent him to the field where he had been. Then when Adam came, she told him of what Cain had done. But Adam grieved and held his peace, and said not a word.

An additional short segment on this same subject is the following:

But as to hard-hearted Cain, Satan came to him by night, showed himself and said unto him, "Since Adam and Eve love thy brother Abel much more than they love thee, and wish to join him in marriage to thy beautiful [twin] sister, because they love him; but wish to join thee in marriage to his ill-favoured [twin] sister, because they hate thee; "Now, therefore, I counsel thee, when they do that, to kill thy brother; then thy sister will be left for thee; and his sister will be cast away."[13]

Besides providing a different explanation for Cain's murder of Abel, this story may suggest some mistakes that parents can make. Showing favoritism to one child with respect to another is unlikely to have desirable consequences, and failure to discipline an errant child can lead to tragedy (Adam "said not a word").

The Garden And The Flood

The biblical explanation for Cain's anger, as quoted above, was that God didn't accept Cain's offering (Genesis 4:2-7). An expanded and modified version of this same explanation is also included in *Adam and Eve*.[14]

> Then on the morrow Adam said unto Cain his son, "Take of thy sheep, young and good, and offer them up unto thy God; and I will speak to thy brother, to make unto his God an offering of corn." They both hearkened to their father Adam, and they took their offerings, and offered them up on the mountain by the altar.
>
> But Cain behaved haughtily towards his brother, and thrust him from the altar, and would not let him offer up his gift upon the altar; but he offered his own upon it, with a proud heart, full of guile, and fraud. But as for Abel, he set up stones that were near at hand, and upon that, he offered up his gift with a heart humble and free from guile.
>
> Cain was then standing by the altar on which he had offered up his gift; and he cried unto God to accept his offering; but God did not accept it from him; neither did a divine fire come down to consume his offering. But he remained standing over against the altar, out of humour and wroth, looking towards his brother Abel, to see if God would accept his offering or not. And Abel prayed unto God to accept his offering. Then a divine fire came down and consumed his offering.[15] And God smelled the sweet savour of his offering; because Abel loved Him and rejoiced in Him. And because God was well pleased with him He sent him an angel of light in the figure of man who had partaken of his offering, because He had smelled the sweet savour of his offering, and they comforted Abel and strengthened his heart.
>
> But Cain was looking on all that took place at his brother's offering, and was wroth on account of it. Then

2. Adam and Eve

he opened his mouth and blasphemed God, because He had not accepted his offering. But God said unto Cain, "Wherefore is thy countenance sad? Be righteous, that I may accept thy offering. Not against Me hast thou murmured, but against thyself."

For the most part, the general flow of this story is much like that of the biblical version. A few extra details on the offering process and its aftermath are provided, but the main difference is in the offerings themselves. In the Bible as quoted above, "Abel was a keeper of sheep, and Cain a tiller of the ground" (Genesis 4:2). Cain's offering was of "the fruit of the ground," while Abel's offering was of "the firstlings of his flock" (Genesis 4:3), and the same was true in *Jasher*. However, in Adam and Eve, as quoted immediately above, these occupations are reversed: Cain raised sheep while Abel raised corn (or wheat).

A further indication of what was wrong with Cain's offering is provided in a report of events after the offerings of Cain and Abel:

> Then they came down from the altar, and went to the cave in which they dwelt. But Abel, by reason of his joy at having made his offering, repeated it three times a week, after the example of his father Adam. But as to Cain, he took no pleasure in offering; but after much anger on his father's part, he offered up his gift once; and when he did offer up, his eye was on the offering he made, and he took the smallest of his sheep for an offering, and his eye was again on it.[16]

2.7 Fallen angels

As a last note here on the subject of the creation of people and angels, it may be noted that sometimes the angels are referred to as sons of God. Sometimes also these angels (or sons of God) fell victim to temptation:

> When men began to multiply on the face of the ground, and daughters were born to them, the sons of God saw that the daughters of men were fair; and they took to wife such of them as they chose. . . . The Nephilim were on the earth in those days, and also afterward, when the sons of God came in to the daughters of men, and they bore children to them. These were the mighty men that were of old, the men of renown (Genesis 6:1-2, 4).

The fallen angels are said to have incurred God's wrath:

> And it came to pass when the children of men began to multiply on the face of the earth and daughters were born unto them, that the angels of God saw them on a certain year of this jubilee, that they were beautiful to look upon; and they took themselves wives of all whom they chose, and they bare unto them sons and they were giants [Nephilim]. . . . And against the angels whom He [God] had sent upon the earth, He was exceedingly wroth, and He gave commandment to root them out of all their dominion, and He bade us to bind them in the depths of the earth, and behold they are bound in the midst of them, and are (kept) separate. And against their sons went forth a command from before His face that they should be smitten with the sword, and be removed from under heaven (Jubilees 5:1,6-7).

This event is also mentioned in the New Testament as one of several examples of God's judgment of the ungodly:

> For if God did not spare the angels when they sinned, but cast them into hell and committed them to pits of nether gloom to be kept until the judgment; if he did not spare the ancient world, but preserved Noah, a herald of righteousness, with seven other persons, when he brought a flood upon the world of the ungodly; if by turning the cities of Sodom and Gomor'rah to ashes he condemned them to extinction and made them an example to those who were to be ungodly; and if he rescued righteous Lot, greatly distressed by the licentiousness of the wicked (for by what that righteous man saw and heard as he lived among them, he was vexed in his righteous soul day after day with their lawless deeds), then the Lord knows how to rescue the godly from trial, and to keep the unrighteous under punishment until the day of judgment, and especially those who indulge in the lust of defiling passion and despise authority (2 Peter 2:4-10).

> And the angels that did not keep their own position but left their proper dwelling have been kept by him in eternal chains in the nether gloom until the judgment of the great day; just as Sodom and Gomor'rah and the surrounding cities, which likewise acted immorally and indulged in unnatural lust, serve as an example by undergoing a punishment of eternal fire (Jude 6-7).

Besides the good angels and the fallen angels mentioned above, the sons of God seem also to have included Satan: "Now there was a day when the sons of God came to present themselves before the Lord, and Satan also came among them" (Job 1:6).

2.8 Polytheism

Polytheism was a recurring problem for the ancient Israelites and their forebears. The first mention of polytheism in these writings occurred in Section 1.5 and concerned the recently discovered empire of Ebla in the area now known as Syria. The thousands of cuneiform tablets that have been discovered there have, among other things, revealed much about the religion of the Eblaites in the third millennium B.C. The official religion as reflected in a creation story would seem to have been a form of monotheism, while the religion of the masses has been referred to as "rank polytheism" having a pantheon of about 500 divinities.

The concept of polytheism arose again at the beginning of this chapter in the creation stories of Genesis (Section 2.2) and Jubilees (Section 2.3). The frequent use of the plural words **us**, **our**, and **we** in association with God could be interpreted to mean that there were considered to be many gods at the time of creation. However, it was suggested in this chapter that the plural forms probably refer to angels acting as a group or acting together with God. Actual polytheism involving the worship of multiple gods seems to have developed later in the biblical story.

Widespread polytheism among the ancestors of the Hebrews is reported in the Bible to have begun while those ancestors were dwelling among the polytheistic inhabitants of Ur of the Chaldees:

> Then Joshua gathered all the tribes of Israel to Shechem, and summoned the elders, the heads, the judges, and the officers of Israel; and they presented themselves before God. And Joshua said to all the people, "Thus says the Lord, the God of Israel, 'Your fathers lived of old beyond the Euphra'tes, Terah, the father of Abraham and of Nahor; and they served other gods (Joshua 24:1-2).

2. Adam and Eve

"Now therefore fear the Lord, and serve him in sincerity and in faithfulness; put away the gods which your fathers served beyond the River, and in Egypt, and serve the Lord. And if you be unwilling to serve the Lord, choose this day whom you will serve, whether the gods your fathers served in the region beyond the River, or the gods of the Amorites in whose land you dwell; but as for me and my house, we will serve the Lord" (Joshua 24:14-15).

Based on the approximate analysis in Section 12.4 and summarized in Table 12.2, Terah the father of Abram may have been born in roughly 1783 B.C.

According to Jubilees, the initial tendency in Ur toward idolatry may have occurred long before Terah. Thus, Reu, the great-grandfather of Abram's father Terah was probably involved in polytheism, and this occurrence was during the earlier presence of ancestors of the Hebrews in Ur:

> And in the thirty-fifth jubilee, in the third week, in the first year thereof, Reu took to himself a wife, and her name was 'Ôrâ, the daughter of 'Ur, the son of Kêsêd, and she bare him a son, and he called his name Sêrôh [also called Serug], in the seventh year of this week in this jubilee. . . . And 'Ur, the son of Kêsêd, built the city of 'Arâ [Ur] of the Chaldees, and called its name after his own name and the name of his father. And they made for themselves molten images, and they worshipped each the idol, the molten image which they had made for themselves, and they began to make graven images and unclean simulacra [statues of gods], and malignant spirits assisted and seduced (them) into committing transgression and uncleanness. And the prince Mastêmâ [apparently Satan] exerted himself to do all this, and he sent forth

other spirits, those which were put under his hand, to do all manner of wrong and sin, and all manner of transgression, to corrupt and destroy, and to shed blood upon the earth. For this reason he called the name of Sêrôh, Serug, for every one turned to do all manner of sin and transgression. And he grew up, and dwelt in Ur of the Chaldees, near to the father of his wife's mother, and he worshipped idols and he took to himself a wife in the thirty-sixth jubilee, in the fifth week, in the first year thereof, and her name was Mêlkâ, the daughter of Kâbêr, the daughter of his father's brother. And she bare him Nahor, in the first year of this week, and he grew and dwelt in Ur of the Chaldees, and his father taught him the researches of the Chaldees to divine and augur, according to the signs of heaven. And in the thirty-seventh jubilee, in the sixth week, in the first year thereof, he took to himsel a wife, and her name was 'Îjâskâ, the daughter of Nêstâg of the Chaldees. And she bare him Terah in the seventh year of this week (Jubilees 11:1, 3-10).

Thus Reu was an ancestor of Terah and may have been born in roughly 1874 B.C. (Table 12.2). The descendants of Terah are of particular interest:

Now these are the descendants of Terah. Terah was the father of Abram, Nahor, and Haran; and Haran was the father of Lot. Haran died before his father Terah in the land of his birth, in Ur of the Chalde'ans. And Abram and Nahor took wives; the name of Abram's wife was Sar'ai, and the name of Nahor's wife, Milcah, the daughter of Haran the father of Milcah and Iscah. Now Sar'ai was barren; she had no child (Genesis 11:26-30).

While Abram was still a child in Ur he is said to have abandoned the idolatry of his family, and he chose only to obey

2. Adam and Eve

the Lord. A few words on Abram's religion-related conclusions may be found in Jubilees. Abram recognized the uselessness of idols, and an appealing story of that recognition follows:

> And the child [Abram] began to understand the errors of the earth that all went astray after graven images and after uncleanness, and his father taught him writing, and he was two weeks of years old, and he separated himself from his father, that he might not worship idols with him. And he began to pray to the Creator of all things that He might save him from the errors of the children of men, and that his portion should not fall into error after uncleanness and vileness (Jubilees 11:16-17).
>
> And it came to pass in the sixth week, in the seventh year thereof, that Abram said to Terah his father, saying, "Father!" And he said, "Behold, here am I, my son." And he said, "What help and profit have we from those idols which thou dost worship, and before which thou dost bow thyself? For there is no spirit in them, for they are dumb forms, and a misleading of the heart. Worship them not: Worship the God of heaven, who causeth the rain and the dew to descend on the earth, and doeth everything upon the earth, and hath created everything by His word, and all life is from before His face. Why do ye worship things that have no spirit in them? For they are the work of (men's) hands, and on your shoulders do ye bear them, and ye have no help from them, but they are a great cause of shame to those who make them, and a misleading of the heart to those who worship them: worship them not."
> And his father said unto him, "I also know it, my son, but what shall I do with a people who have made me to serve before them? And if I tell them the truth, they will slay me; for their soul cleaveth to them to worship them and honour them. Keep silent, my son, lest they slay thee" (Jubilees 12:1-7).

The Garden And The Flood

Eventually, Terah took his family away from Ur of the Chaldees to bring them to the land of Canaan, where they apparently could worship as they pleased. Abram had successfully kept himself from idolatry in contrast to other members of Terah's family. However, the entire family didn't get as far as Canaan:

> Terah took Abram his son and Lot the son of Haran, his grandson, and Sar'ai his daughter-in-law, his son Abram's wife, [and Nahor his son and Milcah his daughter-in-law, his son Nahor's wife,] and they went forth together from Ur of the Chalde'ans to go into the land of Canaan; but when they came to Haran, they settled there (Genesis 11:31).

Also, it will be noted below that the idolatry probably practiced by Abram's brother Nahor in Ur of the Chaldees seems to have been continued by his son Bethuel and his grandson Laban in Haran (Genesis 22:20-24, 28:1-5).

After the death of Terah, Abram took his wife Sar'ai, his nephew Lot, and his servants and cattle and moved southwest from Haran to Canaan:

> Now the Lord said to Abram, "Go from your country and your kindred and your father's house to the land that I will show you. And I will make of you a great nation, and I will bless you, and make your name great, so that you will be a blessing. I will bless those who bless you, and him who curses you I will curse; and by you all the families of the earth shall bless themselves."
>
> So Abram went, as the Lord had told him; and Lot went with him. Abram was seventy-five years old when

he departed from Haran. And Abram took Sar'ai his wife, and Lot his brother's son, and all their possessions which they had gathered, and the persons that they had gotten in Haran; and they set forth to go to the land of Canaan (Genesis 12:1-5).

However, there was a famine in the land, and this caused Abram in about 1727 B.C. to take his family and possessions on a few-year detour into Egypt. The chronology and some of the activities of Abram and his family in Egypt are considered in Chapter 13.

After Abram's return from Egypt to Canaan, he became the father of Ishmael and Isaac, and then Isaac had a family of his own:

> These are the descendants of Isaac, Abraham's son: Abraham was the father of Isaac, and Isaac was forty years old when he took to wife Rebekah, the daughter of Bethu'el the Aramean of Paddan-aram, the sister of Laban the Aramean. And Isaac prayed to the Lord for his wife, because she was barren; and the Lord granted his prayer, and Rebekah his wife conceived. . . . When her days to be delivered were fulfilled, behold, there were twins in her womb. The first came forth red, all his body like a hairy mantle; so they called his name Esau. Afterward his brother came forth, and his hand had taken hold of Esau's heel; so his name was called Jacob (Genesis 25:19-21, 24-26).

> Then Isaac called Jacob and blessed him, and charged him, "You shall not marry one of the Canaanite women. Arise, go to Paddan-aram to the house of Bethu'el your mother's father; and take as wife from there one of the daughters of Laban your mother's brother" (Genesis 28:1-2).

The Garden And The Flood

Through Laban's intrigue Jacob eventually had four wives including Laban's daughter Leah and her maid Zilpah as well as Laban's daughter Rachel and her maid Bilhah (Genesis 29:15-30:13). Similarly, Abram's first two wives had included Sarai (later renamed Sarah) and her maid Hagar (Genesis 16:1-6). Eventually, Jacob sought to take his family and the livestock that he had accumulated in Haran and return to Canaan. But Jacob didn't trust Laban, so he left in secret. This story provides an early example of how polytheism would continue to haunt the Hebrews. Unknown to Jacob, Rachel had stolen Laban's household gods, and Laban especially wanted to have those back:

> And Laban overtook Jacob. . . . And Laban said to Jacob, "What have you done, that you have cheated me, and carried away my daughters like captives of the sword? . . . And now you have gone away because you longed greatly for your father's house, but why did you steal my gods?" Jacob answered Laban, "Because I was afraid, for I thought that you would take your daughters from me by force. Any one with whom you find your gods shall not live. In the presence of our kinsmen point out what I have that is yours, and take it." Now Jacob did not know that Rachel had stolen them (Genesis 31:25, 26, 30-32).

Laban didn't find the idols, and Jacob and his family and possessions were able to continue on to Canaan.

As Jacob and his family made their way south through Canaan, however, the subject of idol worship came up once again:

> God said to Jacob, "Arise, go up to Bethel, and dwell there; and make there an altar to the God who appeared to

2. Adam and Eve

you when you fled from your brother Esau." So Jacob said to his household and to all who were with him, "Put away the foreign gods that are among you, and purify yourselves, and change your garments; then let us arise and go up to Bethel, that I may make there an altar to the God who answered me in the day of my distress and has been with me wherever I have gone." So they gave to Jacob all the foreign gods that they had, and the rings that were in their ears; and Jacob hid them under the oak which was near Shechem (Genesis 35:1-4).

A few years later (c. 1625 B.C.) a famine brought Jacob and his family to Egypt, even as the earlier famine had brought Abram and his family. This later visit lasted until the exodus in 1440 B.C. Unfortunately, in the longer visit the Hebrews may have absorbed some of the idol-worshiping traits of the Egyptians. This is suggested by the people's making and worshiping of the golden calf when Moses seemed to delay in coming down from Mount Sinai. The Egyptians worshiped live Apis bull calves throughout much of their history, and images of such calves are common. This Hebrew idolatry incident will be quoted in more detail in Section 5.5.

Other incidents of idolatry and polytheism are also recorded, and a few of these may be mentioned from later in the history of the Hebrews. A conspicuous example is provided by King Solomon and his wives:

> He had seven hundred wives, princesses, and three hundred concubines; and his wives turned away his heart. For when Solomon was old his wives turned away his heart after other gods; and his heart was not wholly true to the Lord his God, as was the heart of David his father. For Solomon went after Ash'toreth the goddess of the

The Garden And The Flood

Sido'nians, and after Milcom the abomination of the Ammonites. So Solomon did what was evil in the sight of the Lord, and did not wholly follow the Lord, as David his father had done. Then Solomon built a high place for Chemosh the abomination of Moab, and for Molech the abomination of the Ammonites, on the mountain east of Jerusalem. And so he did for all his foreign wives, who burned incense and sacrificed to their gods (1 Kings 11:3-8).

The polytheism introduced by King Solomon did not cease with his death. Following that death, Rehoboam his son reigned in his stead. Then a tax dispute caused the northern kingdom of Israel under the leadership of Jeroboam to separate from the southern kingdom of Judah under Rehoboam. Both kingdoms continued in idolatry:

And Jerobo'am said in his heart, "Now the kingdom will turn back to the house of David; if this people go up to offer sacrifices in the house of the Lord at Jerusalem, then the heart of this people will turn again to their lord, to Rehobo'am king of Judah, and they will kill me and return to Rehobo'am king of Judah." So the king took counsel, and made two calves of gold. And he said to the people, "You have gone up to Jerusalem long enough. Behold your gods, O Israel, who brought you up out of the land of Egypt." And he set one in Bethel, and the other he put in Dan (1 Kings 12:26-29).

The wording of this story is remarkably similar to the story of the golden calf in the exodus account. In that story Aaron is said to have made a golden calf, but from the plural forms used (indicated in boldface in the following quotations) one could wonder whether Aaron, like Jeroboam, had actually made two calves:

> When the people saw that Moses delayed to come down from the mountain, the people gathered themselves together to Aaron, and said to him, "Up make us **gods**, who shall go before us; as for this Moses, the man who brought us up out of the land of Egypt, we do not know what has become of him" (Exodus 32:1).

> [They] said, "These are your **gods**, O Israel, who brought you up out of the land of Egypt!" (Exodus 32:4, 8).

> And Aaron said, "Let not the anger of my lord burn hot; you know the people, that they are set on evil. For they said to me, 'Make us **gods**, who shall go before us; as for this Moses, the man who brought us up out of the land of Egypt, we do not know what has become of him'" (Exodus 32:22-23).

> So Moses returned to the Lord and said, "Alas, this people have sinned a great sin; they have made for themselves **gods** of gold" (Exodus 32:31).

It also seems remarkable that Jeroboam could have considered it to be a good idea to repeat the idol-making project of Aaron that had been condemned by the Lord in the strongest possible terms: "And the Lord said to Moses, 'I have seen this people, and behold, it is a stiff-necked people; now therefore let me alone, that my wrath may burn hot against them and I may consume them; but of you I will make a great nation'" (Exodus 32:9,10). "And the sons of Levi did according to the word of Moses; and there fell of the people that day about three thousand men" (Exodus 32:28).

The Exodus story of the golden calf includes many instructions concerning idolatrous behavior that the Hebrews were ordered not to engage in. Jeroboam's flagrant disobedience to most or all of those instructions could nearly be interpreted as a mocking of the traditional worship practices and the Exodus lessons. Some of the real or potential transgressions at the time of the exodus are listed below, and Jereboam is said explicitly to have committed many of these when he made his own golden calves about five hundred years later:

> You shall have no other gods before me (Exodus 20:3).
>
> You shall not make for yourself a graven image, or any likeness of anything that is in heaven above, or that is in the earth beneath, or that is in the water under the earth (Exodus 20:4).
>
> [Y]ou shall not bow down to them or serve them (Exodus 20:5).
>
> You shall not make gods of silver to be with me, nor shall you make for yourselves gods of gold (Exodus 20:23).
>
> Whoever sacrifices to any god, save to the Lord only, shall be utterly destroyed (Exodus 22:20).
>
> Take heed to all that I have said to you; and make no mention of the names of other gods, nor let such be heard out of your mouth (Exodus 23:13).
>
> [Y]ou shall not bow down to their gods, nor serve them, nor do according to their works, but you shall utterly overthrow them and break their pillars in pieces (Exodus 23:24).

2. Adam and Eve

You shall make no covenant with them or with their gods (Exodus 23:32).

They shall not dwell in your land, lest they make you sin against me; for if you serve their gods, it will surely be a snare to you (Exodus 23:33).

When the people saw that Moses delayed to come down from the mountain the people gathered themselves together to Aaron, and said to him, "Up, make us gods, who shall go before us (Exodus 32:1).

And he received the gold at their hand, and fashioned it with a graving tool, and made a molten calf (Exodus 32:4).

[T]hey have made for themselves a molten calf, and have worshiped it and sacrificed to it (Exodus 32:8).

So Moses returned to the Lord and said "Alas, this people have sinned a great sin; they have made for themselves gods of gold (Exodus 32:31).

[F]or you shall worship no other god, for the Lord, whose name is Jealous, is a jealous God (Exodus 34:14).

You shall make for yourself no molten gods (Exodus 34:17).

Do not turn to idols or make for yourselves molten gods: I am the Lord your God (Leviticus 19:4).

The polytheism introduced by King Solomon also thrived under his son Rehoboam:

> Now Rehobo'am the son of Solomon reigned in Judah. . . . And Judah did what was evil in the sight of the Lord, and they provoked him to jealousy with their sins

> which they committed, more than all that their fathers had done. For they also built for themselves high places, and pillars, and Ashe'rim on every high hill and under every green tree; and there were also male cult prostitutes in the land. They did according to all the abominations of the nations which the Lord drove out before the people of Israel (1 Kings 14:21-24).

Following this non-promising beginning to worship in the divided kingdom, the following reigns varied widely in their attitude toward polytheism.

Elijah was an important prophet in the northern kingdom of Israel during the ninth century B.C. He was instrumental in replacing the worship of the Canaanite god Baal by the traditional worship of the biblical Lord or Yahweh. Baal worship was defended by King Ahab (c. 874/73-853 B.C.[17]) and also especially by his wife Queen Jezebel. Elijah's efforts were continued by those of his disciple Elisha. The southern kingdom of Judah also experienced times of religious reform, but the most important of these times wasn't until the leadership of King Josiah (c. 637/36-605 B.C.[18]):

> In the late 8th century both Judah and Israel were vassals of Assyria. Israel rebelled, and was destroyed c. 722 B.C. Refugees fleeing to Judah brought with them a number of new traditions (new to Judah, at least). One of these was that the god Yahweh, already known and worshiped in Judah, was not merely the most important of the gods, but the only god who should be served. This outlook influenced the Judahite landowning elite, who became extremely powerful in court circles after they placed the eight-year-old Josiah on the throne following the murder of his father, Amon of Judah.

By the eighteenth year of Josiah's reign, Assyrian power was in rapid deline, and a pro-independence movement gathered strength in the court. This movement expressed itself in a state theology of loyalty to Yahweh as the sole god of Israel.[19]

Following the exile the Israelites began increasingly to accept and declare the fact of monotheism that had been emphasized already in the first commandment: "You shall have no other gods before me" (Exodus 20:3).

2.9 Conclusion

In the stories of the Bible, the first human beings were Adam and Eve and their descendants. This chapter has examined several aspects of the lives of these people, and one of the dominant themes could be referred to as sin and punishment. Adam and Eve succumbed to Satan's tempting and were consequently expelled from their paradise in the garden of Eden. Cain was jealous of his brother Abel. That jealousy led to Cain's murder of Abel, and as a result Cain fled from the presence of his parents. Even some of the angels committed sins with humans, and they too were punished. A long-lasting difficulty for the proper biblically-endorsed worship practices of the Hebrews was the persistent recurrence of polytheism.

---

1. *The Oxford Annotated Bible, Revised Standard Version,* Edited by H. G. May and B. M. Metzger (Oxford University Press, 1962), p. 2, comment on Genesis 1:26.
2. *The Book of Jubilees or Little Genesis,* Translated from the Ethiopic text by R. H. Charles, With an introduction

by G. H. Box (The Macmillan Company, 1917). As noted in Section 1.3D of *Patterns of Biblical Chronology,* The *Book of Jubilees* is included in the biblical canon of the Ethiopian Orthodox Church, and it will usually be abbreviated here as Jubilees. This book is also well represented among the Dead Sea Scrolls.
3. *The Book of Yashar,* Translated from the Hebrew and Published by Mordecai Manuel Noah (Hermon Press, New York, 1972), (Cited here and below in the form *Jasher*); *Jasher* 1:7.
4. Heidel, Alexander, *The Babylonian Genesis* (University of Chicago Press, Chicago, 1951), Figure 17.
5. *Adam and Eve,* Translated by S. C. Malan from the Ethiopic edition edited by E. Trumpp, included in *The Forgotten Books of Eden,* Edited by Rutherford H. Platt, Jr. (Alpha House, Inc., 1927).
6. "What's in season: fig," *The Guide Istanbul,* 12 August 2016, http://www.theguideistanbul.com/article/whats-in-season-fig, Web. 19 June 2017.
7. *Jasher* 1:17-20.
8. R. Rogers and O. Hammerstein II, "The farmer and the cowman," from the musical *Oklahoma* (1943).
9. *Adam and Eve,* op. cit., Book 1, 78:1-6, 11-15.
10. *Adam and Eve*, op. cit., Book 1, 77:9. "Cain was fifteen years old, and Abel twelve years old."
11. *Adam and Eve*, op. cit., Book 1, 75:11.
12. *Adam and Eve*, op. cit., Book 1, 74:6-8, 75:12.
13. *Adam and Eve*, op. cit., Book 1, 76:10-11.
14. Adam and Eve, op. cit., Book 1, 78:16-26.
15. The divine fire that consumed Abel's offering is reminiscent of the fire that came down from God to consume Elijah's offering (and the altar of stones) in the time of King Ahab (1 Kings 18:30-40).
16. *Adam and Eve*, op. cit., Book 1, 77:6-7.

17. L. W. Casperson, Patterns of Biblical Chronology (Westbow Press, Bloomington, Indiana, 2012), Table 3.2, p. 50.
18. L. W. Casperson, op. cit., Table 3.4, p. 55.
19. "Book of Deuteronomy," Wikipedia. Web. 6 March 2017.

3. FROM ADAM TO NOAH

"This is the book of the generations of Adam." . . . "When Adam had lived a hundred and thirty years, he became the father of a son in his own likeness, after his image, and named him Seth . . ." (Genesis 5:1,3).

3.1 Introduction

The preceding chapter has summarized some of the activities of people mentioned in the Bible near the time of creation. For chronological purposes, it is important to seek connections between the creation stories and later datable biblical events. It is possible that the named people in even the earliest stories in Genesis were real human beings with ordinary life spans, and the purpose of this chapter is to begin to estimate when they might have lived. This procedure involves using the biblical genealogies to connect the earliest people to datable historical events.

Unfortunately, the genealogies sometimes prove to be difficult to employ with confidence. While the arrangements of the names in these genealogies are highly consistent among versions, the time periods associated with each generation vary widely. Besides uncertainties with the recorded time intervals, it

3. From Adam to Noah

is sometimes difficult to find historical events that are both datable and connectable to the genealogies. In fact, for this period in Bible history only a limited number of significant and potentially datable events are even mentioned. For the named people in the genealogy from Adam to Noah, the only chronological information typically provided in the Bible is their age at the birth of an important child and their age at death (Genesis 5). These indicated ages sometimes seem implausibly high, and there is no way to know with certainty which of them (if any) might be literally correct.

The most obvious early event to consider for dating purposes is the flood associated with Noah. The physical interpretation of that major incident will be of recurring interest and is deferred to following chapters. If the cause or consequences of a major flood episode could be dated by some means and identified with Noah's flood, then there would be a possibility of obtaining an absolute date associated with the life of Noah. Rough estimates of other dates of the patriarchal era could then be attempted using the not-always-consistent biblical genealogies.

For later use, the genealogies from Adam to Noah are reviewed in Section 3.2 considering the Masoretic (or Jewish Masoretic), Samaritan (or Israelite Samaritan), and Septuagint versions of the Pentateuch. A differently formatted genealogy expressed in terms of the sabbatical and jubilee calendars is provided by Jubilees. It is shown in Section 3.3 that the Jubilees genealogy agrees closely with the genealogy of the Samaritan version of the Bible. The possible historical development of the various biblical chronologies for this era is reviewed in Section

The Garden And The Flood

3.4. The Bible-related genealogies of *Jasher* and Josephus's *Antiquities of the Jews* are considered in Section 3.5.

3.2 Biblical genealogy

In Genesis 5-10 the Bible has provided us with a detailed genealogy of the time period from Adam to the beginning of the flood, and the biblical data is summarized in this section. Many of the time intervals in this genealogy seem very large by modern standards, and a further concern is that these periods are not consistent between the various versions of the Bible. The standard biblical numbers are compared in this section, and the numbers from Jubilees will be considered in Section 3.3.

A tabulation of the ages of the antediluvian patriarchs at the births of their sons is given in Table 3.1 according to the Masoretic, Samaritan, and Septuagint versions of the Bible. All of these patriarchs are said to have lived for two or more centuries after the births of their sons. While these years are interesting (though doubtful), the numbers given explicitly in Table 3.1 are the stated ages at the births of their sons. These ages, if meaningful, would be the most important for considering the overall pre-flood patriarchal time interval. In calculating the time period from the creation of Adam to the start of the flood, it should be noted that Noah was said to have been six hundred years old at the time of the flood (Genesis 7:6,11). Including the patriarchs before Noah, the flood should have occurred in the years shown. On the other hand, Shem is said to have been one hundred years old two years after the flood, when Arpach'shad was born (Genesis 11:10). Thus, Shem must have been born when Noah was five hundred and two years old. This result is also stated explicitly in *Jasher*:[1] "And Noah was five hundred and two years old when Naamah [his wife] bare Shem. . . ."

Table 3.1. Biblical chronology of the antediluvian patriarchs from Adam to the beginning of the flood. The numbers shown are the indicated ages of the patriarchs at the births of their sons.

Patriarch	Verses	Masoretic	Samaritan	Septuagint
Adam	Genesis 5:3	130	130	230
Seth	Genesis 5:6	105	105	205
Enosh	Genesis 5:9	90	90	190
Kenan	Genesis 5:12	70	70	170
Mahalalel	Genesis 5:15	65	65	165
Jared	Genesis 5:18	162	62	162
Enoch	Genesis 5:21	65	65	165
Methuselah	Genesis 5:25	187	67	167
Lamech	Genesis 5:28	182	53	188
Noah	Genesis 5:32	502	502	502
Shem	Genesis 11:10	100	100	100
Year of world at flood	Genesis 7:6,11	1656	1307	2242

3.3 Jubilees genealogy

Besides the data shown in Table 3.1, another important resource for information on biblical chronology is Jubilees.[2] As an introduction to the time-keeping system in Jubilees, some comments by Box may be helpful:[3]

> It is obvious that Jubilees is dominated by certain interests and antipathies. It is to a large extent polemical

in character, and its author desires at once to protest against certain tendencies which, in his view, threaten true religion, and to inculcate certain reforms. Incidentally it commends certain religious practices, and endeavours to invest them with enhanced sanctions. In the forefront, as its name (*"The Book of Jubilees"*) suggests,[4] stands the question of the Calendar. It is all important in the author's view that the divinely ordained principle according to which history is divided up by year-weeks (i.e., periods of 7 years) and Jubilees (i.e., periods of 7×7 years) is recognized (cf. Jubilees Chapter 1, verse 26 f.). Accordingly, he gives a history from Creation to Moses, in which the sequence of events is recorded and dated exactly by Jubilee-periods, or portions of such. This leads up to a final section in which the law respecting jubilees and sabbatical years is solemnly enjoined. The writer's aim seems to have been nothing less than a reformation of the Jewish Calendar. The prevailing system has led to the nation "forgetting" new moons, festivals, and sabbaths (and (?) jubilees);[5] in other words, it has produced grave irregularities in the observance of matters which were of divine obligation.

    A cardinal feature of the writer's system is the jubilee-period, which consists of 7×7 (i.e. 49) years. Here we are confronted with a difficulty. The passage in Leviticus (25:8-14) which ordains the observance of the jubilee-year expressly identifies this, in the present form of the text, with the fiftieth year (Leviticus 25:10,11). But it is incredible that the author of our Book would deliberately have violated the express injunctions of the Pentateuch on such a matter, and we are driven to conclude that he had a text before him in which the word "fiftieth" was absent.[6] The wording of verses 8 and 9 is ambiguous, and allows of the explanation that the jubilee-year was the forty-ninth and not the fiftieth. It is quite possible that in verses 10 and 11 "fiftieth" has been added to the text, in the

3. From Adam to Noah

interests of the rival explanation that ultimately prevailed, for, as has been pointed out already, our Book presupposes a text of the Pentateuch that is independent of and earlier than M.T. [Masoretic text]. This explanation suffers from the difficulty that the LXX [Septuagint text] and other ancient versions (including the Samaritan text) support the currently received reading. But it is not improbable that on such a matter the influence of orthodox views may have operated to bring their text of the verses into harmony with the currently accepted theory.[7]

The use in Jubilees of the forty-nine year interval may explain why that book is not included in most Bibles, in spite of the fact that it was found among the Dead Sea Scrolls and was probably well known to the Jews of that era.

As a first step in a consideration of the chronology of Jubilees, it is of interest to compare the Jubilees version of the period from Adam to the flood with the biblical version summarized in Table 3.1. In undertaking such a comparison, one quickly finds that the chronology of Jubilees agrees quite well with the chronology given in the Samaritan Pentateuch. Agreement with the Masoretic and Septuagint versions is much less satisfactory. Table 3.2 shows first the Jubilees chronology from the margins of the Charles translation[8] in its original jubilee-based form. Thus, the third column lists the actual year of the world at which a patriarch's chronologically most important son was born, with the jubilee/sabbatical format converted to years from the creation of the world. For convenience in comparing with biblical data, the fourth column is derived from the third and shows the age of each patriarch at the birth of his son. The fifth or last column shows the biblical age of the

The Garden And The Flood

patriarchs at the births of their sons according the Samaritan Pentateuch, as given previously in Table 3.1.

Table 3.2. Comparison of the Jubilee chronology and the Samaritan chronology for the patriarchs from Adam to the flood.

Patriarch	Verses	Year of world	Jubilees age	Samaritan age
Adam	Jubilees 4:7	130	130	130
Seth	Jubilees 4:11	235	105	105
Enosh	Jubilees 4:13	325	90	90
Kenan	Jubilees 4:14	395	70	70
Mahalalel	Jubilees 4:15	461	66	65
Jared	Jubilees 4:16	522	61	62
Enoch	Jubilees 4:20	587	65	65
Methuselah	Jubilees 4:27	652	65	67
Lamech	Jubilees 4:28	701-707	49-55	53
Noah	Jubilees 4:33	1207	506-500	502
Shem	Jubilees 7:18	1310	103	100
Year of world at flood	Jubilees 5:23	1308	1308	1307

A few further remarks may be appropriate concerning the numbers shown in Table 3.2. The actual Jubilees data for the year of the world in which Lamech's wife gave birth to her son Noah were not specified precisely, and only the date range as shown was given. This uncertainty leads to the correlated ambiguities for Lamech and Noah at the births of their sons. Also, the date at which Shem's wife gave birth to their son Arpachshad is not explicitly stated. Rather, Arpachshad is said in Jubilees 7:18 to have been born two years after the flood, and the flood is said to have begun in 1308. Thus, the year 1310 as shown may have been the author's intention concerning the birthdate.

The discrepancies between the numbers in columns four and five are seen to be very small, and the overall period from Adam to the flood may differ by only about one year between the two versions. In contrast, it may be seen from Table 3.1 that there are many individual differences of a hundred years or more between the Samaritan numbers and those of the Masoretic or Septuagint versions. The overall period from Adam to the flood is more than three hundred years higher in the Masoretic version than in the Samaritan version, while the Septuagint total is more than nine hundred years higher than the Samaritan total. These results suggest that the origins of the chronological elements of the Jubilees and Samaritan documents may in some way be related.

3.4 Genealogy of chronologies

Many suggestions have been made as to the origin and authorship of Jubilees. The results shown in Table 3.2 raise the question of why the Jubilees and Samaritan chronologies for this period should be so similar, and this question has long been of interest. In the simplest interpretations it would seem that either the Jubilees account for this particular period is a chronological embellishment of the Samaritan version (at a time when that version still had forty-nine-year jubilee periods), or alternatively that the Samaritan Pentateuch version adapted the Jubilees chronology for substitution into a version that was otherwise similar to what became the Septuagint or Masoretic Bible. In this regard it may be mentioned that the idea of Jubilees having a Samaritan author was suggested long ago by Beer.[9] Beer's conclusions were not always widely accepted. Thus, for example, it has been said that "Beer, with arguments that are extensive and erudite but not convincing, ascribes it [Jubilees] to a Samaritan author."[10] The apparently close relationship shown above between the chronologies of the Samaritan Pentateuch and Jubilees is consistent with the idea that Jubilees might have been based, at least in part, on an earlier Samaritan version of the Pentateuch. This would also, of course, be compatible with the idea that Jubilees may have had a Samaritan author.

The alternative possibility is that a source for Jubilees might predate the Samaritan Pentateuch. As noted previously in Sections 2.2 and 2.3, Genesis 1:26 and other verses seem to suggest that when God created man he may have had participation by subordinate spirits or angels. On the other hand, in Jubilees 2:2 the close association of God with his angels at creation is noted explicitly.

A related question involves the relationship between the Samaritan Pentateuch (SP), the Septuagint (LXX), and the Masoretic Text (MT). Recent detailed analyses of differences between early manuscripts of these versions and in comparison to the Dead Sea Scrolls have yielded important results:

> This leads to the conclusion that the translators of the LXX had before them, among many of the most ancient manuscripts of the Pentateuch up until their time (the third century B.C.E.), texts that were closer in content to the SP and were likely similar to those found in Qumran Cave 4, written in the same script that Jewish writers called "Samaritan Hebrew." Scholars call these manuscripts "Proto-Samaritan" or "Pseudo-Samaritan." These texts from Qumran, and undoubtedly the earlier

texts that were in the hands of the LXX translators, are the earliest texts of the Pentateuch in use in ancient times that are known today. Thus, the SP presents the earliest known text of the Pentateuch.[11]

In spite of the limitations of our information on the authorship and original date of early Bible-related texts, it is tempting to at least sketch a possible genealogy of some of the variations considered here, and such a sketch is shown below in Table 3.3. It has been suggested in Section 1.2 that the earliest source document for the creation stories (labelled 1. Original Pentateuch in Table 3.3 and identified there with Moses) may not have included the concept that the creation events must be fitted to a seven-day week. Rather, the seven-day sequence may have been added to the creation story in a new version (labelled 2a. Modified Pentatuch in the table) to enhance the antiquity and authority of the seven-day week, together with its sabbath day requirements. Alternatively, Jubilees (labelled 2b. Jubilees in the table) could predate and have influenced the writing of the modified or proto-Samaritan Pentateuch.

Table 3.3. Chronological patterns in the creation stories and in the commandments of Exodus and Leviticus, starting at the top from a possible form ascribed there to Moses.

Version	Creation Stories (Genesis, Jubilees 2)	Commandments (Exodus, Leviticus, Jubilees 50)
1. Original Pentateuch (Moses)		7 day week 7 year sabbatical 49 year jubilee
2a. Modified Pentateuch (Proto-Samaritan)	7 day week	7 day week 7 year sabbatical 49 year jubilee
preceded or followed by		
2b. Jubilees	7 day week 7 year sabbatical 49 year jubilee finished work on sixth day	7 day week 7 year sabbatical 49 year jubilee
3. Later Pentateuchs	7 day week	7 day week 7 year sabbatical 50/49 year jubilee
(Samaritan, Septuagint) (Masoretic)	finished work on sixth day finished work on seventh day	

In a similar way it has been suggested in this section that the author of Jubilees may have taken the seven-year sabbatical period and forty-nine year jubilee period from the Bibles of which he was aware (Proto-Samaritan) and added a discussion of those periods to the creation stories and commandments of his new book (labelled 2b. Jubilees in the table). Later editors may have attempted to replace the forty-nine year period in the Modified Pentateuch with a fifty-year period, as suggested above in Section 3.3.

3. From Adam to Noah

The chronological content in the creation stories indicated in Table 3.3 refers mostly to Genesis, and the commandments refer mostly to Exodus and Leviticus. The idea that the seven-day week might have provided the chronological framework for the creation events may have been introduced with either 2a. Modified Pentateuch (Proto-Samaritan) or 2b. Jubilees. The book Deuteronomy, the last book of the Pentateuch, is usually thought to have been composed during about the reign of Josiah, many centuries after the death of Moses.[12]

The Masoretic text of Genesis differs from earlier versions in suggesting that God finished his work on the seventh day of creation prior to beginning his rest. This discrepancy can be seen from a comparison of the Samaritan and Septuagint versions of Genesis 2:2 with the Masoretic version. The idea of God working on the seventh day was mentioned previously in Section 1.3. The three versions of Genesis 2:2 are written below in the translations by Tsedaka[13] and Brenton[14] with the day numbers in boldface. A related Jubilees account is also included,[15] as well as a statement by Josephus:

Samaritan text (Tsedaka):
And in the **sixth** day Eloowwem completed His work which He had done. And He rested on the **seventh** day from all His work which He had done (Genesis 2:2).

Septuagint text (Brenton):
And God finished on the **sixth** day his works which he made, and he ceased on the **seventh** day from all his works which he made (Genesis 2:2).

Masoretic text (Tsedaka):
And on the **seventh** Elohim finished His work which He

had made; and He rested on the **seventh** day from all His work which He had made (Genesis 2:2).

Jubilees text (Charles):
And he finished all His work on the **sixth** day – all that is in the heavens and on the earth, and in the seas and in the abysses, and in the light and in the darkness, and in everything. And He gave us a great sign, the Sabbath day, that we should work **six** days, but keep Sabbath on the **seventh** day from all work (Jubilees 2:16-17).

Josephus text (Whiston)
Accordingly Moses says, That in just **six** days the world, and all that is therein, was made. And that the **seventh** day was a rest, and a release from the labour of such operations; whence it is that we celebrate a rest from our labours on that day, and call it the sabbath, which word denotes rest in the Hebrew tongue.[16]

The Septuagint text is the translation of the Old Testament used by the earliest Christians and still in use by the Eastern Orthodox churches. The Bibles of most western churches are based on the Hebrew or Masoretic text, which states that God finished his work on the seventh day. This text was subjected to editing as late as the tenth century A.D.:

The standard version of the Old Testament is the Biblia Hebraica Stuttgartensis, published in Stuttgart, Germany, in 1966-77, the fourth edition of a version edited by Rudolf Kittel in 1902. The biblical text is that of the Leningrad Codex, which dates to 1010, making it the oldest complete Hebrew Bible in existence, but it also incorporates notes on variant readings, including some from the Dead Sea Scrolls.[17]

3. From Adam to Noah

It may be noted that the Masoretic creation text of Genesis is also different from the corresponding texts of Exodus:

And on the **seventh** day God finished his work which he had done, and he rested on the seventh day from all his work which he had done (Genesis 2:2).

[I]n **six** days the Lord made heaven and earth, the sea, and all that is in them, and rested the **seventh** day (Exodus 20:11).

[I]n **six** days the Lord made heaven and earth, and on the **seventh** day he rested, and was refreshed (Exodus 32:17).

3.5 Other Genealogies

Besides the genealogies found among the books and versions of the Bible, there are other Bible-related texts that also include variations of the same genealogies. A tabulation of the ages of the antediluvian patriarchs at the births of their sons based on *Jasher* is given in Table 3.4.[18] The spellings of the names of the patriarchs vary slightly between sources, and the spellings used previously in Table 3.1 are retained here. The ages in this table are mostly identical to the corresponding ages summarized in Table 3.1 based especially on the Masoretic and Samaritan versions of the Bible. However, the age for Jared in Table 3.4 (62 years) agrees with the Samaritan version for Jared while being less than in the Masoretic version by one hundred years. On the other hand, the age of Methuselah in Table 3.4 (87 years) is less than in the Masoretic version for Methuselah by one hundred years. The age for Lamech in Table 3.4 (approximately 180 years) is close to the age in the Masoretic version. No value is given by *Jasher* for the age of Shem at the birth of Arpachshad.

Table 3.4. Approximate chronology of the antediluvian patriarchs from Adam to the beginning of the flood according to *Jasher*. The numbers shown are the ages of the patriarchs at the births of their sons.

Patriarch	Verse	Age
Adam	*Jasher* 2:1	130
Seth	*Jasher* 2:2	105
Enosh	*Jasher* 2:10	90
Kenan	*Jasher* 2:15-16	70
Mahalalel	*Jasher* 2:37	65
Jared	*Jasher* 2:37	62
Enoch	*Jasher* 3:1	65
Methuselah	*Jasher* 3:13	87
Lamech	*Jasher* 4:11,13	180
Noah	*Jasher* 5:18	502
Shem	*Jasher* 7:15,19	?
Year of world at flood	*Jasher* 6:13	1454

3. From Adam to Noah

A few further words might be helpful here concerning *Jasher*. While this work has been referred to a few times in this study, elsewhere it is sometimes dismissed as being of no historical value. That it might contain useful material is supported by many arguments, including the fact that a work of this name is mentioned explicitly in the Bible. Thus, the text in Joshua 10:12-13 concerns a battle led by Joshua in which he commanded the sun to stand still until the battle was won. This story is said in the Bible to be found in *Jasher*, and indeed the same text is found in *Jasher* 88:63-64. Similarly, in 2 Samuel 1:18 (KJV) King David quotes from *Jasher* Jacob's command that the children of Judah should be taught the use of the bow. The source for this quote is *Jasher* 56:9.

A further indication of the possible importance of *Jasher* relates to the chronologies that have been reviewed in Section 3.4 above. Those discussions have included the idea that the chronology of the Samaritan Pentateuch may be older and more reliable than the chronologies given in the Masoretic or Septuagint Pentateuchs. With that background, it is of interest now to compare those chronologies to the chronology given by *Jasher* and shown in Table 3.4. It is perhaps significant that the ages of the first seven patriarchs (from Adam through Enoch) at the births of their sons as recorded in the Samaritan Pentateuch are exactly the same as the corresponding ages reported by *Jasher*. This agreement can be interpreted to support the antiquity of both the Samaritan and *Jasher* versions.

Another tabulation of the ages of the antediluvian patriarchs at the births of their sons is given in Table 3.5 based on data from Book 1 of Josephus's *Antiquities of the Jews*.[19] Again, the spellings used previously in Table 3.1 are retained here. The ages in Table 3.5 from Adam to Enoch are in agreement with the corresponding ages in the Septuagint as shown in Table 3.1. The ages for Methuselah and Lamech are in agreement with the Masoretic data of Table 3.1. No explicit value seems to be given by Josephus for the age of Shem at the birth of Arpachshad. The ages in Table 3.5 for the eight patriarchs from Adam to Methuselah are one hundred years greater than the corresponding ages of Table 3.4 given by *Jasher*.

Table 3.5. Approximate chronology of the antediluvian patriarchs from Adam to the beginning of the flood according to Book I of Josephus's *Antiquities of the Jews*. The numbers shown are the ages of the patriarchs at the births of their sons.

Patriarch	Verse	Age
Adam	Josephus 2:3, 3:4	230
Seth	Josephus 3:4	205
Enosh	Josephus 3:4	190
Kenan	Josephus 3:4	170
Mahalalel	Josephus 3:4	165
Jared	Josephus 3:4	162
Enoch	Josephus 3:4	165
Methuselah	Josephus 3:4	187
Lamech	Josephus 3:4	182
Noah	Josephus 3:3, 4:1	500
Shem	Josephus 6:4	?
Year of world at flood	Josephus 3:3	2256

3.6 Conclusion

The purpose of this chapter has been to develop a relative chronology for the time period from Adam to Noah. This process is dependent on the genealogies of the people living during that period of time. Unfortunately, the genealogies available represent only one family line, and the ages of the individuals in that line at the births of their descendants often seem implausibly high and vary between texts.

The chronologies given in Tables 3.1, 3.2, 3.4, and 3.5 may be called relative chronologies because absolute dates have not been suggested. One possibility for obtaining absolute dates would be to find an association between the biblical flood and a datable archeological or geophysical event. The investigation of possible geographical locations and absolute chronologies for the biblical characters and events will continue in the following chapters. Approximate rationalizations of the ages of the patriarchs in these genealogies are developed in Chapter 12.

As considered in Section 3.3 above, a forty-nine year jubilee calendar system has been adapted to a relative chronology for the period from Adam to Noah. That chronology is similar to the one in the Samaritan text of Genesis and differs more substantially from other versions. However, none of the chronologies for this time period seem plausible in all respects, and other approaches will be considered in Chapter 12. On the other hand, the jubilee calendar may provide the best framework available for obtaining many exact dates from the time of the exodus and throughout the following Old Testament era.[20]

---

1. *The Book of Yashar,* Translated from the Hebrew and Published by Mordecai Manuel Noah (Hermon Press, New York, 1972), (Cited here and in other chapters in the form *Jasher*); *Jasher* 5:18.
2. *The Book of Jubilees or Little Genesis*, from the Ethiopic text by R. H. Charles, With an introduction by G. H. Box

(The Macmillan Company, 1917). *The Book of Jubilees* is included in the biblical canon of the Ethiopian Orthodox Church, and it will usually be abbreviated here as Jubilees.
3. R. H. Charles, op. cit.; Introduction by G. H. Box, pp. xv-xvi.
4. "This is obscured by such titles [of Jubilees] as 'the little Genesis,' 'the Apocalypse of Moses,' etc."
5. "Jubilees 6:34; cf. 1:10."
6. "So, R. Leszynsky, *Die Sadduzäer* (Mayer & Müller, Berlin, 1912) pp. 156 ff."
7. "It should be noted that the Talmud (T.B., Ned., 61a) refers to the view (held by R. Jehuda) that the jubilee-period was forty-nine years."
8. R. H. Charles, op. cit.
9. B. Beer, *Das Buch der Jubiläen und sein Verhältnis zu den Midraschim* (Wolfgang Gerhard, Leipzig, 1856); B. Beer, *Noch ein Wort über das Buch der Jubiläen* (Heinrich Hunger, Leipzig, 1857).
10. "Jubilees, Book of," in *The Jewish Encyclopedia: A Descriptive Record of the History, Religion, Literature, and Customs of the Jewish People from the Earliest Times to the Present Day* (Funk and Wagnalls Company, New York, 1904), Volume 7, p. 301.
11. *The Israelite Samaritan Version of the Torah, First English Translation, Compared with the Masoretic Version*, edited and Translated by Benyamim Tsedaka, Coedited by Sharon Sullivan (William B. Eerdmans Publishing Company, Grand Rapids, Michigan, 2013), pp. xxix-xxx.
12. G. von Rad, "Deuteronomy," in *The Interpreter's Dictionary of the Bible: An Illustrated Encyclopedia* (Abingdon Press, Nashville, 1962), Volume 1, A-D, pp. 831-838; p. 836.

13. *The Israelite Samaritan Version of the Torah*, op. cit., p. 7.
14. *The Septuagint*, Translated by L. C. L. Brenton (Samuel Bagster and Sons, Ltd., London, 1851).
15. R. H. Charles, op. cit.
16. F. Josephus, *The Life and Works of Flavius Josephus*, Translated by William Whiston (Holt, Rinehart and Winston, New York), *The Antiquities of the Jews*, Book 1, 1:1.
17. S. M. Miller and R. V. Huber, *The Bible: A History, The making and impact of the Bible*, (Good Books, Intercourse, PA, 2004), "In search of a reliable text," pp. 220-221.
18. *Jasher*, op. cit., Chapters 2-7.
19. F. Josephus, op. cit., Book 1.
20. L. W. Casperson, *Patterns of Biblical Chronology* (Westbow Press, Bloomington, IN, 2012), Section 5.6, "Summary of possible jubilee-related events," pp. 106-108, Table 5.3.

4. RELATIVE TIMES AND PLACES

"In the six hundredth year of Noah's life, in the second month, on the seventeenth day of the month, on that day all the fountains of the great deep burst forth, and the windows of the heavens were opened" (Genesis 7:11).

4.1 Introduction

As a first step in identifying the times and places associated with biblical events between creation and the flood, it might be helpful to review some of the data on these aspects of the Bible stories. Then one could attempt to use studies of history, geography, archeology, geology, etc., to determine the physical locations or times of events that might have been associated with this period. Unfortunately, for such early times only limited potentially useful information is available.

One of the most dramatic physical occurrences in the early chapters of Genesis is the flood of Noah's era. Assuming that there is a factual basis for the stories, the flood would by any reckoning have involved a large geographical area and a large number of people. Thus, one could, for example, look for evidence of a datable high water level at some reasonable

The Garden And The Flood

location and time in the past. One could also review historical records of relevant ancient civilizations, to see whether any suitable garden or flood conditions might have been mentioned or be otherwise discernible. A process of this sort will be initiated in the present chapter, and it will be continued in Chapters 5 to 7.

The relative chronology of the biblical flood and the calendar underlying that chronology are considered in Section 4.2. The possibility that the flood could correspond to a major inundation event of one of the rivers of Mesopotamia is discussed in Section 4.3. Approximate locations for the garden of Eden and the flood are tentatively suggested in Section 4.4. It has long been inferred that the geographical location of the garden of Eden and of the landing place of the ark are not widely separated, and an early map of a postulated arrangement is reviewed in Section 4.5. The ancient data underlying this map and its limitations are considered in Section 4.6.

4.2 The biblical flood
The flood as recorded in the Bible is of interest as a possible aid in determining an absolute chronology for the earliest stories of the Bible. In such a chronology the dates of the events of the first chapters of Genesis would be known at least approximately. Using only the genealogical information reviewed in Sections 3.2 and 3.3, it is clear that one is limited to considering a relative or floating chronology with no absolute dates. Because of disagreements between versions of the Bible and (some would say) the implausibly high ages that sometimes have been indicated, uncertainties with such relative chronologies amount to

4. Relative Times and Places

many centuries. It is in fact because of these issues that the early stories in Genesis are sometimes treated as partially fictional.

The purpose of this section is to begin an effort at understanding the many physical questions associated with the biblical flood in particular. If in some sense the flood was a real event in history, then there would be at least the possibility of finding the date of its occurrence. As a starting point in considering the nature and chronology of the biblical flood, it may be helpful to review some of the biblical texts on this subject:

> In the six hundredth year of Noah's life, in the second month, on the seventeenth day of the month, on that day all the fountains of the great deep burst forth, and the windows of the heavens were opened. And rain fell upon the earth forty days and forty nights. On the very same day Noah and his sons, Shem and Ham and Japheth, and Noah's wife and the three wives of his sons with them entered the ark . . . (Genesis 7:11-13).

> The flood continued forty days upon the earth; and the waters increased, and bore up the ark, and it rose high above the earth. The waters prevailed and increased greatly upon the earth; and the ark floated on the face of the waters. And the waters prevailed so mightily upon the earth that all the high mountains under the whole heaven were covered; the waters prevailed above the mountains, covering them fifteen cubits deep (Genesis 7:17-20).

> But God remembered Noah and all the beasts and all the cattle that were with him in the ark. And God made a wind blow over the earth, and the waters subsided; the fountains of the deep and the windows of the heavens were closed, the rain from the heavens was restrained, and

the waters receded from the earth continually. At the end of a hundred and fifty days the waters had abated; and in the seventh month, on the seventeenth day of the month, the ark came to rest upon the mountains of Ar'arat. And the waters continued to abate until the tenth month; in the tenth month, on the first day of the month, the tops of the mountains were seen.

At the end of forty days Noah opened the window of the ark which he had made, and sent forth a raven; and it went to and fro until the waters were dried up from the earth. Then he sent forth a dove from him, to see if the waters had subsided from the face of the ground; but the dove found no place to set her foot, and she returned to him to the ark, for the waters were still on the face of the whole earth. So he put forth his hand and took her and brought her into the ark with him. He waited another seven days, and again he sent forth the dove out of the ark; and the dove came back to him in the evening, and lo, in her mouth a freshly plucked olive leaf; so Noah knew that the waters had subsided from the earth. Then he waited another seven days, and sent forth the dove; and she did not return to him any more.

In the six hundred and first year, in the first month, the first day of the month, the waters were dried from off the earth; and Noah removed the covering of the ark, and looked, and behold, the face of the ground was dry. In the second month, on the twenty-seventh day of the month, the earth was dry. Then God said to Noah, "Go forth from the ark, you and your wife, and your sons and your sons' wives with you. Bring forth with you every living thing that is with you of all flesh – birds and animals and every creeping thing that creeps on the earth – that they may breed abundantly on the earth, and be fruitful and multiply upon the earth." So Noah went forth, and his sons and his wife and his sons' wives with him. And every beast, every creeping thing, and every bird, everything that moves upon the earth, went forth by families out of the ark (Genesis 8:1-19).

For brevity these selections omit some topics from the Genesis account such as construction details of the ark, descriptions of the animals that were brought inside, and the deaths of the people and animals that remained outside. Concerning the interpretation of the flood, one of the first questions that one might ask involves its magnitude. It will be assumed here that the word "earth" in the story refers in the actual event to a broad area of land and water but not necessarily the whole earth as understood today. The original human author probably would not have known that the earth was approximately spherical or imagined that the highest mountain on earth (Mt. Everest) was covered by the flood to a depth of fifteen cubits (about twenty-five feet). It is very possible, however, that heavy rain with local flooding could have occurred over a large area.

The biblical story recalled above provides several details about the chronology of the flood as experienced by Noah and his family. It may be helpful to place this information in the form of a table:

The Garden And The Flood

Table 4.1. Chronology of the flood in the lifetime of Noah.

Event	Genesis reference	Year	Month	Day
Rain began	7:11	600	2	17
Rain ended	7:12		(3	27)
Ark landed	7:24, 8:3-4		7	17
Mountains visible	8:5		10	1
Raven sent	8:6-7		(11	11)
Dove sent	8:8			
Dove sent again	8:10			
Dove sent again	8:12			
(intercalary days?)				
Covering removed	8:13	601	1	1
Depart from ark	8:14-19		2	27

The information suggested in parentheses is not addressed explicitly in the Bible. The values shown represent a preliminary attempt to estimate the dates for some of the undated events in the table.

4. Relative Times and Places

The first assumption underlying the numbers shown in parentheses is that the months each had a length of exactly thirty days. This idea is suggested by the fact that the rain is said to have begun on the seventeenth day of the second month (Genesis 7:11), while the ark seems to have come to rest one hundred fifty days later on the seventeenth day of the seventh month (Genesis 7:24, 8:3) or exactly five months later.[1] The simplest ancient calendars for which these results would be possible would be non-lunar systems in which each month had exactly thirty days. For such a calendar twelve months would be equivalent to three hundred sixty days, while the actual solar year is approximately three hundred sixty-five and one quarter days. This calendar could be harmonized with the solar year, if that were desirable, by the addition after the twelfth month of five or six intercalary days. Depending on the harmonization adopted, the number of days needed (five or six) could be established by detailed record-keeping and observations to begin the year at, for example, the vernal equinox. The standard civil calendar of the ancient Egyptians always used five intercalary days, so the calendar year drifted away from its starting point by about one quarter of a day per year.[2] Modern calendars have unequal month lengths, with most years of three hundred sixty-five days but every fourth year (leap year) of three hundred sixty-six days (with additional less frequent corrections).

It is not being suggested here that the calendar described above is exactly the one that underlies the Genesis accounts. However, to the extent that the data are historical, this calendar may be adaptable to the limited information that Genesis provides. The rain is said to have begun on the seventeenth day of the second month and to have lasted forty days (Genesis 7:11-12, 7:17). Thus, in Table 4.1 the rain is estimated (as shown by the parentheses) to have stopped on the twenty-seventh day of the third month. This estimation is based on the assumption just mentioned that all months lasted for thirty days.

Similarly, the raven seems to have been sent out forty days after the mountaintops became visible "in the tenth month, on the first day of the month" (Genesis 8:6-7). This date of sending should perhaps be the eleventh day of the eleventh month as shown by parentheses in the table. There is less clarity on the dates associated with the sending out of the dove. There is no explicit information on how long it was after the raven was sent out that the dove was first sent. It is, however, indicated that the second sending of the dove was seven days after the first, and the third sending was seven days after the second. Thus, it might be a reasonable assumption that the first dove was sent out seven days after the raven was sent.[3] According to the Jewish historian Flavius Josephus (first century A.D.), the dove was sent out only once at seven days after the sending of the raven.[4] In the Jubilees version the birds aren't mentioned at all (Jubilees 5:21-32),[5] and neither are they mentioned in *Jasher*.[6] There is also no explicit information on the possible use of intercalary days to adapt the calendar year to the solar year, so it isn't clear how many days were in one year.

4.3 Inundation and the flood

In an effort to identify the chronology of the biblical flood, the location and cause of the flood are also of particular interest. As will be considered in more detail below, Mesopotamia has

4. Relative Times and Places

sometimes been considered as a possible location for the departure of the ark. Thus, it is reasonable first to consider whether flooding is known or possible in the lands associated with Mesopotamia. A partial answer is that some degree of flooding in the vicinity of the Tigris and Euphrates Rivers was a regular annual event, and the reciting of *Enûma elish* (a Babylonian creation story considered in Section 1.4) each spring may have been intended to limit excess damage to buildings or lands caused by the annual floods:

> It is possible that on this occasion the epic was recited as a magic formula against the coming inundation of Babylonia caused by the rise of the Tigris and the Euphrates following the melting of the snows in the mountains of Armenia and Kurdistan; for at the time of these floods it seemed as if the primordial chaotic condition of "water, water everywhere" were to return. ... The recitation of *Enûma elish* presumably reflects the annual battle between Marduk and the watery chaos produced by the spring inundations.[7]

In the context of flooding it is important to understand that in Mesopotamia rainfall is not the primary source of the water needed for growing crops:[8]

> Albert T. Clay[9] called attention to the fact that the rainy season and the overflow of the rivers of Babylonia do not synchronize, and the average fall of rain in Babylonia, amounting to about six inches per year,[10] is too small to be of any consequence; in fact, it is so small that the land would be a desert were it not for the irrigation canals and the inundations. The rivers do not flood in the winter but in the spring, from March to June, following the melting

of the snows on the Zagros and the mountains of Armenia.

This mechanism for water to reach the crops is similar to that in Egypt where there is also very little direct rainfall, except that in Egypt the inundation by the Nile River results from the monsoon conditions to the south of the country[11] whereas the inundation in Mesopotamia is due to the melting of snow to the north. While the inundations are clearly a type of flooding, even an unusually large inundation would probably have been considered a normal and not necessarily unwelcome component of Mesopotamian agriculture. Also, unlike the biblical flood, which is closely identified with rain, the inundations are not in general synchronized with the rainy season. In any case, it seems doubtful that an ordinary weather event would lead to rainfall lasting "forty days and forty nights" as reported in the Genesis account (Genesis 7:12).

In short, it does not appear likely that the normal annual inundations in Mesopotamia could have been identified with the extraordinary flood that is associated with Noah and his family. There are other possible causes for this flood, however, and some of these will be considered in Chapter 7.

4.4 Approximate locations

If Mesopotamia was not directly associated with the flooding described in Genesis and there was a factual basis for the flood stories, then it would remain reasonable to attempt to identify a different geographical setting for the flood and its associated events. It could possibly also be of interest to attempt to infer a location for the garden of Eden. As a preliminary consideration,

4. Relative Times and Places

one can recall again the creation story of Genesis that was the subject of Chapter 1. As noted previously in Section 2.4, after the first man was created he was placed in the garden of Eden. The story suggests a fairly specific location for that garden:

> And the Lord God planted a garden in Eden, in the east; and there he put the man whom he had formed. And out of the ground the Lord God made to grow every tree that is pleasant to the sight and good for food, the tree of life also in the midst of the garden, and the tree of the knowledge of good and evil.
>
> A river flowed out of Eden to water the garden, and there it divided and became four rivers. The name of the first is Pishon; it is the one which flows around the whole land of Hav'ilah, where there is gold; and the gold of that land is good; bdellium and onyx stone are there. The name of the second river is Gihon; it is the one which flows around the whole land of Cush. And the name of the third river is Tigris, which flows east of Assyria. And the fourth river is the Euphra'tes (Genesis 2:8-14).

The first sentence in this quotation should probably be understood to indicate that the garden is east of the location of the author and his expected audience. That sentence also indicates that Adam was placed in the garden, but it is not stated where he came from, i.e., where he was created or born. This subject was mentioned previously in Section 2.4. In any case, the first people in the biblical story lived initially in the garden "of Eden" (Genesis 2:15), or perhaps the garden "in Eden" (Genesis 2:8), or the garden "out of Eden" (Genesis 2:10). The wording isn't entirely clear, but it seems that the author's intent may have been to suggest that the four rivers flowed to the four corners (or quarters) of the earth, or at least in four different directions into much of the locally relevant world (Genesis 2:10-14). Within these verses it is only the Tigris and Euphrates rivers that are easily recognized. The Euphrates is left with no geographical location, perhaps because it was already familiar to the writer's anticipated readership. As it happens, the headwaters of the Tigris and Euphrates Rivers are not far from each other compared to the lengths of these rivers, and thus their headwaters suggest a possible approximate location for the garden.

The idea that the world could be described as having four somewhat distinguishable quarters has a very old history, even outside of the Bible:

> In connection with Ebla's ideology of kingship an interesting suggestion can be made. We remember that Naram-Sin of Akkad [c. 2190-2154 B.C., short chronology] was the first in Mesopotamian history to assume the title 'king of the four regions'. The reference was intended cosmologically rather than geographically, to the four quarters of the world rather than to four particular countries as it was later to be interpreted. . . . [I]t seems pertinent to ask whether precedents are to be found in the culture of Ebla, which was destroyed by Naram-Sin. In fact, they are. Two significant bits of evidence suggest that the quadripartite conception of the world was peculiarly native to Ebla. First, in the palace seals of Mardikh IIB1 there is that frequently recurring figure of the naked hero kneeling and supporting on his head with raised arms the circular symbol composed of four heads. . . . Secondly, in the topography of Protosyrian Ebla both the Archive texts and archaeological exploration show that the city was quadripartite, essentially subdivided into four quarters arranged approximately like the quadrants of a circle, with the Acropolis at its centre. . . . The regular division

4. Relative Times and Places

into four of the city inscribed in an approximately circular perimeter wall is connected with the notion of the city as microcosm, by which the image of the world is reflected in that of the city, understood as centre of the universe.

If the idea of the division of the world into four was really typical of the conception of Ebla, it would then be quite understandable that Naram-Sin, after the destruction of his great western rival, should have assumed the title of 'king of the four regions'. At one blow he would have appropriated from the Eblaite culture the concept foreign to the Mesopotamian world of the division of the cosmos into four, signified by it the re-gathering into one of the civilised world, and symbolised the universality of his political power, henceforth without a rival.[12]

The concept of the earth having four quarters was continued later in Mesopotamia, and another example is from the reign of Ur-Nammu:

> Ur-Nammu's reign [2047-2030 B.C., short chronology] began the Third Dynasty of Ur, also known as the Ur III Period. . . . Ur-Nammu died in battle and was succeeded by his son Shulgi [2029-1982 B.C., short chronology] who completed his father's ziggurat and temple for Nanna at Ur. . . . At some point early in his reign he followed the precedent of Naram-Sin, taking the title "Ruler of the Four Quarters (of the earth)" and declaring himself a god. . . . Shulgi also produced the earliest law code yet known, but unfortunately, it survives only in a poorly preserved later copy. This collection of laws used to be ascribed to Ur-Nammu, but a recently recognized fragment of the prologue makes it clear that Shulgi was its author.[13]

The Garden And The Flood

The idea that the four rivers flowing from the garden of Eden watered the whole earth is suggested also in Adam and Eve: "And Adam and Eve went from before the [western] gate of the garden to the southern side of it, and found there the water that watered the garden, from the root of the Tree of Life, and that parted itself from thence into four rivers over the earth."[14]

Further statements relating to the location of the garden are also given. Thus, it may be recalled that Adam and Eve were expelled from the garden for eating the forbidden fruit:

> Then the Lord God said, "Behold, the man has become like one of us, knowing good and evil; and now, lest he put forth his hand and take also of the tree of life, and eat, and live for ever" – therefore the Lord God sent him forth from the garden of Eden, to till the ground from which he was taken. He drove out the man; and at the east of the garden of Eden he placed the cherubim, and a flaming sword which turned every way, to guard the way to the tree of life (Genesis 3:22-24).

Thus, Adam was sent out from the garden, apparently through the east gate. We are also told explicitly that Adam's eldest son Cain "dwelt in the land of Nod, east of Eden" (Genesis 4:16). It is not necessarily true that Adam or Cain would have continued to live east of the garden for any great period of time. Also, it may be noted in passing that the phrase "East of Eden" introduced in Genesis became the title of a popular book written by John Steinbeck.[15]

Another subject of interest is the location of the flood and in particular the location of the landing place of the ark. As quoted above, after the heavy rain the ark came to rest on "the mountains

4. Relative Times and Places

of Ar'arat" (Genesis 8:4). The area of Ararat (or ancient Urartu or Armenia, as discussed further in the following section) is to the north of Mesopotamia, the region between the Tigris and Euphrates Rivers. The mountains of Ararat might be considered to include the headwaters of both of these rivers. Thus, one can imagine initially that the flood experienced by Noah and his family may have occurred somewhere between Mesopotamia and the mountains of Ararat to the north.

In spite of the above relative information, it would seem that the basic questions concerning the absolute location and mechanism of the flood still remain to be resolved. As noted above in Section 4.3, the annual inundations of the Tigris and Euphrates rivers are not associated with local rainfall, and the usual limited rainfall in that region isn't easily compatible with heavy rain lasting for forty days. It is also difficult to imagine that even a major inundation event along one of these rivers could float the ark in that location for as long as the five months indicated in the Bible (Genesis 7:24, 8:3-4). In that time it seems likely that the ark would have been interacting with the shoreline of the river, if it did not actually run aground. If the ark remained afloat on one of these rivers, it would have washed ever farther from the river's headwaters and perhaps all the way to the Persian Gulf. It would also have been swept ever farther from its biblically reported landing place in the mountains of Ararat. A different and more specific possibility for the location and nature of the flood will be considered starting in Chapter 5. We will continue here by considering a somewhat traditional interpretation for the location of the garden of Eden.

4.5 The Terrestrial Paradise

As discussed above, it would seem that a possible location for both the garden of Eden and the landing place of the ark could be somewhere between Mesopotamia and the mountains of Ararat. Not surprisingly, this proposed location is not a new idea but rather goes back many years. As an illustration, it may be helpful to consider the map in Figure 4.1.[16] This map attempts to identify the ancient location of the garden of Eden and by implication also the approximate landing place of the ark. The map was made by the distinguished mapmaker Emanuel Bowen in the mid-eighteenth century. Bowen served as Royal Mapmaker to both King George II of England and Louis XV of France. It must be emphasized at the outset that this is an early map of the subject region, and its main interest is that it provides an explicit interpretation of the location of the garden of Eden. The map's inclusion here should not be interpreted to suggest that it is considered to be correct in all of its geographical details, such as the locations and current names of many of the places of interest.

4. Relative Times and Places

Figure 4.1 A Map of the Terrestrial Paradise by Emanuel Bowen (1694?-1767). (Reproduced with permission)

The title Terrestrial Paradise is, of course, an alternative name for the garden of Eden, and both names are given on the enclosed area that is the central focus of the map. As indicated previously, the four rivers mentioned in Genesis 2:10-14 as flowing out of the Garden of Eden include the Pishon, Gihon, Tigris, and

Euphrates Rivers. The mapmaker has identified all four of these rivers on his map as flowing out of the Terrestrial Paradise. The Pishon River is here spelled Pison and is equated on the map with the Phasis River of ancient times that flows into the Black Sea through the Kingdom of Colchis. The Gihon River is equated with the Araxes River of Armenian tradition that flows into the Caspian Sea.

There are two major unnamed lakes that appear on the map in or near the Terrestrial Paradise. By comparing with a modern map of this region, the left-most of these lakes, that is largely included in the Paradise region, corresponds to contemporary Lake Van. The lake to the southeast of the Paradise region corresponds to contemporary Lake Urmia. Because of limitations of the geographical data available to Bowen two hundred fifty years ago, the map shown in Figure 4.1 is not as accurate as one could wish. Places of interest are not always located correctly, and also for example the shapes of Lakes Van and Urmia shown here differ significantly from those given on modern maps. Nevertheless, Bowen's map provides a useful illustration of several concepts relating to the Garden of Eden and the flood that have been of interest to many generations of Bible readers.

It should also be acknowledged that, due to limitations of resolution and content, the details of the map are not all easy to read in Figure 4.1. A somewhat clearer expanded view of the central part of this map is given in Section 4.6 below. It is evident from the map (and other more modern and accurate maps) that there are many hills and mountains in the general Paradise region besides Mt. Ararat. Any of these could be regarded as among "the mountains of Ararat" (Genesis 8:4). As such they could be considered to have been possible landing places for Noah's ark. While the precision is limited, it is striking that a single small geographical area as shown on the map includes plausible locations for several of the important events described in the first chapters of the Bible.

For perspective it may be helpful to recall the names that at later dates in history were associated with the area of the Terrestrial Paradise. The biblical name Ararat was originally represented as Urartu and later also as Armenia. A brief introductory summary of these names and the history of Urartu have been given by Piotrovsky:

> For three centuries [ninth through seventh century B.C.] Urartu was a formidable rival to Assyria. Though twice defeated by the Assyrians, the Urartians several times prevailed in this contest, and indeed – though only by a few decades – outlasted their rivals. But posterity dealt harshly with the memory of Urartu. The name was preserved in the Old Testament in the corrupt form "Ararat," which in the Latin version became "Armenia." When the Massoretic writers were vocalising the text of the Bible they inserted the vowel *a* into words which were unknown to them, so that "Urartu" became "Ararat;" and it is only within very recent years that the Qumran scrolls have yielded a form of the name with the semi-vowel w in the first syllable.[17]

The name Ararat (Urartu) appears at several places in the Bible. When some Bible readers think of the name Ararat they might assume that the reference is to the specific mountain of that name. However, as already indicated, Ararat in the Bible seems always to be a reference to the geographical region better called Urartu:

> By the end of the 9th century [B.C.] the power of Urartu had increased to such an extent that Assyria was compelled to recognise its dominance in western Asia. From this period began the political and cultural rise of Urartu, which made it for two centuries the largest state in western Asia, occupying the whole of the Armenian highland area.[18]

Bible verses referring to Ararat include the following:

> [A]nd in the seventh month, on the seventeenth day of the month, the ark came to rest upon the mountains of Ar'arat [Urartu] (Genesis 8:4).

> Then Sennach'erib king of Assyria departed, and went home, and dwelt at Nin'eveh. And as he was worshiping in the house of Nisroch his god, Adram'melech and Share'zer, his sons, slew him with the sword, and escaped into the land of Ar'arat [Urartu]. And Esarhad'don his son reigned in his stead (2 Kings 19:36-37, Isaiah 37:37-38).

> Set up a standard on the earth, blow the trumpet among the nations; prepare the nations for war against her [Babylon], summon against her the kingdoms, Ar'arat [Urartu], Minni [Mannea, south of Lake Urmia], and Ash'kenaz [Scythia]; appoint a marshal against her, bring up horses like bristling locusts (Jeremiah 51:27).

Thus, it might be best to interpret Genesis 8:4 as stating that the ark came to rest somewhere on the mountains of a land that later came to be known as Urartu, the Armenia of Figure 4.1.

4. Relative Times and Places

While Mt. Ararat may be counted as among the mountains of Ararat, one needs to be cautious about interchanging the identification of a particular mountain (or any other object) with the name of a group of mountains (or other objects). As an arbitrary illustration, one might imagine substituting the name Washington for Ararat. Most Americans would understand Mt. Washington to refer to the highest peak in the Northeastern United States (about 6,288 feet). On the other hand, the "mountains of Washington" would mean the mountains in the state of Washington, 3,000 miles to the west. The highest of these is Mt. Rainier at about 14,411 feet.

4.6 Origin of the map

Before continuing with the preceding discussion of Bible-related possibilities for the location of the Terrestrial Paradise, it may be appropriate to say a few words about some of the seemingly less reasonable aspects of the map in Figure 4.1. It is worth enquiring, for example, where Bowen might have obtained some of the detailed geographical information suggested by his map. For easier reading, an expanded version of the Terrestrial Paradise and the region to its south are reproduced in Figure 4.2. More interesting details are shown in the south than in most other areas of the map. In particular, it seems to be Bowen's purpose here to represent graphically several unusual (and not necessarily correct) aspects of the Tigris River.

The Garden And The Flood

Figure 4.2 Expanded version of the Terrestrial Paradise and neighboring region, from the map by Bowen shown in Figure 4.1.

It may be noted that some of the peculiar features of the Tigris River as shown in Figures 4.1 and 4.2 seem to have been described many centuries earlier in the writings of Pliny the Elder (23 A.D. to 79 A.D.) and others. To illustrate this possible relationship, we quote here a part of a chapter devoted to the Tigris River from Pliny's *Natural History*. The locations indicated in Pliny's text are not all relevant for comparison with the map in Figure 4.2, and some of the later text is omitted:

4. Relative Times and Places

The River Tigris.

It is also convenient to say somewhat of the River Tigris itself. It beginneth in the Region of Armenia the Greater, issuing out of a great Source in the Plain. The place beareth the Name of Elongosinè. The River itself, so long as it runneth slowly, is named Diglito; but when it beginneth to be rapid, it is called Tigris, which in the Median language signifieth a Dart. It runneth into the Lake Arethusa, which beareth up all that is cast into it; and the Vapours that arise out of it carry Clouds of Nitre. In this Lake there is but one kind of Fish, and that entereth not into the Channel of the Tigris as it passeth through; as likewise the Fishes of the Tigris do not swim out into the Water of the Lake. In its Course and Colour it is unlike the other: and when it is past the Lake and meeteth the Mountain Taurus, it loseth itself in a Cave, and so runneth under, until on the other side it breaketh forth again in a Place which is called Zoroanda. That it is the same River is evident by this, that it carrieth through whatever was cast into it. After this second Spring, it runneth through another Lake, named Thospites, and again taketh its Way under the Earth through Gutters, and 25 Miles beyond it is returned about Nymphaeum. . . . After this it receiveth out of Media the Coaspes; and so passing between Seleucia and Ctesiphon, as we have said, it poureth itself into the Lakes of Chaldsea, which it replenisheth with Water for the Compass of threescore and ten Miles: which done, it issueth forth, gushing out with a very great Stream, and on the right of the Town Charax is discharged into the Persian Sea. . . .[19]

According to Pliny as quoted above, the Tigris River begins "in the Region of Armenia the Greater," and in Bowen's map the entire Terrestrial Paradise including the headwaters of the Tigris River is in Armenia. The specific location of the source of the

Tigris is said to be a spring in the plain Elongosinè from which it flows into Lake Arethusa. The spring at the origin of the Tigris and the larger lake to its east both appear to be represented on Bowen's map in association with the name Elongosin. The water of Lake Arethusa is said to be so dense that everything floats on it, its color is unusual, and only one kind of fish survives. According to Pliny, the river enters a cave when it arrives at Mt. Taurus, and then it emerges on the other side of the mountain. This circumstance is shown clearly on the map in Figure 4.2. The entering and emerging rivers were claimed to be the same on the grounds that whatever was thrown into the river as it entered the mountain emerged in the river at the other side. According to Pliny, the river then enters another lake called Thospites, and this lake is shown in red on the original map. The river is said then to flow underground for twenty-five miles, and this circumstance is indicated on the map as a gap in the river. According to Pliny, the river returns to the surface near the city of Nymphaeum, and perhaps this city could be associated with Ninphae on the map. In the following text (omitted in the quotation), Pliny describes several other points of interest that are not discernible on the map. He does eventually mention that the river flows between Seleucia and Ctesiphon, as evident on the map, before finding its way to the "Persian Sea."

From the above discussion it would appear that the Tigris River as shown on the map by Bowen has many of the same strange properties as reported centuries earlier by Pliny. For both Pliny and Bowen, the Tigris River flows through two lakes and two underground channels. Other early authors report some of the same claims as Pliny. Strabo states that the river flows through

4. Relative Times and Places

only one lake before disappearing into the ground. He agrees that, as with Pliny's Lake Arethusa in the quotation above, the water in this lake is charged with nitre and supports only one kind of fish. In general, however, the names of the lakes and other geographical features vary between the ancient and not so ancient authors, and even the identities of the lake(s) mentioned have not always been clear. This situation has been remarked on by Lynch and others:

> How strange it seems that at the end of the nineteenth century one should be engaged in exploring and mapping this fine country, one of the fairest and most favoured of the Old World! How should we be able to explain, still less to justify, the circumstance to some visitor from another planet? It lies about in the centre of the land area of our hemisphere; the climate is bracing, water is abundant, the sun is warm. Yet it is so little known to the more civilised peoples that their travellers journey thither with the aid of a compass through districts which are now deserts, but which are well capable of supporting the races that are highest in the human scale. The case would appear to have been much the same during the period of the expansion of Greek culture and of the later and beneficent sway of Rome. The knowledge displayed of these regions by representative writers like Strabo, Pliny and Ptolemy is, to say the best of it, vague and fabulous.[20]

As mentioned above, there seems to be some uncertainty about the names and even the reality of some of the features in the vicinity of the Terrestrial Paradise. Thus, Lake Van, which is unnamed in Bowen's map, is often today equated with Lake Thospites in Pliny's report. Also, Lake Arethusa is said by Pliny to be dense, oddly colored, and toxic to most fish because of its salt load, with the Tigris River flowing beside or through it. On the original colored map only Lake Thospites is red, perhaps suggesting toxicity, and the Tigris River apparently flows beside this lake without mixing. Several of these remarkable features are in disagreement with geographical reality, while others are not. An updated interpretation of some of these features will be included in Chapter 5.

In short, neither the map of Bowen nor the ancient writings of Strabo, Pliny, and Ptolemy provide a reliable description of the lakes and rivers of the land formerly known as Armenia. The map is of interest nevertheless, because it indicates an early level of awareness of the geography related to the land of Eden and the mountains of Ararat. These places are of central importance in understanding the settings of the earliest stories in the Bible.

4.7 Conclusion

The first three chapters of this book have sought to clarify the texts and their meaning for the biblical time period from the creation to the flood. The establishment of a relative chronology for this period in the form of genealogies was also attempted. As it happens, inferring the actual calendar date of the biblical flood is a process that can't be approached in isolation. One can't ask when the flood occurred without at the same time pursuing answers to the questions of what caused the flood and where it happened. The general idea of location can be separated initially into the two questions of identifying the launching point of the ark in the flood story and also its landing point. This chapter has first reviewed the biblical texts relating to the flood and their internal relative chronology. Concerning location, the most

4. Relative Times and Places

natural assumption is that the personalities in the flood story may have begun their adventures at a place not far distant from their ancestral origin in the story, namely, the garden of Eden. According to the Bible, the garden of Eden was in the vicinity of the headwaters of four rivers, two of them being the Tigris and the Euphrates. These rivers and their headwaters are well-known, and thus one has a possible general location for the first floating of the ark.

The Bible also informs us that as the flood was receding "the ark came to rest upon the mountains of Ararat" (Genesis 8:4). Happily, the location of the mountains of Ararat coincides well with the location of the headwaters of the Tigris and Euphrates Rivers, and this coincidence supports the idea that the ark builders lived somewhere near the garden of Eden. The coincidence also encourages the idea that the flood might not have been a global event, for an ark drifting on a world-wide ocean for an extended period of time would be unlikely to land almost where it started. It remains then to seek for an interpretation and chronology for an ark-floating flood in the identified and more restricted geographical region.

One natural interpretation is that the flood is somehow related to the well-known annual flooding that occurs along the Tigris and Euphrates Rivers. This interpretation was tentatively excluded in Section 4.3 on the grounds that the usual flooding in Mesopotamia is due to snow-melt far upstream and occurs at the wrong season compared to the timing of the very limited local rainfall. Also, such a flood occurs annually; but, even larger than usual, it might not be special enough to be the basis for the population-destroying flood described in the Bible. Finally, a

The Garden And The Flood

flooding of the rivers would tend to carry an ark away from the mountains of Ararat rather than towards them. It would seem then that the Genesis flood might have some physical interpretation other than the flooding of a river. An early map and ancient texts were considered in an effort to clarify the relevant geography. Chapters 5 to 7 will be concerned more specifically with seeking a suitable flood and its chronology in the vicinity of both Eden and Ararat.

---

1. A. Heidel, *The Gilgamesh Epic and Old Testament Parallels* (The University of Chicago Press, Chicago, 1963), p. 247.
2. J. Finegan, *Handbook of Biblical Chronology, Principles of Time Reckoning in the Ancient World and problems of Chronology in the Bible* (Princeton University Press, Princeton New Jersey, 1964), pp. 23-24, Paragraphs 40 and 41.
3. A. Heidel, *The Gilgamesh Epic and Old Testament Parallels*, op. cit., p. 247.
4. F. Josephus, *The Life and Works of Flavius Josephus*, Translated by William Whiston (Holt, Rinehart and Winston, New York), *The Antiquities of the Jews*, Book 1, 3:5.
5. *The Book of Jubilees or Little Genesis*, Translated from the Ethiopic text by R. H. Charles, With an introduction by G. H. Box (The Macmillan Company, 1917). As noted in Section 1.3D of *Patterns of Biblical Chronology*, The *Book of Jubilees* is included in the biblical canon of the Ethiopian Orthodox Church, and it will usually be abbreviated here as Jubilees.
6. *Book of Yasher*. Translated from the Hebrew and Published by Mordecai Manuel Noah (Hermon Press,

New York, 1972), (Cited here and in other chapters in the form *Jasher*); *Jasher* 6:1-40.
7. A. Heidel, *The Babylonian Genesis, The Story of Creation*, Second Edition (The University of Chicago Press, Chicago, 1951), p. 17.
8. Ibid., p. 98.
9. "A. T. Clay, *The Origin of Biblical Traditions* (New Haven, Conn. 1923), pp. 75-78. See also George A. Barton's remarks in the *Journal of the American Oriental Society*, Volume 45, p. 27 f. (1925), and Clay's rejoinder, ibid., p. 141."
10. "See M. G. Ionides, *The Régime of the Rivers Euphrates and Tigris* (E. & F. N. Spon, Ltd., London, 1937), especially pp. 24-36."
11. L. W. Casperson, *Patterns of Biblical Chonology* (Westbow Press, Bloomington, IN, 2012), Section 20.6, "The famine context," pp. 512-516; and Section 22.3, "Darkness and famine (gloom and doom)," pp. 548-552.
12. P. Matthiae, *Ebla, An Empire Rediscovered* (Doubleday & Company, Inc., Garden City, New York, 1981), pp. 223-224.
13. W. H. Stiebing, Jr., *Ancient Near Eastern History and Culture*, Second Edition (Pearson, Longman, New York, 2009), pp. 80-82.
14. *Adam and Eve*, Translated by S. C. Malan from the Ethiopic edition edited by E. Trumpp, included in *The Forgotten Books of Eden*, Edited by Rutherford H. Platt, Jr. (Alpha House, Inc., 1927); Book 1, 9:2.
15. J. Steinbeck, *East of Eden* (The Viking Press, New York, 1952).
16. E. Bowen (1694?-1767), *A Map of the Terrestrial Paradise*. Available at http://www.armenica.org/cgi-bin/armenica.cgi?806774033515528=1=3==Historical%20maps==1=3=AAA,Web. 23 Feb 2018.

17. B. B. Piotrovsky, *The Ancient Civilization of Urartu*, Translated from the Russian by J. Hogarth (Cowles Book Company, Inc., New York, N.Y. 10022), 1969), p. 13.
18. Ibid., p. 51.
19. *Pliny's Natural History in Thirty-Seven Books*, Translated by Dr. Philemon Holland (George Barclay, Leicester Square, 1847-48), Book 6, Chapter 27; pp. 142-143.
20. H. F. B. Lynch, *Armenia Travels and Studies*, Volume 2 (Longmans, Green, and Co., London, 1901), pp. 39-40.

5. URARTU AND LAKE VAN

"Set up a standard on the earth, blow the trumpet among the nations; prepare the nations for war against her [Babylon], summon against her the kingdoms, Ar'arat [Urartu], Minni, and Ash'kenaz" (Jeremiah 51:27).

5.1 Introduction

Considerations in the previous chapters have suggested that the garden of Eden and the landing place of the ark were probably both in the region that came to be known as Ararat (Urartu) and then later as Armenia. These conclusions were based primarily on interpretations of stories from Genesis. An early map suggesting a possible location for the land of Eden was discussed in Sections 4.5 and 4.6. The purpose of this chapter is to explore further biblical and nonbiblical sources that could provide support for these conclusions.

One of the most familiar images that one might call to mind concerning the garden of Eden involves the concept of Adam and Eve choosing to eat of the tree of the knowledge of good and evil rather than obeying God. The imagery of this story may have survived for centuries and found its way into the art of Urartu. Several examples of this phenomenon are considered in Section

5.2. A newer map of Lake Van is introduced in Section 5.3, and some of the properties of this lake are compared to those quoted by the ancient geographers. The book *Adam and Eve* seems to suggest a geography of the land of Eden, and this subject is explored in Section 5.4.

5.2 Remembrances of Eden

The kingdom of Urartu (Ararat) may be imagined to correspond approximately to the later Armenia as discussed in Section 4.5 and shown on the maps in Figures 4.1 and 4.2. Thus, the land of Eden and the mountains of Ararat are both near the middle of Urartu. The history of Urartu will be of recurring interest in this chapter, and a few items of possible relevance may be mentioned at the outset. Among the most important symbols of the garden of Eden are the so-called tree of life and the tree of the knowledge of good and evil (Genesis 2:15-17). Representations of what are thought to be one or the other of these trees are not uncommon on archeological artifacts from Urartu. For example, Urartian helmets from the eighth century B.C. that were found during excavations at the ancient city of Teishebaini are sometimes decorated in the front with multiple sacred trees.[1] A picture of such a helmet has been published by Piotrovsky,[2] as has a close-up view of a sacred tree.[3] The trees on the helmet are faced on both sides by winged or flightless human-like forms reaching out to the tree with one hand while holding some object with a handle in the other hand. It isn't obvious what these humanoids are doing, but perhaps they are guarding the trees while picking fruit and placing it in baskets. Arching above the sacred trees are what appear to be several large snakes. It is possible to imagine that

5. Urartu and Lake Van

this ancient stylized symbolism on a helmet, in what has often been considered to be the vicinity of the garden of Eden, reflects in some way an intimidating memory of Urartian snakes and the threat that they represent.

The symbolism of trees and snakes was also included in Urartian correspondence. In commenting on the varying quality of letters found at Teishebaini, Piotrovsky has said the following:

> We get an entirely different impression from the letters addressed to the viceroy at Teishebaini by the Urartian kings and high officials in the capital. They are well written, in a handsome cuneiform script, and are stamped at the end with the royal cylinder seal, which bears a representation of the tree of life with fantastic animals on either side.[4]

An example of one of these letters with seal has also been published by Piotrovsky.[5] The trees mentioned above could also perhaps represent fig trees. That there were such trees in the garden of Eden is indicated in the biblical story, as mentioned previously in Section 2.5:

> Then the eyes of both [Adam and Eve] were opened, and they knew that they were naked; and they sewed fig leaves together and made themselves aprons.
> And they heard the sound of the Lord God walking in the garden in the cool of the day, and the man and his wife hid themselves from the presence of the Lord God among the trees of the garden. But the Lord God called to the man, and said to him, "Where are you?" And he said, "I heard the sound of thee in the garden, and I was afraid, because I was naked; and I hid myself" (Genesis 3:7-10).

The Garden And The Flood

In this story Adam and Eve are said to have made clothing of fig leaves, and then they hid among the trees. Thus, it is natural to imagine that the trees among which they were hiding could have included fig trees.

The tree near the center of the picture with the seal is faced on both sides by two-legged winged creatures, and these creatures also appear to be picking fruit into baskets. One could imagine that the fantastic creatures could represent some version of the biblical cherubim. In fact, the idea of cherubim picking figs in the garden of Eden already appears explicitly in *Adam and Eve* in a story that reports events after Adam and Eve were expelled from the garden:

> Then God commanded the cherub, who kept the gate of the garden with a sword of fire in his hand, to take some of the fruit of the fig-tree, and to give it to Adam.
> The cherub obeyed the command of the Lord God, and went into the garden and brought two figs on two twigs, each fig hanging to its leaf; they were from two of the trees among which Adam and Eve hid themselves when God went to walk in the garden, and the Word of God came to Adam and Eve and said unto them, "Adam, Adam, where art thou?"
> And Adam answered, "O God, here am I. When I heard the sound of Thee and Thy voice, I hid myself, because I am naked."[6]

This report states clearly that Adam and Eve were hiding among the fig trees. The resemblance between the biblical account, the *Adam and Eve* version, and the above image from the Urartian cuneiform tablet suggests that all three of these records may, at least in part, have had a common origin in ancient Urartu. As

5. Urartu and Lake Van

suggested in Figures 4.1 and 4.2, the garden of Eden may have been located in the land that came to be called Urartu and Armenia, and Armenian figs are a popular variety today. As a bonus, we may also now have an early pictorial concept for the appearance of cherubim.

Similar but much earlier imagery is to be found on cylinder seals from Ebla. The impression of one of these seals has been published previously by Matthiae.[7] The symbolism on that seal could also possibly relate to scenes from the garden of Eden. The uncoiled snake in the impression seems to be accompanied by anthropoid beings and trees, and the seal is said to date from about 2000 to 1800 B.C. In this context one might recall also the similar cylinder seal impression shown in Figure 2.1. That seal has been said to date from the slightly earlier period of about 2200-2100 B.C. These earlier dates are shortly after the date of the garden of Eden stories that will be inferred in Table 12.4.

In about 300 A.D., Armenia became the first nation in the world to adopt Christianity as its official religion. As Christians, the people would have maintained an interest in any garden of Eden symbolism. A less ancient (915-921 A.D.) representation of Adam and Eve and their temptation was recorded as one of a series of remarkable bas-relief friezes on an outside wall of the Armenian Church of the Holy Cross on Akhtamar, the second largest island of Lake Van. This lake is represented by the unnamed blob on the lefthand side of the Terrestrial Paradise shown previously in Figure 4.2. The friezes include life-size human and animal figures illustrating subjects from Bible history. Of particular interest here is the depiction of "Adam and Eve, standing naked beside the tree of life, and, a little further, the serpent tempting Eve. The treatment of the human form is primitive and almost barbarous, recalling the Romanesque."[8]. A picture of these scenes is shown in Figure 5.1.[9] It may be noted that in this representation of the temptation the serpent is seen as a snake with legs. Significant damage to the figures had already occurred before this picture was taken in about 1898. After the picture was taken, the church was closed by the Ottoman Turks, and further damage to the bas-reliefs due to gunshots and vandalism occurred during and after the Armenian genocide. Thus, later pictures of these scenes are of less visual interest. An adjoining monastery built in 653 A.D. now also has been destroyed.

5. Urartu and Lake Van

Figure 5.1. Garden of Eden frieze on the north wall of the church at Akhtamar, including Adam and Eve at the tree of life, and the serpent tempting Eve at the tree of the knowledge of good and evil.

Archeological findings have suggested that the garden of Eden may have been in the area including modern day Lake Van that came to be known as Urartu (Ararat) and then as Armenia. Thus it may be significant that, of the many Bible-related pictures that could have been chosen for the outside walls of the Akhtamar church, significant space was given to the Genesis story of the temptation of Adam and Eve in the garden of Eden. One could imagine that the familiarity of that particular story in the local lore of Urartu may have made it an important subject for inclusion.

5.3 Lake Van today

Much of Sections 4.5 and 4.6 was devoted to a map of the Terrestrial Paradise as a place in Armenia that may have included the past location of the garden of Eden and perhaps also the landing place of the ark. Many of the detailed geographical features of this map incorporate information from the Bible and from the classical geographers of the Greco-Roman world. The information provided by the classical authors and consequently by the map isn't always accurate and thus doesn't agree in all points with modern geographical reality.

A subject of special interest in this study is the lake now known as Lake Van. It was formed by an eruption of the major volcano called Nemrut (or Nimrod, etc.) blocking water flowing from the Van basin in Armenia. That eruption is thought by geologists to have occurred a few hundred thousand years ago. As suggested in Section 4.5, this lake may be associated with the garden of Eden.

5. Urartu and Lake Van

For various reasons it is appropriate here to introduce a more modern map of Lake Van, and such a map is in Figure 5.2.[10] By comparison it is clear that the representation of Lake Van on Bowen's map in Figures 4.1 and 4.2 is of limited accuracy. Particularly striking in Figure 5.2 is the existence of a large shallow bay, the Ercis (or Arjish, etc.) Gulf on the north side of the lake, which is not hinted at in the earlier map. The existence of this arm of the lake may help in interpreting the multiple lakes reported by Pliny.

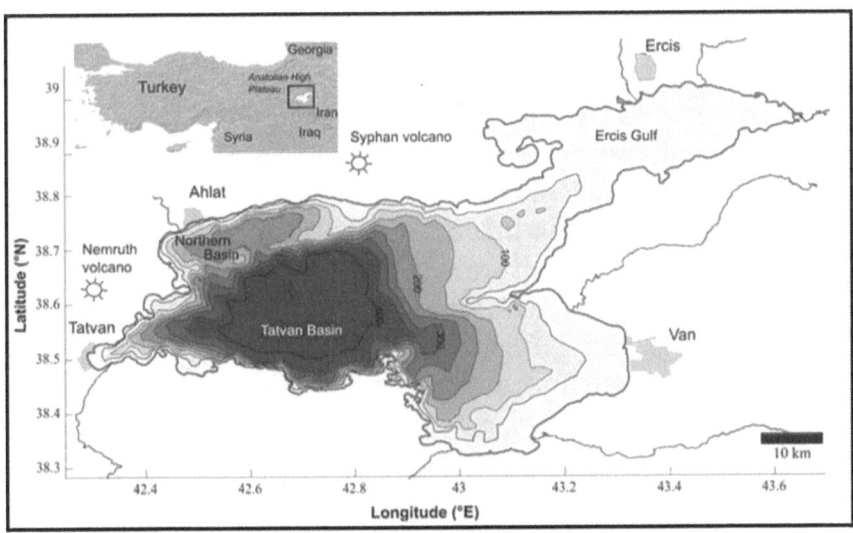

Figure 5.2. Bathymetric map of modern Lake Van, including also the neighboring cities of Ahlat, Ercis, Tatvan, and Van, and the sites of the two volcanoes Nemruth (Nimrod, etc.) and Syphan (Sipan, etc.). Current water depth in meters is indicated. (Reproduced with permission)

Lynch has commented on the possible relationship of Lake Van to the reports of the earlier classical authors:

The Garden And The Flood

> The mention of two lakes by Pliny need not perplex us over-much; for his [Pliny's] Arethusa no doubt denotes the Arjish arm of Lake Van, and his Thospites the principal body of water with the city of Van, the Dhuspas of the cuneiform inscriptions [also Thuspa], upon its eastern shore. Ptolemy, on the other hand, entangles the subject still further by separating the lakes of Areesa – no doubt the Arethusa of Pliny – and Thospitis by four degrees of longitude. This geographer does not give us any indications as to the properties of the lake waters; but he tells us that the Tigris is partly a river of Armenia and that its sources constitute Lake Thospitis. The position which he assigns to the town of Artemita – which is probably the modern Artemid – is further evidence that in speaking of Lake Areesa or Arsissa he was in fact referring to Lake Van. One cannot help concluding that his Thospitis with the town of Thospia was actually the self-same sheet of water.
>
> I think it is plain that the names Thopitis, Thospites, Arsene, Arethusa, and Areesa or Arsissa, are all applied to the great basin with the two immemorial cities, Dhuspas – the modern Van – and Arjish. Moreover, I should be surprised to learn that any lake exhibiting the same properties had been discovered in the belt of mountains south of Lake Van in which the present sources of the Tigris are found.[11]

According to Lynch and others, Pliny was incorrect in associating the lakes Arethusa and Thospites with the headwaters of the Tigris River. Thospites is likely to correspond to the deepest basin of what is now called Lake Van with its ancient associated city of Dhuspas. Ancient Arethusa is probably identifiable with the more shallow Arjish arm of the lake. Thus, it may be supposed that the map given in Figure 5.2 was approximately valid in classical times except that with Lynch's

5. Urartu and Lake Van

interpretation the main body of the lake including the area labelled Tatvan Basin would have been referred to by Pliny as Lake Thospites. Likewise the Ercis Gulf would have been known then as Lake Arethusa.

It is also possible, of course, that some significant changes may have occurred in the configuration of these lakes between the time of Pliny and the present. Even greater changes may have occurred since the earlier time that might be associated with the garden of Eden. Thus, the water level of the surface of the lakes has varied substantially over the centuries, and these changes will be discussed in Chapter 7. Erosion could have affected the shoreline and the now-broad channel between the lakes. Sedimentation could cause a gradual rise in the floors of the lakes. Tectonic forces in this volcanic region could also have caused significant changes in the lake floor and shoreline.

Other statements by the ancient writers have been confirmed by modern explorers:

> The account given by Strabo of the cleansing properties of the lake is thus confirmed in a striking manner. Indeed, the bather issues from his swim as though his limbs had been rubbed with soap – but with a soap of extremely agreeable quality, leaving a velvety feeling upon the skin. The great buoyancy of the waves enhances the pleasure of such exercise, and they are at once pellucid and sparkling under the ruffle of the breeze. On the other hand they are most unpleasant to the taste. The colour of the sheet of water cannot be given in a single word; and indeed it varies with extraordinary range of scale. A cobalt [blue] of great brilliancy is perhaps the most normal hue; but a certain milky paleness is seldom quite absent, becoming

> invested at morning and evening with an infinite number of delicate tints.
>
> Only one kind of fish is found in Lake Van, resembling a large bleak. But, often as I have bathed, I have never seen one gliding through the water, or surprised a shoal while following the shore. It is possible that they adhere to the estuaries of the rivers, up which they make their way in large numbers to spawn during the season of spring freshets.[12]

This excerpt confirms in rapid order the cleansing properties of the lake, the density or buoyancy of the water, and the fact that only one kind of fish lives there. All of these properties were known to the Greco-Roman authors. The excerpt also illustrates a modern author's joy of swimming in Lake Van. It may be noted that the only fish known to live in the lake is today known as the pearl mullet.

As a reminder, we may recall again a few lines of Pliny quoted previously in Section 4.6:

> It [the Tigris River] runneth into the Lake Arethusa, which beareth up all that is cast into it; and the Vapours that arise out of it carry Clouds of Nitre. In this Lake there is but one kind of Fish, and that entereth not into the Channel of the Tigris as it passeth through; as likewise the Fishes of the Tigris do not swim out into the Water of the Lake. In its Course and Colour it is unlike the other: and when it is past the Lake and meeteth the Mountain Taurus, it loseth itself in a Cave, and so runneth under, until on the other side it breaketh forth again in a Place which is called Zoroanda.[13]

These sentences and those of Strabo mentioned above support the identifications made by Lynch. The brilliant blue color reported

5. Urartu and Lake Van

by Lynch is probably similar to the well-known blue color of the waters of Crater Lake in Oregon. Both Lake Van and Crater Lake are endorheic lakes (lakes with no outlet stream), and there is relatively little suspended matter to diminsh the clarity of their waters. Oswald has also commented on the color of the water of Lake Van:[14] "The lake is doubtless of great depth, for its waters appear an intense indigo-blue when seen from above." It may be remarked here that one of the tributaries of the Tigris River does indeed flow through a tunnel in semi-agreement with the above quotation of Pliny. This circumstance will be considered in more detail in Section 10.5.

The city of Van at the eastern end of the main body of the lake of the same name has been the most important settlement on the shore of the lake for thousands of years. It has often been referred to as a garden city because of its beautiful flowers, vegetable gardens, vineyards, grain fields, and orchards of fruit and nut trees. Its citadel has served as a readily defendable fortress through much of the city's history:

> Of the various sites which one might select upon the shores of the lake of Van, none would present as great advantages for a populous and self-contained settlement as that of the city from which it derives its name. . . . Of the beauty of the site it would not be possible to speak too highly; but I tremble to provoke in my English reader a nausea of descriptive writing. The Armenians have a proverb which is often quoted: Van in this world and paradise in the next. The comparison might be justified under happier human circumstances, the perversity of man having converted this heaven into a little hell. Its aptness may be recognised during the course of a walk in the neighourhood, or from the standpoint of the rock

which supports the citadel. In the north across the waters is outspread an Italian landscape – a Vesuvius or an Etna, with their sinuous surroundings, on an Asiatic scale. Nearer at hand and fully exposed, the long barrier of the Kurdish mountains recalls the wildest scenery of the Norwegian coast. From the city herself as from the extremities of the wide basin, the short, sharp ridge of Varag is seen with pleasure to the eye, lifted some 4500 feet above the waters, and, at evening, reflecting the sunset in the most varied hues. The lake is not sufficiently large to separate these various objects by distances which preclude under ordinary conditions the simultaneous enjoyment of the beauty of all from a single shore. And it is large enough to spread at their feet with all the qualities of the ocean – the depth and vastness and changing surface of the high seas.[15]

The human misery implied above would have been well known to Lynch's contemporary readers. It concerns the lot of the native Armenians who at the time of his publication in 1901 had recently experienced multiple massacres at the hands of the Ottoman Turks with a death toll of many tens of thousands.[16] Beginning in 1915, just a few years later, most remaining Armenians, about 1.5 million of them, were massacred by the Turks in one of the largest genocides of modern times.[17] Also, most of the ancient Armenian churches and monasteries were destroyed or converted into mosques. Destruction of the cultural heritage of the Armenians in their former lands has continued to the present, including the leveling in 2005 of a medieval-era cemetery that had contained thousands of majestic funerary monuments.[18] The past and recent perpetrators of these events have denied that they ever occurred.

5. Urartu and Lake Van

The well-known former beauty of Lake Van and its settlements is evident from previous quotations, and it is striking that the Armenian proverb considers Van to be the closest place to paradise on earth. This is of interest in the present context because, as suggested already in Figures 4.1 and 4.2, the Lake Van region is sometimes thought to have been included in the original Terrestrial Paradise. It is tempting to imagine that the proverb quoted above could recall a historical reality in which the garden of Eden actually existed in the vicinity of Lake Van.

5.4 Geography of Eden

As discussed in Section 4.4, the Bible provides only limited information concerning the locations of the garden of Eden and the landing place of Noah's ark. In approximate terms these locations would seem to be in the areas indicated on the maps shown in Figures 4.1 and 4.2. We start here by summarizing very briefly the biblical results that have already been discussed. The garden of Eden is said to have included the headwaters of four rivers (Genesis 2:10). From the names of two of these, the Tigris River and the Euphrates River, a general location in Urartu can be inferred for the site of the garden. As implied already in the maps, the land of Eden might sometimes represent a broader area than just that corresponding to the garden itself.

Concerning the landing place of the ark, the Bible is also not as specific as one could wish: "[T]he ark came to rest upon the mountains of Ar'arat" (Genesis 8:4). The "mountains of Ar'arat" don't refer to a particular mountain. All that can be inferred from this statement is that the landing place was somewhere in the generally mountainous area formerly called Urartu. The dominant geographical feature of Urartu was the lake (or sea) now known as Lake Van, and this lake has been referred to as "the ancient nucleus of Urartu."[19] Thus, in early times the environs of this lake may have had some association, real or imagined, with the headwaters of the rivers and with Noah's flood.

Though the biblical texts are limited, there are other sources of geographical information. One interesting and possibly useful source is *Adam and Eve*, quoted previously in Sections 2.6, 2.7, 4.4, and also in Section 5.2 above. Those quotations involved especially considerations of people and events in the garden. Several further quotations that refer to the geography of the surrounding land of Eden are included here. For convenience, references to *Adam and Eve* in the remainder of this chapter include the indication of book number (1 or 2), chapter, and verse at the end of each quoted section rather than at the end of the chapter. The subjects are arranged mostly by orientation with respect to the garden of Eden:

A. Clear sea to the north

> And to the north of the garden there is a sea of water, clear and pure to the taste, like unto nothing else; so that, through the clearness thereof, one may look into the depths of the earth. And when a man washes himself in it, becomes clean of the cleanness thereof, and white of its whiteness... (1 *Adam and Eve* 1:2-3).

> Then Satan, the hater of all good, said unto Adam, "O Adam, I am an angel of the great God; and, behold the hosts that surround me. God has sent me and them to take thee and bring thee to the border of the garden northwards; to the shore of the clear sea, and bathe thee

5. Urartu and Lake Van

and Eve in it, and raise you to your former gladness, that ye return again to the garden" (1 *Adam and Eve* 28:4-5).

It appears from these quotations that there was a body of water to the north of the garden of Eden that had amazing clearness and cleansing properties, and a possible interpretation of this sea will be suggested below.

Assuming that the garden of Eden was localized to somewhere on the map shown in Figure 5.2, it follows that a portion of the Ercis (or Arjish or Ardjish) arm of the lake, could be considered north of the garden, as indicated in the quotations. As noted in the previous section, however, there may have been significant geological and hydrological changes over the several millennia since any plausible date for the garden of Eden events of interest. Such changes could limit the use of the map in Figure 5.2 to draw excessively detailed conclusions about lake or garden geography in the distant past.

Regarding the clarity of the water in the lake a few comments can be made. As discussed in Section 5.3, the Ercis arm of the lake would probably have been known to Pliny as Lake Arethusa. Today the water of the lake may be considered to be very clear, though often blue in appearance as noted by Lynch. The statement that because of the water's clearness "one may look into the depths of the earth" is also of interest. In a local tradition the clearness of the lake sometimes permits visibility of an underwater bridge that had crossed the shallow gulf (or the corresponding river) when the water level was lower.[20]

The cleansing properties noted by Strabo and again by Lynch in modern times are consistent with the statement in Adam and Eve. The cleansing agent suggested by Strabo is carbonate of soda or washing soda.[21] There is, however, one respect in which the description in Adam and Eve disagrees with the current reality of Lake Van. In one of the above quotations from Adam and Eve the sea of water is said to be "pure to the taste." Strabo's opinion concerning the taste of the water wasn't so favorable: "It contains soda, and it cleanses and restores clothes; but because of this ingredient the water is also unfit for drinking."[22] Lynch's opinion of the taste of the waters as quoted in Section 4.3 also wasn't favorable: "[T]hey are most unpleasant to the taste."[23] It is difficult to imagine that water which cleaned like washing soda could ever taste pure. A possible resolution of this seeming discrepancy is that the Bendimahi River enters the Arjish arm (or Ercis Gulf) of Lake Van at approximately the lake's most northeastern point, and other significant rivers enter this gulf as well. Thus, at lower levels the water in the gulf could once have been pure enough for drinking. On the other hand, at present higher water levels and as noted above it is only in the rivers or their estuaries that fish may now be found.

B. Rocks to the west

> Again, also, because God is merciful and of great pity, and governs all things in a way He alone knows – He made our father Adam dwell in the western border of the garden, because on that side the earth is very broad. And God commanded him to dwell there in a cave in a rock. . . (1 *Adam and Eve* 1:8-9).

> And when they came to the opening of the [western] gate of the garden, and saw the broad earth spread before them, covered with stones large and small, and with sand, they feared and trembled, and fell on their faces, from the

5. Urartu and Lake Van

fear that came upon them; and they were as dead. Because – whereas they had hitherto been in the gardenland, beautifully planted with all manner of trees – they now saw themselves, in a strange land, which they knew not and had never seen (1 *Adam and Eve* 2:2-3).

"What is this overhanging ledge of rock to shelter us, compared with the mercy of the Lord that overshadowed us? What is the soil of this cave compared with the gardenland? This earth, strewed with stones; and that, planted with delicious fruit-trees?" (1 *Adam and Eve* 4:6-7).

After their expulsion from the garden, Adam and Eve are said here to have lived in a cave under a cliff in a very rocky area to the west of the garden. Other texts in *Adam and Eve* imply that they and some of their descendants continued for many years to live in or near this cave. The possibility of living in a cave also occurs at later places in the biblical texts. Several centuries after Adam and Eve, Lot and his two daughters lived for a time in a cave (Genesis 19:30). In later times of distress, Israelites also hid themselves in caves (Judges 6:2; 1 Samuel 13:6, 22:1, 24:3; 2 Samuel 23:13; 1 Kings18:4, 18:13, 19:9, 19:13; 1 Chronicles 11:15, Psalm 57:1, 142:1; Hebrews 11:38) or contemplated the possibility of doing so (Isaiah 2:19-21; Ezekiel 33:27). Members of many other ancient societies are also known to have lived in caves, and much of what is known about such people comes from the study of what they left behind as art and artifacts in sheltered cave environments. It is good to remember, however, that only a tiny percentage of ancient humans could ever have lived in the relatively small number of habitable caves. Also, those that did so might not be representative of their societies as a whole. They could instead have been too poor to afford ordinary daylight housing, or they could have been refugees as Lot and his daughters were, or they could have been hunters needing temporary shelter while on food gathering expeditions.

The idea of living in a cave is not so familiar in modern times, and long-term cave occupants are sometimes known as troglodytes. It is of interest here, however, that people at least as recently as the nineteenth century A.D. have continued to live in caves in approximately the same location as Adam and Eve in the stories under discussion. In those stories the cave of Adam and Eve may have been located on the slopes of Nemrut volcano, west of the suggested site of the garden of Eden. This could correspond approximately to the location of the town of Akhlat, at the northwest corner of Lake Van:[24] "At Akhlát old town the inhabitants have taken up their abode in holes and chambers excavated in these soft volcanic rocks."

The rocky and sandy nature of the terrain in the vicinity of the cave of Adam and Eve is suggestive of a not long dormant volcanic landscape:

> From Akhlát to Tadván, the wide plain, rising towards Nimrúd Dágh [Mt. Nemrut] on the west, is wholly composed of friable grey tuff, which produces a very light, rich soil for grain. Pieces of pumice are abundantly contained in it. . . . Near Zerákh the sand is thickly strewed with large lumps of pumice, black scoriae, and obsidian, containing glassy felspar. The immense quantity of volcanic products here observable is astonishing. . . .[25]

5. Urartu and Lake Van

The towns of Akhlát and Tadván mentioned in this quotation may be seen as Ahlat and Tatvan on the map given previously as Figure 5.2.

C. Very high mountain to the north

> But when they came to the mountain to the north of the garden, a very high mountain, without any steps to the top of it... (1 *Adam and Eve* 28:10).

There were said to be two mountains in the vicinity of the garden. Adam and Eve climbed "the mountain to the north of the garden, a very high mountain, without any steps to the top of it." Perhaps the mention of the absence of steps should be understood to mean only that there was no established trail. On the other hand, the mention of steps could also be a prelude to the discussion of steps associated with some other mountain, and this will be considered in sub-section D below.

The "very high mountain" to the north of the garden could possibly be a reference to Mt. Sipan (or Syphan, Suphan, etc.) volcano indicated in Figure 5.2. Mt. Sipan is an extinct volcanic peak over 13,000 feet high. The name of this mountain is sometimes associated with the landing place of the ark:

> The Subhan, or Sipan Dagh, a magnificent conical peak covered with snow, rises abruptly from the plain, N. of lake Van. According to the tradition, Noah's ark, floating on the waters of the deluge, struck against its top, when the patriarch, congratulating himself on his escape, exclaimed Subhanu-Ilah, "Praise be to God;" and hence the name.[26]

The Garden And The Flood

Of course, many other identifications have also been suggested for the landing-place of the ark. Mt. Sipan has an advantage of including an explanation for the name of the landing place. A landing on the slopes of this volcano is also compatible with the idea of the ark floating on the waters of a then recently augmented Lake Van.

D. Not-so-high mountain to the west

> And when they had come out of the cave they went up the mountain to the west of the garden (1 Adam and Eve 34:2).

Thus, Adam and Eve also climbed "the mountain to the west of the garden." This mountain was presumably less high than the "very high mountain" to the north. The mountain to the west may also be the holy mountain mentioned later, on which their cave was situated. The actual name of this mountain today could be Mt. Nemrut (or Nemruth, Nimrod, etc.) volcano. Nemrut is a dormant (or currently inactive) volcanic peak over 9,000 feet high. It is clear that the summit of Mt. Nemrut would seem to be much lower than that of Mt. Sipan at 13,000 feet, especially considering that the surface elevation of Lake Van is about 5,600 feet. If these identifications of the mountains were correct, then Adam and Eve and some of their descendants might have lived near or on the sometimes-active Nemrut volcano. This volcano has continued to erupt intermittently in historical times.

As mentioned in sub-section C above, the mountain to the north of the garden may be a reference to the volcanic peak Sipan Dagh. It may be recalled that the mountain to the north of the garden was "a very high mountain, without any steps to the top of

5. Urartu and Lake Van

it." Emphasizing that the mountain was without any steps could be understood as a lead-in to the mention of a not-so-high mountain that did have steps. Mt. Nemrut would seem to be a reasonable candidate for this mountain. It has already been suggested that the cave dwelling of Adam and Eve may have been on the side of this mountain, which was not so tall as Sipan Dagh. As it happens, the road (or trail) between Lake Van and the summit crater of Mount Nemrut was narrow and treacherous:

> On the west of A'd-el-Jíwáz, the road is carried 200 or 300 feet above the level of the Lake, which washes the base of the travertin mountain. The rock is frequently cut in steps for the passage of beasts of burden, and is extremely slippery and dangerous. A narrow irregular alluvial plain, formed by the retiring of the mountains from the Lake towards the north, is then crossed.[27]

The steps in this trail suggest that it might be the implied reference for the mention in *Adam and Eve* of a higher mountain without steps. If this were true, it could be interpreted to suggest that the current (or recent) trail with steps along the side of Mount Nemrut was already present at the time of the writing of *Adam and Eve*.

E. New geography with eastern and western borders

> And Adam said unto Eve, "Since we know not what there is to the westward of this cave, let us go forth and see it to-day." Then they came forth and went towards the western border. . . . And God said unto Adam, "O Adam, what seekest thou on the western border? And why hast thou left of thine own accord the eastern border, in which was thy dwelling-place? Now, then turn back to thy cave,

The Garden And The Flood

and remain in it, that Satan do not deceive thee, nor work his purpose upon thee (1 *Adam and Eve* 52:13, 53:4-5).

In this quotation the geographical references seem to have changed. The cave is now said to be on the "eastern border" of their newly understood more westerly homeland, whereas formerly the cave was considered to be west from the garden of Eden. Adam and Eve could now consider going toward the "western border" of their seemingly western-shifted region of interest.

5.5 Down from the mountain

The world with which Adam and Eve were now interacting (in the account given in *Adam and Eve*) proved to be a fairly hostile environment. They occupied an area of very rocky terrain, and their home was in a cave. The geographical hints given previously could be interpreted to suggest that they were living on the slopes of Mount Nemrut volcano. They were later encouraged to go down the west side of the mountain:

> Then the Word of God came and said unto him, "O Adam, go down to the westward of the cave, as far as a land of dark soil, and there thou shalt find food." And Adam hearkened unto the Word of God, took Eve, and went down to a land of dark soil, and found there wheat growing, in the ear and ripe, and figs to eat; and Adam rejoiced over it (1 *Adam and Eve* 66:7-8).

Thus Adam and Eve were sent from their inhospitable cave environment to find a fertile land for farming. Later Cain may have farmed the same land and then settled there with his wife. After he murdered his brother, Cain is said to have married his twin sister, and the two of them moved down the west side of the mountain (Mt. Nemrut in this interpretation): "He then went down to the bottom of the mountain, away from the garden [of Eden], near to the place where he had killed his brother. And in that place were many fruit trees and forest trees. His sister bare him children, who in their turn began to multiply by degrees until they filled that place" (2 *Adam and Eve* 1:7-8). A few generations later some of the descendants of Cain's brother Seth may also have moved down from the western side of the same mountain (2 Adam and Eve 20). This western settlement of the children of Eden will be discussed further in Section 6.5.

The mountain to the west of the garden may sometimes also be referred to as the holy mountain (2 *Adam and Eve* 16:5; 19:8; 20:11,15,26-27,29-30,33,36-37; 21:4-7; 22:6,10). The representation of this mountain seems to suggest that it had a tradition of volcanic activity. The terrain was rocky, and the mountain was sometimes unapproachable: "[T]he stones of that holy mountain were of fire flashing before them, by reason of which they could not go up again" (2 *Adam and Eve* 20:33). What is perhaps the same holy mountain will be referred to again in Section 6.6, as a seemingly similar eruption event was described by the prophet Ezekiel.

It may be remarked that the story of the sins of the descendants of Adam and Eve at the base of the holy mountain (perhaps Mt. Nemrut) has parallels in the sins of later Israelites in the vicinity of other holy mountains. The *Adam and Eve* story includes reports such as the following:

> They began to go down from the Holy Mountain one after another, and to mix with the children of Cain, in foul fellowships (2 *Adam and Eve* 19:8).
>
> And when they looked at the daughters of Cain, at their beautiful figures, and at their hands and feet dyed with colour, and tattooed in ornaments on their faces, the fire of sin was kindled in them. . . , until they committed abomination with them (2 *Adam and Eve* 20:31,32).
>
> And God was angry with them, and repented of them because they had come down from glory, and had thereby lost or forsaken their own purity or innocence, and were fallen into the defilement of sin (2 *Adam and Eve* 20:34).

A similar story involves the sinning of the Israelites at the base of a different holy mountain (Mt. Sinai or Horeb) after the exodus from Egypt. That this was also considered a holy mountain or mountain of God may be seen, for example, in the following verses: Exodus 3:1, 4:27; 18:5; 19:3,17; 24:13; 31:18; 32:1; Deuteronomy 1:6; 9:10; Judges 5:5; 1 Kings 19:8. The golden calf episode in the Bible at the base of this holy mountain includes the following report:

> When Moses didn't come back down the mountain right away, the people went to Aaron. "Look," they said, "make us a god to lead us, for this fellow Moses who brought us here from Egypt has disappeared; something must have happened to him." "Give me your gold earrings," Aaron replied. So they all did – men and women, boys and girls. Aaron melted the gold, then molded and tooled it into the form of a calf. The people exclaimed, "O Israel, this is the god that brought you out of Egypt!" When Aaron saw how happy the people were

about it, he built an altar before the calf and announced, "Tomorrow there will be a feast to Jehovah!" So they were up early the next morning and began offering burnt offerings and peace offerings to the calf idol; afterwards they sat down to feast and drink at a wild party, followed by sexual immorality. Then the Lord told Moses, "Quick! Go on down, for your people that you brought from Egypt have defiled themselves, and have quickly abandoned all my laws. They have molded themselves a calf, and worshiped it, and sacrificed to it, and said, 'This is your god, O Israel, that brought you out of Egypt.' "...

When Joshua heard the noise below them, of all the people shouting, he exclaimed to Moses, "It sounds as if they are preparing for war!" But Moses replied, "No, it's not a cry of victory or defeat, but singing." When they came near the camp, Moses saw the calf and the dancing, and in terrible anger he threw the tablets to the ground, and they lay broken at the foot of the mountain. He took the calf and melted it in the fire, and when the metal cooled, he ground it into powder and spread it upon the water and made the people drink it. Then he turned to Aaron. "What in the world did the people do to you," he demanded, "to make you bring such a terrible sin upon them?" "Don't get so upset," Aaron replied. "You know these people and what a wicked bunch they are. They said to me, 'Make us a god to lead us, for something has happened to this fellow Moses who led us out of Egypt.' Well, I told them, 'Bring me your gold earrings.' So they brought them to me and I threw them into the fire, and ... well ... this calf came out!" When Moses saw that the people had been committing adultery – at Aaron's encouragement, and much to the amusement of their enemies – he stood at the camp entrance and shouted, "All of you who are on the Lord's side, come over here and join me." And all the Levites came (Exodus 32:1-8,17-25 TLB).

Thus, in both the Eden and Sinai stories, the people of God are said to have engaged in sexual immorality near a holy mountain. Similar conduct may also have occurred during the reign of King Josiah at the Temple Mount in Jerusalem: "And he broke down the houses of the male cult prostitutes which were in the house of the Lord, where the women wove hangings for the Ashe'erah" (2 Kings 23:7).

5.6 Conclusion

It has long been considered possible that the garden of Eden and the landing place of the ark may both have been in the same general area that later became known as Urartu. Several consequences of this possibility have been considered here. Archeological artifacts from Urartu and Armenia include symbolism that is suggestive of the garden of Eden, and descriptions in some ancient nonbiblical texts also seem to be compatible with biblical stories relating to Eden. This same geographical area still supports lush gardens and ancient serpent-related traditions. The volcanic nature of the terrain is in agreement with Bible-related records of the landscape including actual datable eruptions. Lake Van itself is an unusual body of water that was referred to by several ancient geographers.

One of the references used several times in this chapter is *Adam and Eve*. That book is a composite work that includes stories that may have been written over an extended period of time. Of particular interest have been some of the earliest events relating to Adam and Eve, their activities, and their immediate descendants. Not considered here are other Bible-related stories in *Adam and Eve* that seem to extend even to New Testament

5. Urartu and Lake Van

events. The origin of many of these stories is difficult to discern. Volcano-related happenings mentioned in *Adam and Eve* have been considered briefly, and similarities between these stories and the biblical writings identified with Ezekiel will be considered further in Section 6.6.

A major purpose of the first part (about sixty-nine chapters) of the first book of *Adam and Eve* seems to be to describe the various adventures and misadventures of the title characters before the arrival of their children. The second book deals mostly with the activities of the descendants of Adam and Eve. The discussions here have provided several tentative identifications of geographical sites related to the garden of Eden. Some of these identifications will be considered further in Chapter 6. One should be careful not to attach too much significance to the detailed contents of the books of *Adam and Eve*, as many of the stories contained in them seem implausible and lacking in independent verification. However, it is perhaps fair to say that the more basic geographical aspects of these descriptions would seem to be compatible with the location of the garden of Eden being somewhere in or near the land previously identified with the Terrestrial Paradise, possibly close to the lake now called Lake Van and especially its shallower Ercis (or Arjish, etc.) Gulf. Adam and Eve are said to have moved from the garden of Eden to a cave home, and that home may have been in the vicinity of Mt. Nemrut volcano. It is also not unreasonable to suppose that early inhabitants, including perhaps descendants of Adam and Eve, may from time to time have had encounters with an active volcano. This possibility will be considered further in Chapter 6. It seems unnecessary to document in detail here, but the then future flood event is also mentioned several times in *Adam and Eve* (1 *Adam and Eve* 53:7, 2 *Adam and Eve* 8:10-13, 17:15, 21:10, 22:5).

---

1. B. B. Piotrovsky, *The Ancient Civilization of Urartu*, Translated from the Russian by J. Hogarth (Cowles Book Company, Inc., New York, N.Y., 1969), p. 160.
2. Ibid., p. 149, Plate 95. See also the caption on p. 214.
3. Ibid., p. 148, Plate 94. See also the caption on p. 214.
4. Ibid., p. 158.
5. Ibid., p. 73, Plate 34. See also the caption on p. 210.
6. *Adam and Eve*, Translated by S. C. Malan from the Ethiopic edition edited by E. Trumpp, included in The Forgotten Books of Eden, Edited by Rutherford H. Platt, Jr. (Alpha House, Inc., 1927), pp. 3-81; Book 1, 36:1-3.
7. P. Matthiae, *Ebla, An Empire Rediscovered*, Translated by D. Holme (Doubleday & Company, Inc., Garden City, New York, 1981), from the page following p. 224.
8. H. F. B. Lynch, *Armenia Travels and Studies*, Volume 2 (Longmans, Green, and Co, London, 1901), p. 132.
9. Ibid., Figure 143, on the page following p. 132.
10. Adapted from C. Glombitza, M. Stockhecke, C. J. Schubert, A. Vetter, and J. Kallmeyer, "Sulfate reduction controlled by organic matter availability in deep sediment cores from the saline, alkaline Lake Van (Eastern Anatolia, Turkey)," *Frontiers in Microbiology*, Volume 4, Article 209 (July 2013), pp. 1-12, Figure 1, p. 2.
11. H. F. B. Lynch, op. cit., p. 42.
12. Ibid., p. 46.
13. *Pliny's Natural History in Thirty-Seven Books*, Translated by Dr. Philemon Holland (George Barclay, Leicester Square, 1847-48), Book 6, Chapter 27; pp. 142-143.
14. F. Oswald, *A Treatise on the Geology of Armenia*, Thesis accepted by the University of London for the Degree of

Doctor of Science (Published by the author at Iona, Beeston, Notts, 1906), p. 118.
15. H. F. B. Lynch, op. cit., pp. 38-39.
16. W. J. Wintle, *Armenia and its Sorrows* (Andrew Melrose, London, 1896).
17. "The forgotten holcaust: The Armenian massacre that inspired Hitler," *The Daily Mail* (London), 11 October 2007.
18. S. Maghakyan, "Sacred stones silenced in Azerbaijan," *History Today*, Volume 57, Number 11, pp. 4-5 (November 2007).
19. B. B. Piotrovsky, op. cit., p. 199.
20. W. K. Loftus, "On the geology of portions of the Turko-Persian frontier, and of the districts adjoining," *The Quarterly Journal of the Geological Society of London*, Volume 11, pp. 247-344 (1855); p. 319.
21. *The Geography of Strabo, Published in eight volumes*, Translated by Horace Leonard Jones (Loeb Classical Library, Harvard University Press, Cambridge Massachusetts, 1928), Volume V, Book 11, Chapter 14, Section 8, p. 327; H. F. B. Lynch, op. cit., p. 40.
22. *The Geography of Strabo*, op. cit.
23. H. F. B. Lynch, op. cit., p. 46
24. W. K. Loftus, op. cit., p. 322.
25. Ibid., p. 323.
26. *A Handbook for Travellers in Turkey: Describing Constantinople, European Turkey, Asia Minor, Armenia, and Mesopotamia*, Third Edition, (John Murray, Albemarle Street, London, 1854), p. 273.
27. W. K. Loftus, op. cit., p. 322.

6. LAND OF EDEN

"Have the gods of the nations delivered them, the nations which my fathers destroyed, Gozan, Haran, Rezeph, and the people of Eden who were in Tel-assar?" (Isaiah 37:12).

6.1 Introduction

Previous chapters have introduced the idea that there was once a land known as Eden, and that land included the garden of Eden, the origins of four rivers, and the landing place of the ark. Several specific stories relating to Eden have been discussed including the creation accounts, the temptation of Adam and Eve, the murder of Abel, and others. Particular attention was paid to chronological implications.

More may be said about the land of Eden and its inhabitants, and a few additional stories are considered here. One of the interesting inhabitants was the serpent. That creature, motivated by Satan, played a central role in the temptation story, and several aspects of snakes in the land that became Armenia and in Bible-related literature are reviewed in Section 6.2. After their disobedience, Adam and Eve were expelled from the garden of Eden. Their new home and the homes of their descendants, introduced in Chapter 5, are considered further in Section 6.3.

6. Land of Eden

Two not-so-familiar biblical regions associated with the early descendants of Adam and Eve are Togarmah and Telassar. The person Togarmah was a great-great-grandson of Noah, and the location of the large area called Togarmah is reviewed in Section 6.4. The place Telassar may be more localized. It is sometimes associated with the "children of Eden," and its probable location is considered in Section 6.5. Several references to the land of Eden are given by Ezekiel, and his hints of a volcanic eruption are of particular interest. These references are examined in Section 6.6. Section 6.7 reviews known eruptions of the Mt. Nemrut volcano, and it is suggested that one of those eruptions may have led to the implications of volcanism noted by Ezekiel and other bible-related authors.

6.2 The serpent

A particularly significant non-human participant in the garden of Eden stories is, of course, the serpent. Thus, it could be of interest to know whether any ordinary snakes might have been living at the high altitude and with the short summers of, for example, Lake Van. The surface elevation of the lake is 5,640 feet, or more than a mile above sea level. As it happens, however, snakes do live near Lake Van, and they have done so since ancient times. Of special interest in this context is the town of Akantz and perhaps the cliff behind and above it.[1] Akantz is also called Ercis and is located near the shore of Lake Van at the north end of the lake. Its position may be seen on the map in Figure 5.2 of the previous chapter.

The Garden And The Flood

Akantz is closely associated with a famous dwelling place of snakes and lizards:[2]

> Close by is the cave where a nest of serpents or large lizards are reputed to have lodged since immemorial times, and have been seen by modern travellers.[3] The place is described as being situated about two miles east of Akantz near the road to Haidar Bey.[4] Messrs. Belck and Lehmann, who have visited Akantz since I was there, were brought some objects in bronze, of which one represented a serpent, and another contained cuneiform characters. They were found by the natives among the ruins of this Zernak.[5]

The bronze serpent just mentioned was described only briefly by Belck and Lehmann, and an approximate translation of their writing on this subject is as follows:

> Excavations in the ruin hill of Sirnakar [Zernak] near Ardjisch on the north shore of Lake Van appear especially promising. We have been brought objects from a variety of places there with the invitation to dig. With the current famine and lack of employment, people are overjoyed when we start to dig anywhere. Among other things, they brought us a bronze snake of about 0.7 m length, coiled several times and having turquoise eyes. . . . We intend to dig in the spring after the snow melts there.[6]

Presumably, snakes and lizards have long lived near Lake Van. A more colorful description of a visit to the cave mentioned above was provided by Müller-Simonis, and an abbreviated and approximate translation of his report follows:

6. Land of Eden

The new Ardjîch [Ercis], or to be more precise, Agantz [Akantz], is a small town built at about one hour and a half away from the lake and is surrounded by a line of villages. The place is an important stopping point on the way from [the city of] Van to Erzerum. . . . [A friend there] assured us of the marvellous accounts that we had heard concerning the Ilan-Dagh, or mountain of the snakes [near the northernmost point of Lake Van]. This mountain, very famous everywhere in the country, should probably instead be called a hill. It dominates the path from Haidar Bey to Agantz, and we spent the morning there investigating it.

Natural caves open at the foot of the rocks, and in the cracks of these caves large snakes may be found. They are prisoners there and cannot leave, but one sees them moving. They have been there from time immemorial, and it is their presence that gave its name to the mountain.

Brrr... I do not like snakes, but they are worth seeing!

This morning we will visit these famous caves. They are about half an hour from Agantz. The volcanic rocks form a kind of great broken wall, dominating the path. The king Sarduris II [764-735 B.C.] carved there two tablets which have cuneiform inscriptions dating from his reign. A third tablet is uninscribed, . . . or perhaps the inscription was removed at an earlier time.

Close to these tablets a small cave opens where one can enter by bending over. Basically, the rock has a long slit, and it is clear that a gray mass fills it. The guide touches this mass with his rod causing it to contract and move! Thus the snakes are not a myth! I always had (I must acknowledge my weaknesses) a nervous fear of snakes, and I need all my courage to examine this phenomenon more closely. These are not serpents, it is a group of crawling animals with stubby legs and short tails. Their heads are turned away from the slit, making it impossible to determine the species of these animals.

They are certainly reptiles (lizards) and they are probably harmless. Their length seems to be 30 to 40 centimeters. It would have been easy to get more information by pulling one of these animals out of its hiding-place, but none of our men wanted to try to do so. I felt only one desire: to get out of there as quickly as possible. Tradition has it that these animals are trapped in this fault. . . . Much more likely, these animals simply have for generations chosen their winter quarters in this cave, and thus have given the name Ilan-Dagh, mountain of the snakes, to this hill since the earliest times.[7]

On this particular visit lizards were encountered rather than the famous snakes. A Bible reader might be inclined to ponder whether the existence of such reptiles could be related in any way to the occurrence of a serpent (with and without legs) in the Genesis story of the garden of Eden. As seen in previous chapters, that garden is often suggested to have been near Lake Van. It may be recalled that, as noted in Section 5.2, snakes also appeared on at least the helmets and cylinder seals of Urartu. One could ponder as well whether the bronze serpent mentioned above might have resembled in any way (appearance, purpose, creation date, etc.) the bronze serpent made by Moses in the wilderness and mounted on a pole for the healing of bites by venomous snakes (Numbers 21:6-9). That serpent later was adopted by the Israelites as an idol (2 Kings 18:4).

After the disobedience of Adam and Eve, they and their descendants were condemned to experience many forms of pain and suffering. The serpent was also cursed:

The Lord God said to the serpent,
"Because you have done this, cursed are you above all

6. Land of Eden

cattle, and above all wild animals; upon your belly you shall go, and dust you shall eat all the days of your life. I will put enmity between you and the woman, and between your seed and her seed; he shall bruise your head, and you shall bruise his heel" (Genesis 3:14-15).

Thus, the serpent would thenceforth go on his belly in contrast to cattle and most wild animals. This presumably is an indication that the reference is to animals originally having four legs. In the words of Josephus, this aspect of God's punishment of the serpent was the following: "And when he [God] had deprived him of the use of his feet, he made him to go rolling all along, and dragging himself upon the ground."[8] The natural assumption is that in the temptation story one particular serpent, possessed by Satan, and presumably also its descendants were changed from being more like lizards to being snakes. Given that snakes and lizards may have existed together in the vicinity of the garden of Eden "from time immemorial," such a transformation, real or mythical, might not have seemed incomprehensible to the human inhabitants of the garden.

The interpretation of the serpent as possibly being originally a lizard-like creature is not reflected in all of the ancient imagery. There, the creature which sometimes has been regarded as tempting Adam and Eve often resembles a snake. If these are actually images referring to the Adam and Eve story, then the snake symbolism perhaps should be considered a simplification reflecting the appearance of the serpent after it had been condemned to slither on the ground and in the absence of information on the appearance of the serpent's original legs if any.

The Garden And The Flood

On the other hand, Josephus indicated clearly that the serpent originally did have feet. It should also be recalled that the tenth century bas-relief representation of the Adam and Eve temptation story on the outside wall of the former Akhtamar church (Figure 5.1 of the previous chapter) includes a picture of the serpent tempting Eve. It is notable that, in the scene showing the temptation of Eve, the serpent does in fact have legs and feet.

There are of course many different kinds of snakes, and a person is free to wonder about the snake descendants of the serpent in the garden of Eden story. One possibility is that they might be something like the snakes that were said to have resided in the region visited by Müller-Simonis a century ago. Pending more information on that, one might also consider any descriptive hints in Bible-related literature concerning the temptation event and its consequences. For this purpose it may be of interest to consider the colorful descriptions in *Adam and Eve*:

> Then Adam and Eve came out at the mouth of the cave, and went towards the garden. But as they drew near to it, before the western gate . . . , they found the serpent that became Satan coming at the gate, and sorrowfully licking the dust, and wriggling on its breast on the ground, by reason of the curse that fell upon it from God. . . . [A]fter it had become venomous, by reason of God's curse, all beasts fled from its abode, and would not drink of the water it drank; but fled from it. When the accursed serpent saw Adam and Eve, it swelled its head, stood on its tail, and with eyes blood-red, did as if it would kill them.[9]

This story provides a down-to-earth interpretation of God's somewhat mysterious and poetical-sounding curse as reported in the biblical account in Genesis 3:14-15, quoted above. Besides being condemned to crawl on its belly, the serpent's descendants and the descendants of Adam and Eve were destined for an unending conflict with each other. The above quotation from *Adam and Eve* provides the explicit suggestion that as part of the curse the serpent had become venomous. In a confrontation with a venomous snake a person might try to trample the snake's head ("he shall bruise your head"), its most vulnerable body part and the site of its poison carrying apparatus. At the same time the snake might try to bite or inject venom into a person's heel or ankle ("you shall bruise his heel"), the person's most easily reachable body part during an attempted trampling. Thus the biblical version of the curse (Genesis 3:14-15) may be imagined once to have included the idea of the serpent being made venomous, as in the *Adam and Eve* text. This concept is also included in Josephus's version:[10] "[God] inserted poison under his [the serpent's] tongue, and made him an enemy to men; and suggested to them, that they should direct their strokes against his head, that being the place wherein lay his mischievous designs towards men, and it being easiest to take vengeance on him that way."

The idea of snakes biting heels is also not otherwise foreign to the Bible, and even horses were understood to be susceptible: "Dan shall be a serpent in the way, a viper by the path, that bites the horse's heels, so that his rider falls backward" (Genesis 49:17). This imagery should already be familiar to those who have watched Western movies and seen what happens when a cowboy's horse is startled by a rattlesnake.

In the *Adam and Eve* serpent story quoted above, the author seems to be suggesting a particular kind of venomous snake. The snake is said to have stood on its tail and swelled its head, and its eyes were said to be "blood-red." The cobra is of course well known for raising the front part of its body and spreading its hood, making it appear to have a large head and be more threatening in appearance. Images of cobras also often show them with red eyes made of ruby, red glass, or today even red light-emitting-diodes. According to *Adam and Eve*, the serpent was eventually carried by the wind to India: "And a wind came to blow from heaven by command of God that carried away the serpent from Adam and Eve, threw it on the sea shore, and it landed in India."[11] Perhaps the ancient author imagined the serpent in the garden to be the ancestor of the archetypal Indian cobra, a kind of snake with which he may have been familiar and which is said now to kill about 10,000 people per year. Also, concerning the idea that the serpent "stood on its tail," it may be noticed that the snake of Figure 2.1 is mostly extended vertically, so this symbolic snake doesn't have its body coiled for balance in the familiar manner of the cobra. It may be recalled, however, that the bronze snake acquired by Belck and Lehmann (quoted previously in this section) was said to have been "coiled several times."

6.3 The new home

After succumbing to temptation by the serpent, Adam and Eve were driven from the garden of Eden. Given the previously discussed information on the location of the ancient "mountain of snakes," one could imagine that the subsequent home of Adam

6. Land of Eden

and Eve might possibly have been nearby, presumably to the west of the garden, as discussed in Section 5.4B.

The following quotation is an approximate translation of the text immediately preceding the text of Müller-Simonis given above. Thus, these words describe the landscape in the vicinity of the snake cave as the visitors approached it from the east:

> We walked along the north shore of the [Ercis branch of the] lake. As I said, the Bendimahi-Tchaï [River] forms an important geological separation; we crossed it and then entered the volcanic region, including Sipan Dagh [a beautiful conical volcano of about 13,314 feet, almost as high as the highest summits in the forty-eight contiguous United States] at its center. The volcanic nature of the terrain and the southern sloping exposure at midday give this lake shore a much warmer climate than the south shore. The snow has almost completely disappeared; it is no longer winter, but seems to be still autumn.
>
> Everywhere the partridges call, and a hunter would be in a land of plenty.
>
> We first crossed the country of Arnis, where several villages lie at the foot of the mountain. Then we crossed a river, and a half hour later we arrived at the village of Haidar Bey, perched on a hill. Walking at a good pace, we soon caught up with our luggage.
>
> To our amazement, the Ardjîch [Ercis] plain that extends to our feet is covered with thick snow. Its valley is undoubtedly more open to the winds from the North and is therefore colder.[12]

The main reason for including this report is to emphasize the dramatic volcanic nature of the lands near the snake cave and to its west. As quoted in the previous section, "Natural caves open at the foot of the rocks, and in the cracks of these caves, large snakes may be found."[13] When exiled from the garden of Eden, Adam and Eve would have needed to find a sheltered place to live. As with the snakes since ancient times, one possibility might have been to locate a suitable cave in the volcanic rock.

Volcanic caves occur in several types. One classic form is the lava tube, which may occur in a lava flow as the surface cools and solidifies while the lower levels continue to flow. Such flows tend to be unstable against tube formation because the fastest moving regions are replenished most quickly from the hot lava source. The slower moving areas tend to become the coolest and solidify first. When the source of the hot lava is exhausted, the remaining fluid lava may drain from the system leaving a network of lava tubes sometimes extending over distances of miles.

Rift caves can form in lava that is subject to stress during cooling or due to tectonic forces after solidification. Erosional caves can form beneath and especially at the edges of lava flows due to the effects of wind and water on softer underlying material. The nature of the caves described by Müller-Simonis isn't so clear from his few words on the subject, but their occurrence at the "foot of the rocks" and perhaps near the lakeshore would suggest that erosion may have been a significant factor.

There is not much information concerning the home of Adam and Eve after they left the garden of Eden, and the information that exists should be regarded as of uncertain reliability. As mentioned in Section 5.4B, Adam and Eve are said to have moved into a cave in what seems to have been a volcanic landscape. This is mentioned because, as discussed in the previous section, snakes and lizards that were reminiscent of the serpent in the garden of Eden story have long been known to live, or at least spend their winters, in sun-warmed caves on the north shore of Lake Van. Caves are also possible in the vicinity of Mt. Nemrut volcano to the west of the imagined location of the garden of Eden. It would seem possible that early human inhabitants of the same region may sometimes have found such caves to be tolerable for their own dwellings. As a modern example, people of this area have continued at least as recently as the nineteenth century A.D. to excavate their own cave homes, as mentioned in Section 5.4B.

The Mt. Nemrut volcano is also a location of particular interest as a site of possible volcanic eruptions in Bible times. The behavior of this volcano would intermittently have become well known to the inhabitants of Urartu and its surrounding civilizations. A more recent encounter may have been the basis for a text ascribed to the prophet/priest Ezekiel (c. 622 B.C. to 570 B.C.) and discussed below in Section 6.6.

6.4 Togarmah

A connection between the early biblical personalities and the ancestors of the Armenians (Urartians) occurs in the stories about the first Armenian king Hayk or Haik:

> According to the earliest Armenian accounts, written sometime between the fifth and eighth centuries AD, the Armenian people are the descendants of Japheth, a son of Noah. After the ark had landed on Mt. Ararat, Noah's family settled first in Armenia and, generations later, moved south to the land of Babylon. The leader of the

Armenians, Haik, a descendant of Japheth, unhappy with the tyranny and evil in Babylon, rebelled and decided to return to the land of the ark. The evil Bel, leader of the Babylonians, pursued Haik. In the ensuing war, good conquered evil when Haik killed Bel and created the Armenian nation. Haik became the first Armenian ruler and his descendants (*Hai* or *Hay* [pronounced high] the Armenian word for "Armenian") continued to lead the Armenians until King Paruir, a descendent of Haik, formed the first kingdom of Armenia and had to face the mighty Assyrian foe.

This legend, probably as old as Mesopotamian legends, including that of Gilgamesh, not only blends historical facts with fable but manages also to place the Armenians in a prominent position within the biblical tradition. Noah, after all, was "the second Adam" and his descendants were chosen and blessed by God to repopulate the earth.[14]

The connection between the Armenians and the family of Noah occurred via Togarmah:

> The people known as Armenians to Darius and to classical writers have always been accustomed to prefer the name of their reputed progenitor, Hayk, the son of Togarmah, great-grandson of Japhet. They call themselves the Hayk or children of Hayk. They believe that their ancestor emigrated from Babylon in a northwesterly direction and ultimately arrived upon the shores of Lake Van. They style the line of their primeval kings the Haykian dynasty, and they relate in a fabulous manner the early struggles of this dynasty with the Assyrian Power.[15]

The biblical genealogy referred to in this quote is found in Genesis 10:1-3: "These are the generations of the sons of Noah,

6. Land of Eden

Shem, Ham, and Japheth; sons were born to them after the flood. The sons of Japheth: Gomer, Magog, Madai, Javan, Tubal, Meshech, and Tiras. The sons of Gomer: Ash'kenaz, Riphath, and Togar'mah." Thus Hayk, the son of Togarmah, would be a great-grandson of Japheth, son of Noah. This genealogy is also repeated in 1 Chronicles 1:4-6.

That the descendants of Togarmah might be associated with the people of Urartu is also supported by other biblical references. In a lamentation over Tyre, many of the places with which Togarmah interacted were discussed by Ezekiel. One of those places was Tyre: "Javan, Tubal, and Meshech traded with you; they exchanged the persons of men and vessels of bronze for your merchandise. Beth-togar'mah [house of Togarmah] exchanged for your wares horses, war horses, and mules" (Ezekiel 27:13-14). The statement that Togarmah exported horses, war horses, and mules is compatible with the high regard in which Urartu held its horses. The distinction in Ezekiel between ordinary "horses" and "war horses" may also be significant:

> Among the animal bones discovered in considerable quantity on Karmir-Blur [Teishebaini, Urartu] were whole skeletons and individual bones belonging to horses of two types, a large breed and a small one. The bronze harness-pieces and bits found by the excavators, with inscriptions in the names of Menua, Argishti and Sarduri [kings of Urartu], belonged to the larger breed of horse, for which western Asia was famous. In a field near Van a stone was found with a cuneiform inscription saying: "Menua, son of Ishpuini, says: 'From this place the horse named Artsibi (the eagle), ridden by Menua, leapt 22 cubits.' " Only a large horse could be imagined to achieve a jump

of this length (about 37 feet). In one of the store-rooms of Teishebaini containing various bronze articles was found a splendid figure of a horse's head which had once decorated the shaft of a chariot; and bronze helmets and quivers were frequently ornamented with representations of chariots and horsemen.[16]

Ezekiel also prophesied against Togarmah and other nations:

> The word of the Lord came to me:
> "Son of man, set your face toward Gog, of the land of Magog, the chief prince of Meshech and Tubal, and prophesy against him and say, Thus says the Lord God: Behold, I am against you, O Gog, chief prince of Meshech and Tubal; and I will turn you about, and put hooks into your jaws, and I will bring you forth, and all your army, horses and horsemen, all of them clothed in full armor, a great company, all of them with buckler and shield, wielding swords; Persia, Cush, and Put are with them, all of them with shield and helmet; Gomer and all his hordes; Beth-togar'mah from the uttermost parts of the north with all his hordes – many peoples are with you" (Ezekiel 38:1-6).

The location of Togarmah coincides closely with the land formerly known as Armenia:

> Togar'mah, a son of Gomer, of the family of Japheth, and brother of Ashkenaz and Riphath (Genesis 10:3). His descendants became a people engaged in agriculture, breeding horses and mules to be sold in Tyre (Ezekiel 27:14). They were also a military people, well skilled in the use of arms. Togarmah was probably the ancient name of Armenia.[17]

6. Land of Eden

A relationship between the words Togarmah and Armenia has also been suggested:

> According to the national traditions, the Armenians are the descendants of Haik, the son of Togarmah, the grandson of Japhet. The historians of the country frequently call the people by the appellation of "Torgomian Doon," the house of Togarmah, and the same expression is used by the prophet Ezekiel [Beth-togar'mah, Ezekiel 38:6]. This tradition is regarded as highly probable by the best authorities.
>
> Professor Delitzsch, in his *Commentary on Genesis*, says: "The Armenians regarded Thorgom (Togarmah), the father of Haik, as their ancestor; and even granting that the form of the name Thorgom was occasioned by Thorgama of the Septuagint, still the Armenian tradition is confirmed by *Tilgarimmu* being in the cuneiform inscription the name of a fortified town in the subsequent district of Melitene, on the southwestern boundary of Armenia."
>
> Professor Rawlinson, whose authority on such subjects no one will dispute, says: "Grimm's view that Togarmah is composed of two elements – *Toka*, which in Sanskrit means a tribe or race, and Armah (Armenia) – may well be accepted. The Armenian tradition which derives the Haikian race from Thorgom, as it can scarcely be a coincidence, must be regarded as having considerable value. Now, the existing Armenians, the legitimate descendants of those who occupied the country in the time of Ezekiel, speak a language which modern ethnologists pronounce to be decidedly Indo-European; and thus, so far, modern science confirms the scriptural account."[18]

6.5 Telassar

Besides Togarmah, another place with a bearing on these subjects is Telassar. This seems to be a place in Eden or a place occupied by people from Eden, and it is mentioned at two places in the Bible. In the following report King Sennacherib of Assyria is threatening to destroy Jerusalem:

> And he [Sennacherib] heard say concerning Tirhakah king of Ethiopia, He is come forth to make war with thee. And when he heard it, he sent messengers to Hezekiah, saying, Thus shall ye speak to Hezekiah king of Judah, saying, Let not thy God in whom thou trustest deceive thee, saying, Jerusalem shall not be given into the hand of the king of Assyria. Behold, thou hast heard what the kings of Assyria have done to all lands by destroying them utterly; and shalt thou be delivered? Have the gods of the nations delivered them which my fathers have destroyed, as Gozan, and Haran, and Rezeph, and the children of Eden which were in Telassar? Where is the king of Hamath, and the king of Arphad, and the king of the city of Sepharvaim, Hena, and Ivah? (Isaiah 37:9-13 KJV, similar to 2 Kings 19:9-13).

Thus "the children of Eden" had been living in Telassar. By association, this place may have been somewhere in the vicinity of the city of Haran, which had been the temporary home of Abraham and his family many centuries earlier (Genesis 11:31-32). In one case we find the phrase "children of the east" where "children of Eden" might have been expected: "Then Jacob went on his journey, and came to the land of the children of the east" (Genesis 29:1, ASV). Jacob had arrived at his ancestral home of Haran, where Abraham and his family had previously arrived from the east.

6. Land of Eden

A dictionary definition of Telassar is the following:

> Telas'sar (*Assyrian hill*) is mentioned in 2 Kings 19:12 and in Isaiah 37:12 as a city inhabited by "the children of Eden," which had been conquered and was held in the time of Sennacherib, by the Assyrians. It must have been in western Mesopotamia, in the neighborhood of Harran and Orfa.[19]

A more detailed discussion of the location of Telassar is the following:

> This city, which is referred to by Sennacherib's messengers to Hezekiah, is stated by them to have been inhabited by the "children of Eden." It had been captured by the Assyrian king's forefathers, from whose hands its gods had been unable to save it. Notwithstanding the vocalization, the name is generally rendered "Hill of Asshur," the chief god of the Assyrians, but "Hill of Assar," or Asari (a name of the Babylonian Merodach), would probably be better.
>
> As Telassar was inhabited by the "children of Eden," and is mentioned with Gozan, Haran, and Rezeph, in Western Mesopotamia, it has been suggested that it lay in Bit Adini, "the House of Adinu," or Betheden, in the same direction, between the Euphrates and the Belikh [River, on which is located Haran]. A place named Til-Assuri, however, is twice mentioned by Tiglath-pileser IV (Ann., 176; Slab-Inscr., II, 23), and from these passages it would seem to have lain near enough to the Assyrian border to be annexed. The king states that he made there holy sacrifices to Merodach, whose seat it was. It was inhabited by Babylonians (whose home was the Edinu or "plain"; see Eden). Esarhaddon, Sennacherib's son, who likewise conquered the place, writes the name *Til-Asurri*, and states that the people of Mihranu called it *Pitanu*. . . .[20]

The Garden And The Flood

Thus, Telassar probably would have been downhill and southwest from the garden of Eden, assuming that the garden was in Urartu near Lake Van, as suggested earlier in Section 5.4A. One could perhaps imagine identities for these "children of Eden" and various ways in which they might have arrived at the vicinity of Haran. For example, one could speculate that Abraham and his extended family could themselves be identified as the children of Eden (or the children of the east) referred to in the above quotations. Abraham has an explicit biblical family tree that traces his ancestry back to Adam and Eve in the garden of Eden, and he is reported to have moved with his family from "Ur of the Chalde'ans" to Haran (Genesis 11:31). As foreigners in Haran they might have been given a genealogy-based identification as children of Eden. Using a chronology postulated previously, that move would have occurred in the eighteenth century B.C.[21]

On the other hand, it is also possible that some of the children of Eden arrived at the place later called Telassar more directly from Eden and at a much earlier time. The books of *Adam and Eve* purport to explain how Adam and Eve came to know of the good land below and to the west of their cave home in the seemingly volcanic area that may have been associated with the Nemrut volcano:

> Then the Word of God came and said unto him, "O Adam, go down to the westward of the cave, as far as a land of dark soil, and there thou shalt find food." And Adam hearkened unto the Word of God, took Eve, and went down to a land of dark soil, and found there wheat growing, in the ear and ripe, and figs to eat; and Adam rejoiced over it. Then the Word of God came again to

Adam, and said unto him, "Take of this wheat and make thee bread of it, to nourish thy body withal." And God gave Adam's heart wisdom, to work out the corn [wheat] until it became bread. Adam accomplished all that, until he grew very faint and weary. He then returned to the cave; rejoicing at what he had learned of what is done with wheat, until it is made into bread for one's use.[22]

Then Adam and Eve took of the corn, and made of it an offering, and took it and offered it up on the mountain, the place where they had offered up their first offering of blood.[23]

Then they both stood up that night and prayed to God. And when it was morning they went out, and went down westward of the cave, to the place where their corn was, and there rested under the shadow of a tree, as they were wont.[24]

Adam and Eve also tells us that Cain and his wife were the first of the children of Eden to actually move to the good land that Adam and Eve had been told of:

As for Cain, when the mourning for his brother was ended, he took his sister Luluwa and married her, without leave from his father and mother; for they could not keep him from her, by reason of their heavy heart. He then went down to the bottom of the mountain, away from the garden, near to the place where he had killed his brother. And in that place were many fruit trees and forest trees. His sister bare him children, who in their turn began to mutiply by degrees until they filled that place.[25]

After the death of Adam and of Eve, Seth severed his children and his children's children, from Cain's children. Cain and his seed went down and dwelt westward, below the place where he had killed his brother Abel. But Seth and his children, dwelt northwards upon the mountain of the cave of Treasures, in order to be near to [the body of] their father Adam. . . . But because of their own purity, they were named "Children of God."[26]

Then Genun [descendant of Cain] said to them [descendants of Seth] from down below, "Go to the western side of the mountain, there you will find the way to come down.". . . Enoch at that time was already grown up, and in his zeal for God, he arose and said, "Hear me, O ye sons of Seth, small and great – when ye transgress the commandment of our fathers, and go down from this holy mountain – ye shall not come up hither again for ever." But they rose up against Enoch, and would not hearken to his words, but went down from the Holy Mountain.[27]

It would appear from the above quotations that two groups of descendants of Adam and Eve may be said to have come down from the Holy Mountain at the western edge of Eden and moved into the more fertile land below. The first group consisted of Cain and his wife, to be joined a few years later by the disobedient children of Seth. Before their disobedience these descendants of Seth seem to have been called "children of God." Either of these groups, or both of them combined, may have been referred to by Isaiah many years later as "children of Eden which were in Telassar" (Isaiah 37:12 KJV). With these interpretations, a very approximate chronology to be developed in Section 12.3 could place the beginning of the migration of the children of Eden to Telassar at a date of roughly 2200 B.C.

It may seem doubtful that the descendants of Adam and Eve could have maintained their cultural identity as "children of

6. Land of Eden

Eden" for more than a thousand years until the time of Isaiah. On the other hand, the Hebrews originally from the northern kingdom of Israel and the southern kingdom of Judah have, at least in part, kept their identities as Samaritans and Jews for about twenty-five hundred years until the present.

6.6 Eden and Ezekiel
The preceding sections of this book have sometimes referred to the land of Eden as an approximate synonym for Urartu (Ararat) and Armenia. Within Eden may have been the garden of Eden, the headwaters of several rivers, the lake now known as Lake Van, and the landing place of the ark. An early graphical representation of the land of Eden was shown previously in Figures 4.1 and 4.2. Several possible indications relating to the geography of Eden are also included in the first and second books of *Adam and Eve*.

The land or people of Eden were mentioned several times by the prophet/priest Ezekiel. It is striking that the word Eden as a place-name is used more times in the book of Ezekiel than in any other book of the Bible. Thus the place Eden is named seven times in the following six verses: Ezekiel 27:23, 28:13, 31:9, 31:16, 31:18(2), and 36:35. One might have expected that Eden would be mentioned much more often in Genesis, but there it is used only six times. To complete this inventory, the place Eden is also named twice in Isaiah and once each in 2 Kings, Joel, and Amos. Thus, in comparison with other biblical authors, Ezekiel seems to have been very interested in Eden.

Ezekiel also provides an indication of volcanic activity in the vicinity of Eden, and the relevant verses are the following:

Moreover the word of the Lord came to me: "Son of man, raise a lamentation over the king of Tyre, and say to him, Thus says the Lord God:

"You were the signet of perfection (Hebrew obscure), full of wisdom and perfect in beauty. You were in Eden, the garden of God; every precious stone was your covering, carnelian, topaz, and jasper, chrysolite, beryl, and onyx, sapphire (or *lapis lazuli*), carbuncle, and emerald; and wrought in gold were your settings and your engravings (Hebrew uncertain). On the day that you were created they were prepared. With an anointed guardian cherub I placed you (Hebrew uncertain); you were on the holy mountain of God; in the midst of the stones of fire you walked. You were blameless in your ways from the day you were created, till iniquity was found in you. In the abundance of your trade you were filled with violence, and you sinned; so I cast you as a profane thing from the mountain of God, and the guardian cherub drove you out from the midst of the stones of fire. Your heart was proud because of your beauty; you corrupted your wisdom for the sake of your splendor. I cast you to the ground; I exposed you before kings, to feast their eyes on you. By the multitude of your iniquities, in the unrighteousness of your trade you profaned your sanctuaries; so I brought forth fire from the midst of you; it consumed you, and I turned you to ashes upon the earth in the sight of all who saw you. All who know you among the peoples are appalled at you; you have come to a dreadful end and shall be no more for ever" (Ezekiel 28:11-19).

In this lamentation over the deceased king of Tyre, the king is said initially to have been in "Eden, the garden of God," with many precious stones and a guardian cherub. He was on "the holy mountain of God" and walked among the "stones of fire." But when he became sinful, he was thrown from the mountain of

God, driven from the stones of fire, humiliated, and burned to ashes. Perhaps the most important point initially is that stones of fire were to be found on the holy mountain of God in Eden. As it happens, this is reminiscent of the volcanic landscape near the cave home of Adam and Eve on the mountain to the west of the garden of Eden. In Sections 5.4 and 5.5 the holy mountain was identified tentatively as Mt. Nemrut. Thus, it is natural to imagine that the holy mountain of God mentioned by Ezekiel could also have been the dormant (intermittently active) Nemrut volcano.

The references to what might be interpreted as volcanic effects on the mountain of God are very similar between the texts of Ezekiel and *Adam and Eve*:

> In the midst of the stones of fire you walked (Ezekiel 28:14).

> I cast you as a profane thing from the mountain of God, and the guardian cherub drove you out from the midst of the stones of fire (Ezekiel 28:16).

> I brought forth fire from the midst of you; it consumed you, and I turned you to ashes upon the earth in the sight of all who saw you (Ezekiel 28:18).

> But after they had thus fallen into this defilement, they returned by the way they had come, and tried to ascend the Holy Mountain. But they could not, because the stones of that holy mountain were of fire flashing before them, by reason of which they could not go up again.[28]

The stones of fire in the stories of Ezekiel and *Adam and Eve* will be discussed further in the following section.

The Garden And The Flood

The role of the cherub in Ezekiel's account is somewhat similar to the role of the cherubim in the Genesis story of the guardians of Eden: "He [God] drove out the man; and at the east of the garden of Eden he placed the cherubim, and a flaming sword which turned every way, to guard the way to the tree of life" (Genesis 3:24). Similar stories of the cherubim also appear in *Adam and Eve*:

> And the cherub who guarded the garden was standing at the western gate, and guarding it against Adam and Eve [whose cave home was to the west of the garden], lest they should suddenly come into the garden. And the cherub turned round, as if to put them to death; according to the commandment God had given him.
>
> When Adam and Eve came to the eastern border of the garden – thinking in their hearts that the cherub was not watching [the eastern gate] – as they were standing by the gate as if wishing to go in, suddenly came the cherub with a flashing sword of fire in his hand; and when he saw them, he went forth to kill them. For he was afraid lest God should destroy him if they went into the garden without His order. And the sword of the cherub seemed to flame afar off.[29]

In short, the texts of Ezekiel and *Adam and Eve* have much in common including cherubim, a holy mountain, and unusual references to what appear to be volcanic effects. It is perhaps not unreasonable to speculate on how these common features came to be present in what would otherwise be considered independent sources. One could, for example, consider the possibility of some measure of common authorship. It will be suggested in Section 6.7 that Ezekiel may have lived immediately after a volcanic event on Mt. Nemrut in the region that could be called the land of

6. Land of Eden

Eden. Thus he would have been very familiar with the circumstances of a typical eruption on the holy mountain. However, no suitable eruption event has yet been suggested for Eden at an earlier time that might be between the Adam and Eve stories and the flood. Thus, a possible interpretation could be that Ezekiel himself or some other writer of an appropriate time and place may have been aware of eruptions on Mt. Nemrut. If so, such a writer could have had a role in creating at least the volcanism descriptions that came to be included in the stories of *Adam and Eve*.

It may be helpful here to say a few words about the definition of the place-name Eden, as this definition seems to have changed as the human family of the Bible developed. Initially the terms Eden or the garden of Eden seem to have referred to actual physical places where Adam and Eve first lived in the Genesis creation story. After they disobeyed God, Adam and Eve were expelled from the garden. We then might infer, mostly from *Adam and Eve* and various geographical hints, that the garden may have been located near the Ercis Gulf of Lake Van as shown in Figure 5.2. After they left the garden, Adam and Eve are said to have occupied a cave home that may have been in the vicinity of Mt. Nemrut volcano. They began farming downhill and farther west from their cave home. Their son Cain and then some of the descendants of their son Seth moved down to the western region that they had begun farming.

The first stories just mentioned seem significantly dependent on the not widely employed history given in *Adam and Eve*. But then other biblical sources become useful, and the land of Eden (as opposed to the smaller garden of Eden) expanded still farther

The Garden And The Flood

westward. As noted in Section 6.5 above, the children of Eden seem to have migrated at least as far west as Telassar in northern Mesopotamia (Isaiah 37:12), near Haran where Abram and his family settled when they first arrived from Ur of the Chaldeans (Genesis 11:31).

At the time of the prophet Amos (c. 750 B.C.[30]) the house of Eden may have extended as far west as Syria: " 'I will break the bar of Damascus, and cut off the inhabitants from the Valley of Aven, and him that holds the scepter from Beth-eden [house of Eden]; and the people of Syria shall go into exile to Kir,' says the Lord" (Amos 1:5). Later, in his allegory of the cedar, Ezekiel (c. 622-570 B.C.) pronounced a punishment on pharaoh king of Egypt. That punishment would be like that reserved for a cedar tree among the other great trees of Eden in the land of Lebanon:

> Son of man, say to Pharaoh king of Egypt and to his multitude:
> "Whom are you like in your greatness? Behold, I will liken you to a cedar in Lebanon, with fair branches and forest shade, and of great height, its top among the clouds.
> . . .
> Thus says the Lord God: When it [the tree] goes down to Sheol I will make the deep mourn for it, and restrain its rivers, and many waters shall be stopped; I will clothe Lebanon in gloom for it, and all the trees of the field shall faint because of it. I will make the nations quake at the sound of its fall, when I cast it down to Sheol with those who go down to the Pit; and all the trees of Eden, the choice and best of Lebanon, all that drink water, will be comforted in the nether world. They also shall go down to Sheol with it, to those who are slain by the sword; yea, those who dwelt under its shadow among the nations shall perish. Whom are you thus like in glory and in greatness

among the trees of Eden? You shall be brought down with the trees of Eden to the nether world; you shall lie among the uncircumcised, with those who are slain by the sword.

This is Pharaoh and all his multitude, says the Lord God" (Ezekiel 31:2-18).

Thus, the land of Eden may then have included a band of territory extending from Lake Van westward through northern Mesopotamia, Syria, and as far as Lebanon.

The references just indicated include deprecations against the children of Eden, and this treatment would be consistent with the much earlier negative prophecies against the children of Cain and some of the children of Seth. An example from *Adam and Eve*, quoted previously, is the following:

> Enoch at that time was already grown up, and in his zeal for God, he arose and said, "Hear me, O ye sons of Seth, small and great – when ye transgress the commandment of our fathers, and go down from this holy mountain – ye shall not come up hither again for ever." But they rose up against Enoch, and would not hearken to his words, but went down from the Holy Mountain.[31]

Sometimes prophecies were more favorable: "This land that was desolate has become like the garden of Eden" (Ezekiel 36:33-36). And sometimes they were not: "The land is like the garden of Eden before them, but after them a desolate wilderness" (Joel 2:1-3). The contrast between these two prophecies is striking. By the time of such prophesies, long after the flood, the original garden of Eden may no longer have been a known real place. Nevertheless, its memory and name could still be used as a metaphor for a place of security and prosperity.

6.7 Eruptions in Eden

It is not always possible to identify with confidence the occurrence, size, or date of a volcanic eruption in the distant past. Nevertheless it might be of some value first to consider a well-known eruption of the Nemrut volcano in the not-so-distant past (1441 A.D.). A contemporary note in the Armenian Chronicles of 1401-1450 A.D. includes the following.

> In 1441 [A.D.] a great sign took place, for the mountain called Nemrud, which lies between Kelath and Bitlis, suddenly began to rumble like heavy thunder. This set the whole land into terror and consternation, for one saw that the mountain was rent asunder to the breadth of a city; and from out of this cleft flames arose, shrouded in dense, whirling smoke, of so evil a stench that men fell ill by reason of the deadly smell. Red-hot stones glowed in the terrible flames, and boulders of enormous size were hurled aloft with peals of thunder. Even in other provinces men saw all this distinctly.[32]

It would seem that Ezekiel's description of the holy mountain in Eden (Section 6.6) and the eyewitness accounts of a much later eruption of the Nemrut volcano (above) are similar to each other, including the volcano's hurling of fiery lava bombs that people might attempt to walk between or find themselves burned up by.

The next question is whether there might have been an eruption of the Nemrut volcano at about the time of the activities of the prophet Ezekiel. The eruption history of Nemrut, as best it is known at present, is available on a web page maintained by the Smithsonian Institution. The eruptions that have occurred more recently than 5000 B.C. are listed in Table 6.1.[33]

6. Land of Eden

Table 6.1. Known eruptions of the Nemrut volcano more recent than 5000 B.C.

> [1692 Apr 13 A.D., Uncertain]
> 1650 Oct 27 A.D.
> 1597 A.D. (in or before)
> 1441 A.D.
> 1402 A.D. (?)
> 1111 A.D.
> 531 B.C. (?)
> 657 B.C. ± 25 years
> 787 B.C. ± 25 years
> 1396 B.C. (?)
> 1662 B.C. (?)
> 4055 B.C. ± 50 years
> 4321 B.C. (?)
> 4615 B.C. (?)
> 4849 B.C. (?)
> 4938 B.C. ± 75 years

The important point in Table 6.1 is that Nemrut volcano is known to have erupted in about 657 B.C. ± 25 years. Uncertainties are large, but this eruption may slightly predate the estimated birthdate of Ezekiel, which as mentioned above was about 622 B.C. If this were so, Ezekiel could have learned about the eruption of Nemrut and its effects from residents of Babylon and other captives with him. Then he might have been able to write eloquently about the recent and dreadful demise of the sinful king of Tyre. It may not be possible to determine with certainty the identity of the king of Tyre that is being referred to. A preliminary candidate for this role could, for example, be Baal I of Tyre, who is believed to have reigned during the first half of the seventh century B.C.[34] Thus, this king could have died at the time of an eruption in about 657 ± 25 B.C.

6.8 Conclusion

Many Bible-related events took place in a land that may be loosely identified as Eden. In the earliest Genesis accounts the garden of Eden is prominent, and this garden seems likely to have been located near modern-day Lake Van. After Adam and Eve were expelled from the garden, they and some of their descendants may have migrated westward into a much larger land of Eden that eventually could have been imagined to extend as far as Lebanon. This chapter has indicated some of the textual suggestions for such an expansion. The evidence also suggests that eruption activities of Mt. Nemrut volcano may have influenced some of the early historical writings relating to the land of Eden.

1. H. F. B. Lynch, *Armenia Travels and Studies* (Longmans, Green, and Co., London, 1901), Volume 2, pp. 25-30.
2. Ibid., p. 28.
3. "See especially Müller-Simonis, *Du Caucase au Golfe Persique*, Paris, 1892, p. 393."
4. "Müller-Simonis, op.cit., pp. 292 and 555."
5. *"Verhandlungen der B. G. für Anthropologie*, 1898, Heft VI, p. 591. These travellers add yet another name to the supposed ruins, viz. that of Sirnakar."
6. W. Belck and C. F. Lehmann, "Entdeckungen in Armenien," *Verhandlungen der Berliner Gesellschaft für Anthropologie, Ethnologie und Urgeschichte,* Redigirt von Rudolf Virchow (Verlag von A. Asher & Co., Berlin, 1898), pp. 568-592; p. 591. (translated by the author)
7. P. Müller-Simonis and H. Hyvernat, *Du Caucase au Golfe Persique à travers l'Arménie, le Kurdistan et la Mésopotamie* (Université Catholique d'Amerique, Washington, D.C., 1892), pp. 292-294. (translated by the author)
8. *The Life and Works of Flavius Josephus*, Translated by William Whiston (Holt, Rinehart and Winston, New York), *Antiquities of the Jews*, Book I, 1:4.
9. *Adam and Eve*, Translated by S. C. Malan from the Ethiopic edition edited by E. Trumpp, included in The Forgotten Books of Eden, Edited by Rutherford H. Platt, Jr. (Alpha House, Inc., 1927); Book1, 17:1-2,6; 18:1.
10. *The Life and Works of Flavius Josephus*, op. cit., Book I, 1:4.
11. *Adam and Eve*, op. cit., Book 1, 18:9.
12. P. Müller-Simonis and H. Hyvernat, op. cit., pp. 291-292. (translated by the author)
13. P. Müller-Simonis and H. Hyvernat, op. cit., p. 292.
14. G. A. Bournoutian, *A Concise History of the Armenian People* (From Ancient Times to the Present), Fifth Edition (Mazda Publishers, Inc., Costa Mesa, California, 2006), p. 16.

15. H. F. B. Lynch, op. cit., p. 70.
16. B. B. Piotrovsky, op. cit., pp. 156-157.
17. "Togar'mah," in William Smith, *A Dictionary of the Bible: Comprising its Antiquities, Biography, Geography, Natural History and Literature Smith's Bible Dictionary* (Henry T. Coates & Co., Philadelphia, 1884), p. 705.
18. W. J. Wintle, *Armenia and its Sorrows*, Third Edition (Andrew Melrose, London, 1896), pp. 22-23.
19. "Telas'sar," in William Smith, op. cit., pp. 677-678.
20. T. G. Pinches, "Telassar," in *The International Standard Bible Encyclopaedia*, James Orr, General Editor (The Howard-Severance Company, Chicago, 1915), Volume 5, p. 2925.
21. L. W. Casperson, *Patterns of Biblical Chronology*, (Westbow Press, Bloomington, IN, 2012), Section 24.5, "Revised chronology," pp. 609-610.
22. *Adam and Eve*, op. cit., Book 1, 66:7-10.
23. *Adam and Eve*, op. cit., Book 1, 66:10.
24. *Adam and Eve*, op. cit., Book 1, 69:10.
25. *Adam and Eve*, op. cit., Book 2, 1:6-8.
26. *Adam and Eve*, op. cit., Book 2, 11:1-2,4.
27. *Adam and Eve*, op. cit., Book 2, 20:21,29-30.
28. *Adam and Eve*, op. cit., Book 2, 20:33.
29. *Adam and Eve*, op. cit., Book 1, 54:3-5.
30. "Amos," *The Interpreter's Dictionary of the Bible, An Illustrated Encyclopedia* (Abingdon Press, Nashville, 1962), Volume 1, A-D, pp. 116-121; p 118.
31. *Adam and Eve*, op. cit., Book 2, 20:29-30.
32. F. Oswald, *A Treatise on the Geology of Armenia*, Thesis accepted by the University of London for the Degree of Doctor of Science (Published by the author at Iona, Beeston, Notts, 1906), p.154.
33. "Nemrut Volcano," Smithsonian Institution, *National Museum of Natural History, Global Volcanism Program*, Database Search, March 26, 2014.

34. A. S. Kapelrud, "Tyre," *The Interpreter's Dictionary of the Bible, An Illustrated Encyclopedia* (Abingdon Press, Nashville, 1962), Volume 4, R-Z, pp. 721-723; p. 722.

PART 2

CAUSES AND EFFECTS OF THE FLOOD

7. LAKE VAN AND THE FLOOD

"The waters prevailed and increased greatly upon the earth; and the ark floated on the face of the waters" (Genesis 7:18).

7.1 Introduction

Assuming that the flood of Noah's time was an actual historical event, one might hope that the causes or consequences of that event would have left identifiable geological, archeological, or other non-biblical historical records. Such records could possibly provide methods for determining the date of the flood. This idea has occurred several times in previous chapters, but only the possibility that the flood could be associated with an inundation caused by flooding rivers was considered in any detail. However, as discussed in Section 3.3, that interpretation of the flood seems doubtful. The biblical flood story doesn't mention any rivers. Also, the inundation in Mesopotamia is due to melting of winter snows near the headwaters of the Tigris and Euphrates Rivers, and in contrast to the biblical story the timing of the melting is not the same as the timing of the very limited local rainfall. In addition, the ark is said to have come to rest on the mountains of Ararat or Urartu (Genesis 8:4), whereas flooding of the Tigris

7. Lake Van and the Flood

and Euphrates Rivers would have carried anything floating ever farther away from the mountains of Ararat.

As suggested by the title of this chapter, the emphasis here will be on the idea that the flood was somewhat localized and may have occurred as a result of an increased water level of Lake Van itself. The surface of the lake is more than a mile above sea level and may be said to be located in the mountains of the land that was formerly called Urartu. Thus, an ark floating on Lake Van would necessarily have landed on the mountains of Ararat, consistent with the biblical account. With this interpretation it would, however, remain to explain a rapid rise in the level of the lake at a time that would be compatible with any discernible physical or historical evidence of the biblical flood.

According to the Bible, there were two causes for the rising water level. The first cause mentioned for the flood was the release of the "fountains of the deep," and this concept is considered in Section 7.2. The other cause was an exceptional rainfall lasting for many days, and this possibility is discussed in Section 7.3. As it happens, both flooding by underground waters and intense multi-day rainfall are events that can be imagined to have occurred in the vicinity of Lake Van. Based on observations in modern times, the possibility of the water level of Lake Van rising in a few days to a level that could float a large boat is considered in Section 7.4. In Section 7.5 information from the ancient past is examined to see if there is any specific evidence for a sufficient rise in the lake level at a time consistent with the approximate date of the flood. Causes of lake level changes are considered in Section 7.6, and the possible consequences of such changes for the kingdom of Urartu are discussed in Section 7.7.

7.2 Fountains of the deep

The Bible provides only somewhat limited information concerning the mechanism of the "flood of waters:"

> Noah was six hundred years old when the flood of waters came upon the earth (Genesis 7:6).
>
> [O]n that day all the fountains of the great deep burst forth, and the windows of the heavens were opened. And rain fell upon the earth forty days and forty nights (Genesis 7:11-12).
>
> The flood continued forty days upon the earth; and the waters increased, and bore up the ark, and it rose high above the earth. The waters prevailed and increased greatly upon the earth; and the ark floated on the face of the waters (Genesis 7:17-18).

Thus, the cause of the flood is said here to have been the bursting forth of the fountains of the great deep and the opening of the windows of heaven. It seems clear enough that the opening of the windows of heaven is a metaphor for heavy rain. However, it isn't so clear what might be meant by "the fountains of the great deep." In any case the result was that the ark floated.

Eventually the waters receded:

> But God remembered Noah and all the beasts and all the cattle that were with him in the ark. And God made a wind blow over the earth, and the waters subsided; the fountains of the deep and the windows of the heavens were closed, the rain from the heavens was restrained, and the waters receded from the earth continually (Genesis 8:1-3).

7. Lake Van and the Flood

Again, there is here a reference to both "the fountains of the deep and the windows of the heavens."

One reasonable interpretation for the words "on that day all the fountains of the great deep burst forth" is that there might have been a large volume of water beneath the surface of the earth that was suddenly somehow freed and brought to the surface. If this water were anywhere in the catchment basin of Lake Van, it could have contributed to a rise in the level of the lake. For example, a substantial earthquake could be imagined to fracture the ground, release subterranean water, and perhaps cause a subsidence of the land surface in the vicinity of the water release. Such effects could lead to the floating of an ark that initially was situated on ground near the shore of the lake.

It will be noted below in Section 7.4 that changes in the level of Lake Van are even today a common occurrence. Thus, the idea of preparing in advance a boat for survival in the event of a rapid rise in lake level might not have seemed a particularly foreign concept. This would have been especially true if Noah and his family were residents of a low-lying plain on the north side of the lake or on one of the islands located in the lake. That a rapid rise in lake level could be due to "fountains of the deep" may not sound especially likely, but some of the effects mentioned above may have been seen in the same vicinity in modern times. Thus, the effects of an earthquake near Mt. Ararat in 1840 A.D. may be of interest. An almost contemporary report of that earthquake and its aftermath are quoted here:

> An earthquake, which changed in a few moments the entire aspect of the country in the neighbourhood of Mount Ararat, commenced on the 20th of June (Russian,

Causes And Effects Of The Flood

or old style), 1840, at about 45 minutes after six in the evening. Repeated, but intermittent, shocks, which seemed to come from the mountain, gave to the earth a movement resembling waves, which continued for about two minutes. The first four and most formidable shocks, which were accompanied by a subterraneous sound, proceeding in the direction of east-north-east, have left on the summits of hills and bottoms of valleys within the range of the agitation, traces which will not soon disappear, and which the eye of the scientific observer will recognise after many ages shall have passed away. It was at the same time observed, that numerous rents or fissures took place on the banks of the Araxes and the Karassu [rivers], from the source of the latter to its confluence with the Arpatchai, on all the spots where the banks of those rivers are somewhat elevated. These fissures, which were parallel to the course of the streams, ploughed the earth to a distance of a verst [a Russian measure of distance equivalent to 3500 feet or 0.6629 mile or 1.067 kilometers] from the beds of the rivers, and, in accordance with the movement given to the soil by the shocks, they were seen every moment to open and shut. There also occurred a great number of violent explosions from the bottoms of holes like little craters, which, opening and shutting in the same way as fissures, spouted out torrents of water, and cast up immense quantities of pebbles and gravel. The waters of the Araxes were so violently agitated, that they rose over both banks, now causing inundation, then sinking again into the centre of the bed, which thus appeared concave. According to the accounts of the people of the country, several parts of the river became dry, while at other parts the body of the stream rose to a great height, making a noise like the sound of boiling water. During these awful moments terror and desolation reigned everywhere to a great distance around Mount Ararat. The Persian town of

7. Lake Van and the Flood

Maku, and Baiazeth, the chief town of a Turkish pashalic, also suffered from the earthquake. Its successive shocks convulsed in a few minutes the earth as far as Shusa and Tabris on the one side, and to Tiflis on the other. But its ravages extended chiefly over the Russian territory. The ancient and venerable monument of St. James, and the village of Acorhi, with its 200 houses and 1,000 inhabitants, situated on the skirts of Mount Masis, at the foot of the Great Ararat, were entirely destroyed by the immense masses of rock which were detached from the summit of the mountain, and by the colossal glaciers accumulated during an incalculable series of ages in that region of eternal snow, which those rocks brought with them in their descent. In the cantons of Erivan, Sharur, Nakhichevan, and Ordubat, nearly all the houses were entirely destroyed. The influence of the earthquake on all the wells within the districts of Erivan and Nakhichevan was very remarkable. In the canton of Nakhichevan upwards of thirty springs were dried up for some time; some continued even several days after the catastrophe to yield only thick and whitish-coloured water; others, on the contrary, became more abundant than they had previously been; and in the vicinity of several of the latter new springs made their appearance. It is difficult to afford any idea of the scene which presented itself in the narrow valley of Acorhi. The masses of rock, ice, and snow, detached by the first shock from the summit of the Ararat and its lateral points, were thrown, at one single bound, from a height of 6,000 feet to the bottom of the valley, where they lay scattered over an extent of seven versts.[1]

It is clear from this example that a major earthquake near Mt. Ararat could have several important consequences for the storage and movement of water on and beneath the surface of the earth.

As seen in the above quotation, these consequences include at least the following:

> Opening and closing of fissures near the banks of rivers
> Explosions from the opening and closing of craters away from rivers
> Spouting by craters of "torrents of water" and "immense quantities of pebbles and gravel"
> Erratic level changes in rivers causing both inundations and dry stream beds
> Streams rising "to a great height, making a noise like the sound of boiling water"
> "Immense masses of rock" and "colossal glaciers" falling "from a height of 6,000 feet to the bottom of the valley"
> Drying of numerous springs, and flows changing to muddy water
> Enhancement of numerous springs, and appearance of new ones

The result of such events could perhaps be referred to loosely as the bursting forth of "the fountains of the great deep."

The above descriptions were interpreted by their author as being consequences of a major earthquake that occurred near Mt. Ararat in the year 1840 A.D. It should be noted that on approximately the same date there is now known to have been an eruption of the Mt. Ararat volcano. Table 7.1 includes a listing of known eruptions of this volcano that have occurred more recently than 5000 B.C.[2] The letters *VEI* represent the volcanic explosivity index, and the number 3 implies a relatively modest eruption. It follows that a much earlier earthquake could have occurred at the time of Noah and perhaps in the vicinity of Lake Van. As with the modern earthquake just considered, there may

7. Lake Van and the Flood

have been hydrological effects of such an earthquake that could have been described in Genesis as the opening of the fountains of the deep.

Table 7.1. Known eruptions of the Ararat volcano more recent than 5000 B.C.

> 1840 Jul 2 A.D., $VEI = 3$
> [1783 A.D., uncertain]
> [1450 A.D., uncertain]
> 550 B.C.
> 2450 B.C. $\pm$ 50 years

The general idea in most of the preceding paragraphs of this section is that there could be water under the surface of the earth. That such underground water is very common and sometimes abundant is already evident by the occurrence of natural springs and numerous successful drilled wells. Some of that underground water could be released to the surface from time to time by inevitable natural processes, and earthquakes provide a well-established mechanism for such releases. The resulting water could provide at least a partial explanation for the fountains of the deep that were said to have contributed to the flood of Noah's era.

One obvious source for underground water is rainfall that soaks into the earth and accumulates over a long period of time. If that water collects in underground cavities, it could be

Causes And Effects Of The Flood

available for rapid release in the event of ground-fracturing earthquakes. Other possibilities for the origin of large volumes of underground water have also been proposed. In some locations ice caves store massive amounts of ice underground. Many such caves have been found, and for example some in Eastern Washington have existed for thousands of years.[3] The insulated volumes of ice that they contain could, at least in principle, be sources of underground water that might be available for long periods of time.

7.3 Exceptional rainfall

According to the biblical record, heavy rainfall was an important contributor to the flood. A natural preliminary question is whether any known and documented rainfall of any time or place, if it had occurred at Lake Van and continued for many days, might have been sufficient to cause an ark-floating flood. As a partial answer to this question, a few record rainfalls are indicated in Table 7.2:[4]

7. Lake Van and the Flood

Table 7.2. Greatest recorded rainfall for various time intervals.

Time	Total (meters)	Average Rate (meters/day)	Forty-day total (meters)
1 minute	0.0312	44.928	1797.12
1 hour	0.3050	7.320	292.80
1 day	1.8250	1.825	73.00
2 days	2.4930	1.247	49.86
3 days	3.9290	1.310	52.39
4 days	4.8690	1.217	48.69
5 days	4.9790	.996	39.83
6 days	5.0750	.846	33.83
7 days	5.4000	.771	30.86
8 days	5.5100	.689	27.55
9 days	5.5120	.612	24.50

Thus, if a record average rainfall rate that has been sustained for a week at one location on earth could have been sustained for forty days at Lake Van, then a total rainfall there of over 30 meters would have been the result ($5.4 \times 40/7 = 30.86$).

It is evident from Table 7.2 that the highest rainfall rates are unlikely to be sustained for many days in sequence. A plot of the highest official rainfall recorded for various integer numbers of days is included in Figure 7.1, based on the data shown in Table 7.2. The curve in the plot represents the approximate fit of a logarithmic function to the tabulated data. Thus, the equation governing the approximated rainfall can be written

$$y = 6 \times \log(x), \tag{7.1}$$

where y represents the cumulative rainfall in meters after $x-1$ days.

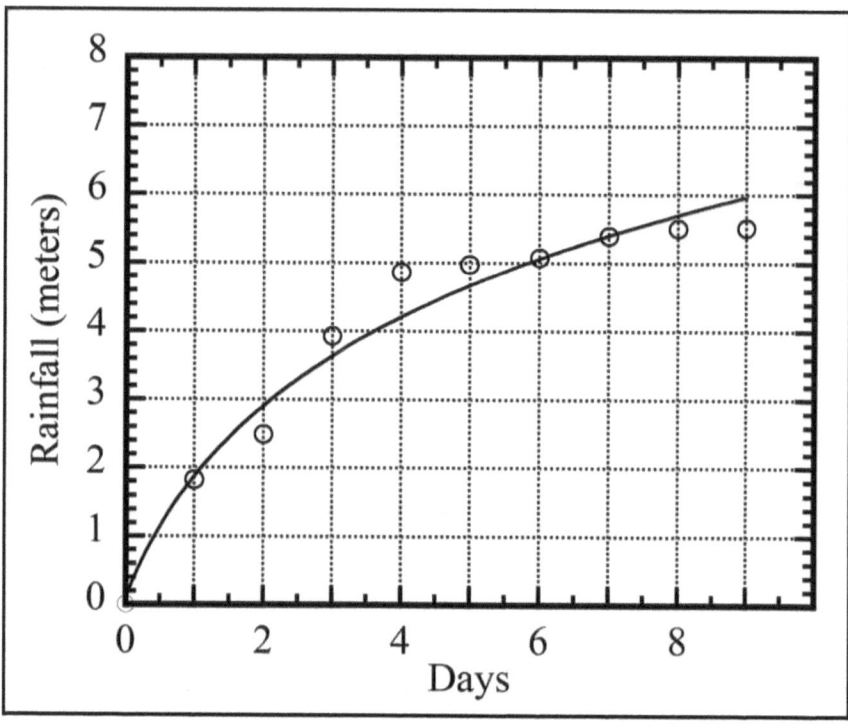

Figure 7.1. Highest official cumulative rainfall for various periods of time, represented by the circles.

The observational data for maximum rainfall in a given time period listed in Table 7.2 is provided for a rainfall period of up to nine days. Assuming that Equation (7.1) provides a reasonable statistical representation of the underlying physics, it might be of interest to use that equation directly to predict the maximum rainfall that might occur anywhere on earth in a forty day period (i.e., $x-1=40$ or $x=41$). If Equation (7.1) were a valid

7. Lake Van and the Flood

representation of the maximum rainfall over a given time interval, then one could estimate that the maximum forty-day rainfall would be roughly $y = 6 \times \log(41) = 9.68$ meters.

It is, of course, not only the rainfall that strikes the surface of Lake Van that could cause flooding, but rather the rainfall that lands in the entire catchment basin of the lake. Thus, some of the rainfall would be carried into the lake by streams and rivers. By one estimate, the lake has a surface area of 3522 km² at present, while the drainage basin covers about 16,096 km².[5] Thus, for modest increases in the lake level, those increases would be equal to the rainfall (averaged over the basin) multiplied by about $16,096/3522 = 4.57$. Thus, a rainfall of about 9.68 meters as in the above estimation would result in a lake level rise of about $9.68 \times 4.57 = 44.24$ meters. Actually, the rise would be somewhat less than this value, because as the level rises the surface area of the lake increases. It should also be noted that the level couldn't increase more than about 90 meters, because at that height the water would begin to spill over the edge of the drainage basin (as it exists today). The altitude today of the lake surface is about 1647 meters above sea level, while the threshold at which water would flow from the basin is 1736 meters above sea level.[6] A water level increase of far less than 44.24 meters would be adequate to float a moderate-size ark if, as seems reasonable, it were built on one of the plains along the shore of the lake or one of its islands. Nevertheless, a conventional heavy rainfall seems unlikely to have been responsible for the flood of Noah's time.

The above calculation does suggest that a forty-day rainfall at the highest rate that could be considered possible anywhere on earth would, if it occurred in the region surrounding Lake Van, raise the level of the lake more than enough to float the ark. The obvious obstacle with this interpretation is that none of the rainfall rates discussed above occurred anywhere near Lake Van, and there is no reason to expect that they might have done so due to routine weather events of the past. The record rainfalls just discussed generally occur in association with localized tropical storms closer to the equator. Furthermore, it will be suggested below that during the biblical flood event, exceptional rainfall rates may have occurred simultaneously over a wide area of the globe.

The above conclusions would leave as suitable occasions of exceptional rainfall less routine events including volcanic eruptions and possibly astrophysical encounters. Eruptions are often associated with heavy rainfall, and some eruptions have been large enough to affect weather over much of the northern hemisphere. While encounters with comets or asteroids are not common, it may be noted that many comets consist largely of water ice. A leading theory of cometary structure is called the icy conglomerate model (also known as the dirty snowball model).[7] Astrophysical events will be considered further in Section 9.2

7.4 Flooding of Lake Van

Several aspects of the biblical and other texts that have been considered here would seem to suggest that at least the garden of Eden and the landing place of the ark may have been in the vicinity of Lake Van. These tentative conclusions lead to the further possibility that the ark in its drifting was actually floating on Lake Van. As noted already (in Section 5.3) the lake is large enough that in terms of wind and waves its behavior would be more like a sea than a typical lake: "And it is large enough to spread at their [the observers'] feet with all the qualities of the ocean – the depth and vastness and changing surface of the high seas. It is about six times as large as the lake of Geneva, having an area of some 1300 square miles."[8]

In order for Lake Van to be a viable contender for the body of water on which Noah's ark floated, it would be helpful if the level of the lake were known to be variable. In other words, if the level of the lake were known to rise during times of relatively heavy rain (and perhaps bursting forth of "the fountains of the great deep"), then during a severe storm a boat built near the shore of the lake might suddenly find itself to be floating. An important part of the miracle of the story is that God was said to have told Noah in advance that he would "bring a flood of waters upon the earth" (Genesis 6:17). Even today the long-range prediction of volcanic eruptions, earthquakes, and major storms would usually be considered difficult or impossible.

Independent of the ability to predict, it is certainly true that the level of Lake Van has varied dramatically during both modern and prehistoric times. One relatively recent level rise with lasting consequences occurred in 1838 and the years immediately following. Several sentences from the report of an observer are reproduced here:

The Lake of Van

It is not generally known that a remarkable rise has of late years taken place in the water of the Lake of Van; and, as this phænomenon comes under the consideration of the geologist, I give here the results of my observations and inquiries on the subject. . . .

According to the statement of the natives on its shores, the water of the Lake, owing to some unaccountable cause, began gradually to increase in the year 1838 or 1839. For some time it fluctuated a good deal, but at the end of twelve months it had gained considerably upon the land wherever there was a shelving shore, the depth being increased nearly two yards. During the second and third years the rise continued, until the increased depth had reached ten or twelve feet. Many towns and villages upon the Lake were surrounded, destroyed, or deserted, and the effects are still apparent on the north shore, where it is less rocky than elsewhere. Having in three years attained its standard height at twelve feet, it so remained, slightly rising and falling until 1850, from which time the natives say there has been a considerable and gradual subsidence of the water.

The effects of this phænomenon are best seen at Ardish, a town formerly containing 5000 inhabitants, with a castle, two mosques, a Christian church, and several caravanseraïs. . . . In 1841 the whole town was completely surrounded, partially under water, and entirely deserted by its inhabitants. . . . All accounts, however, agreed as to the rapid encroachment of the lake upon the town in 1838 and 1839; and that the inhabitants were driven out during the two years following; the foundations of the houses gave way, and fresh water failed them.

At the time of my visit the water reached up to the very base of the town and castle walls on the west and east sides, and within ten paces of the large mosque on the south side. . . . There was, however, undoubted evidence of its having been much higher, for within the walls in various places were large pools of stagnant water; the ground in several parts was saturated with salt and overgrown with saline plants. The quantity of land gained

7. Lake Van and the Flood

by the lake is enormous. . . . A like fate befell many other villages along the shores. Iskella, a small fishing village, a mile and a half from Van, is now half deserted. The islands in the Lake, on the authority of the Armenian Bishop at Chijis Monastery, were in a similar manner gained upon by the rise of water.

There is a tradition that the Lake of Van now covers what was formerly a spacious plain, studded with villages and gardens, – that the Bend í Máhi Sú [river] and the two streams of Ardish met and formed one large river about midway between Ardish and Bitlis, – that a short way below their junction there was a large bridge (which is now sometimes seen at the bottom of the Lake by the boatmen), – that at some distance below the bridge the river suddenly disappeared through a large hole, near which was a powerful salt-spring, – that by some sudden convulsion the hole became closed, and the accumulated streams, having no outlet, gradually formed the Lake of Van. . . .

Similar oscillations in the levels of lakes have been observed in America and other portions of the world; but I am not aware of any on so great a scale, and where the effects have been so felt by the natives on the shore.

The rise of the Lake of Van is probably due to some change of climate for a succession of years, such as the fall of a greater quantity of snow in the mountains, or a deficiency in the usual evaporation. This is undoubtedly the most reasonable explanation; but it must not at the same time be forgotten, that climatic alternations in those regions are exceedingly regular and uniform. . . .

Again, it is difficult to conceive how four small streams alone, supplied by melting snow, should have sufficient influence to elevate twelve feet such a large body of water as the Lake of Van; though they may have a local effect at their mouths.

Another solution which presents itself, is the existence of intermittent springs in the bottom of the Lake, bursting forth at long intervals. We are not, however, aware of results on so great a scale produced by such a cause.[9]

According to his report, the geologist Loftus visited the vicinity of Lake Van in 1852, a few years after a significant and unexpected rise in the water level of the lake beginning in 1838. The level increased for two or three years but had already begun to fall before Loftus's arrival. A particular area of his interest seems to have been near the town of Ardish (or Arjish, or now Ercis), at almost the northernmost point of the lake. The total rise in the level of the lake was about twelve feet, causing major problems for the inhabitants. Many towns near the lake were "surrounded, destroyed, or deserted" as a result of this flood. Even the ancient town of Ardish with its population of about five thousand people had to be permanently abandoned together with its historical castle. The town was rebuilt on higher ground in 1841 and is indicated as Ercis in the map shown in Figure 5.2. While some areas on shore became surrounded by the encroaching water, previously existing islands were diminished in size.

Of particular interest are Loftus's reports and musings about the ancient history and causes of such variations in the level of the lake. He mentions a tradition that the level had once been much lower than at the time of his visit. According to that tradition, the shallow Arjish Gulf had then been reduced to a gathering of its present feeder streams into a single river flowing toward the southwest and out of the gulf. In Section 7.7 it will be noted that between about 849 B.C. and 675 B.C. the level of Lake

7. Lake Van and the Flood

Van was about 150 m below its present value. As considered in Section 5.4A, a bridge that had spanned this combined river was said sometimes to be visible to boatmen beneath the waters of the gulf. The average depth of the gulf today is about twenty meters.[10] From a consideration of the bathymetric map given in Figure 5.2, a location for such a bridge would be easy enough to imagine spanning an Ercis Gulf that was narrowed to a single river. The reported disappearance of the river into a hole in the ground and its subsequent emergence into Lake Van is reminiscent of Pliny's discussion of the headwaters of the Tigris River, quoted in Section 4.6. Apparently independent of such local traditions, Lynch arrived at a similar interpretation of Pliny's writings on this subject:

> The mention of two lakes by Pliny need not perplex us over-much; for his [Pliny's] Arethusa no doubt denotes the Arjish [Ercis] arm of Lake Van, and his Thospites the principal body of water [Lake Van] with the city of Van, the Dhuspas of the cuneiform inscriptions [also Thuspa], upon its eastern shore. . . . The mention of two lakes by Pliny and Ptolemy may point to a former isolation of the Arjish arm.[11]

Depending on the degree of ancient isolation of Lake Arethusa from the principal part of Lake Van, as inferred by Lynch above, the water level of Lake Arethusa could have been much more sensitive to changes in local rainfall than the modern combined Lake Van system. The surface area of Lake Arethusa could also have been more sensitive to water input than the more steeply-shored portions of Lake Van. Loftus was obviously very interested in the cause of the rise in water level of Lake Van. He

217

first offered what he considered to be the most natural explanation for this rise, but he seemed not to be fully convinced of the correctness of this explanation:[12] "The rise of the Lake of Van is probably due to some change of climate for a succession of years, such as the fall of a greater quantity of snow in the mountains, or a deficiency in the usual evaporation." Two concerns he had with this explanation were indicated:

> 1. "[B]ut it must not at the same time be forgotten, that climatic alternations in those regions are exceedingly regular and uniform. . . ."
> 2. "Again, it is difficult to conceive how four small streams alone, supplied by melting snow, should have sufficient influence to elevate twelve feet such a large body of water as the Lake of Van. . . ."

Loftus then continued by suggesting a completely different interpretation for the rise in the water level:[13] "Another solution which presents itself, is the existence of intermittent springs in the bottom of the Lake, bursting forth at long intervals." This suggestion is remarkable in the present context. The description of the biblical flood begins with the following words: "[O]n that day all the fountains of the great deep burst forth. . . ." (Genesis 7:11). If "springs in the bottom of the Lake" are considered to be the same as "fountains of the great deep," then these two descriptions are essentially identical. In both cases the springs/fountains are said to have burst forth. This identity is remarkable because there is no other indication that Loftus had in mind the biblical flood, and there is no indication in the Bible that the flood was associated in any way with a lake.

7. Lake Van and the Flood

In the present study it is suggested that Lake Van, considered by Loftus in interpreting the flood of the 1800s, may be the very same lake that at a much earlier time could have been responsible for the biblical flood. Independent of that possibility, it is tempting to imagine that the rapidly rising lake levels between 1838 and 1840 had some connection with the earthquakes and volcanic eruption of Mt. Ararat in 1840 A.D., discussed in Section 7.2. Those geological events caused the spouting of "torrents of water" and the appearance of new springs. These springs (and others on the floor of Lake Van) could have been recent representations of the bursting forth of "all the fountains of the great deep" (Genesis 7:11), as occurred at the time of Noah's flood.

It is perhaps of interest here to consider again the chronology of the rising water level reported by Loftus as quoted above. The water is said to have encroached on the town of Ardish in 1838 and 1839, and by 1841 the town had been abandoned. Eruptions may be preceded by the filling from below of a magma chamber or chambers that eventually feed magma to the volcano. This filling leads to deformation or swelling of the ground in the vicinity of the volcano, depending on the configuration of the chambers. In the case of Mt. Ararat, such swelling could in principle have been associated with increases in the water level of Lake Van prior to the eruption in 1840. Records of known eruptions of Mt. Ararat are indicated in Table 7.1.

Major eruptions are often associated with extended periods of heavy rainfall. Thus, a volcanic eruption and earthquakes at the time of Noah could account for both the rain and the underground water sources that are reported in the story of Noah's flood.

Causes And Effects Of The Flood

These water sources might have raised the lake level and thereby floated the ark while flooding surrounding low-lying lands. Unfortunately, data concerning the earliest eruptions of Mt. Ararat and other volcanoes are not necessarily complete. Thus, it isn't possible to say with certainty whether the major climate event associated with Noah's flood might have resulted from volcanic activity.

The consequences of the rising water level of Lake Van discussed above can be emphasized by the loss of Arjish and its fortress as well as the loss of other ancient fortresses adjacent to the lake:[14]

> [Arjish] was known to Marco Polo (thirteenth century) as one of the three greatest cities of Armenia; and at the commencement of the sixteenth century it formed one of the seven fortresses which encircled the lake of Van.[15] In the summer of 1838 it was still peopled; but in the winter of that year the waters of the lake rose, until in 1841 they had attained an increase of some 10 or 12 feet. The foundations of the houses gave way and the supply of fresh water failed.[16] Arjish was evacuated by its reduced population, and is at the present day not tenanted by a single soul.

With respect to the other encircling fortresses, Lynch has remarked as follows."

> The three old fortresses of Akhlat, Adeljivas and Arjish all bear testimony to a considerable rise in the level of the lake since the days when they were built. The walls of the first two on the side of the water have either fallen in or are being slowly undermined. Arjish has been permanently abandoned by its inhabitants. Immemorial

villages, like that of Kizvag between Akhlat and Tadvan, are being menaced by the latest periodical increase, which seems to have commenced about 1895.[17]

Thus, three of the original seven fortresses had been abandoned or destroyed as a result of the rising waters, and other villages were "being menaced" as of Lynch's writing in 1901.

Like Loftus, Lynch was also interested in the possible causes of the level changes in Lake Van, and a few of his comments on this subject are quoted here:

> A feature which has occupied considerable attention . . . is the fluctuation in level of these Armenian lakes [Van, Gökcheh, and Urmi]. There can be no doubt that they are all three subject to more or less pronounced periodical changes; and various reasons have been assigned. . . . If I venture to join in the discussion I would submit the suggestion that we should for convenience group the phenomena under two heads. Temporary variations should be distinguished from any differences of a more permanent nature the existence of which it may be possible to prove.
>
> The same phenomenon of a rise in level was apparent on the margin of the large lake in the crater on Nimrud. There the brushwood, representing the growth of many years, was submerged; and much had already perished from want of sustenance. All the evidence points to the fact that such changes are of a temporary nature, and that a period of increase is followed by one of decline. The most probable explanation is that they are due to climatic conditions, which, it is well known, are variously operative over cycles of years. In the absence of any observatory in these countries this question is largely a matter of surmise or, at best, of inference. The existence of such periodical fluctuations may be regarded as having

been established; it remains to consider the changes of a more permanent order. . . . Much the same conditions as at the present day appear to have prevailed during the historical period – a vast sheet of water, deep and translucent, dammed up by the volcanic barrier at its westerly extremity. I think there can be no doubt that the permanent tendency of this sheet of water has been to rise in level. . . . The evidence which may be collected upon the shores of Lake Van all points in the same direction of a progressive upward tendency.[18]

Lynch's observation that the level of Lake Van is also subject to periodic increases is supported by modern studies, and the lake level variations are found to closely parallel changes in sunspot numbers. Based on the study of a thirty-year period (1944-1974, only three solar cycles), the following inferences have been drawn:

With a one-year delay the lake level rises with increasing sunspot number, suggesting that there may be a causal relationship between them. At times of high solar activity, weather conditions tend to become more unstable with more rain, less evaporation and higher humidity, factors that promote a lake level rise.[19]

Solar cycle 13 began in 1890, so according to the above quotation the lake level would have begun to increase in about 1891. Thus the increase may have started slightly earlier than Lynch's suggested date of "about 1895."

Significant variations of the lake level, both positive and negative, are known to have occurred in prehistoric times. For the recent historical period the exceptional increase during about 1838-1840 has been mentioned as well as the periodic increases

7. Lake Van and the Flood

noted by Lynch and now associated with sunspot cycles. In addition to these events, there seems to be a long-term upward trend:

> Much the same conditions as at the present day appear to have prevailed during the historical period – a vast sheet of water, deep and translucent, dammed up by the volcanic barrier at its westerly extremity. I think there can be no doubt that the permanent tendency of this sheet of water has been to rise in level.[20]

More recent measurements seem to confirm this trend, and high accuracy in these measurements is now possible. As an example of the general trend, there was an overall increase in water level by about two meters between 1944 and 2002.[21]

The long-term increases in the lake level noted already by Lynch in 1901 and continued into the twentieth century, as just indicated, may be among the many consequences of world-wide industrialization beginning especially in the nineteenth century. Probably the best-known of these consequences is global warming with its associated changes in weather patterns. Such changes may be the cause of the increased rainfall that was likely responsible for the recent increases in lake level. In these chronological studies the duration of climate changes due to a volcanic eruption in 1912 was less than it would have been if the atmosphere had not become anthropogenically polluted.[22] In a report from 1910, determinations of the smallest visible phase of the moon yielded results that might be consistent with degraded visibility in comparison with earlier times.[23]

Of particular interest here could be implications considered previously that at some times in the ancient past the Ercis gulf to the north of the main body of Lake Van might have appeared to Pliny and others as a separate lake that drained through an underground channel into the main lake. Besides short term variations, such as those associated above with the eruption of Mt. Ararat in 1840 A.D., the water level and area of the Ercis gulf seem generally to have been increasing. Thus, since at least the beginning of the sixteenth century, the fortress of Arjish near the present Ercis gulf was one of the seven fortresses that encircled Lake Van. Rising water levels of the lake, including especially the Ararat events described above, caused the fortress to be permanently abandoned; and other towns and fortresses continue to be threatened by water.

Some qualitative insight into the increasing water level and surface area of the Ercis gulf might possibly be obtained from consideration of early texts and maps. It has already been mentioned in Section 5.3 that in the time of Pliny the Elder the gulf may have been better described as a small lake (Lake Arethusa). Figure 4.2 is a map of the Terrestrial Paradise drawn in the 1700s. The resolution of that map is limited, but in comparison with the bathymetric map of Figure 5.2, it appears that in Figure 4.2 Lake Arethusa has been entirely omitted. In a later map prepared by Loftus and published in 1855,[24] the Ercis gulf is clearly drawn and appears somewhat similar to the gulf as shown on the map of Figure 5.2. A possibly important difference, however, is that in the later 2013 map the Erics gulf appears to have encroached further onto the plain northeast of the main lake and the channel between the lake and the gulf has grown substantially wider. These would appear to be indications of the continuing increase in water level, and that increase could be a

7. Lake Van and the Flood

result of continuing climate change. It can be imagined that at an earlier time the narrowed channel might have been crossed by a bridge as in the local traditions.

7.5 Ancient lake levels

The previous discussions have mostly been concerned with the water level of Lake Van over the past few centuries. To increase the relevance of these considerations for the time of the biblical flood, it would be helpful to find information about the water level in the lake for times several thousand years in the past. Large changes have occurred, but it may not be possible to obtain the detailed information that would be necessary to clearly identify any flood-related level changes. In any case the level of Lake Van has varied dramatically during prehistoric times:

> The lake level was at its highest at 72 m above present at the height of the last ice age about 18,000 yr B. P. [before present]. A dramatic drop to over 300 m below present occurred about 9500 yr B. P., with an equally dramatic rise around 6500 yr B. P. These sudden variations . . . left distinctive imprints on the sedimentary, geochemical and pollen record.[25]

Level changes continue to occur, but such dramatic changes as quoted above haven't occurred in historic times.

A few authors have also estimated changes in the level of Lake Van during times of interest in this study. Landmann, et al. have discerned three recessions in the level of the lake that are at least somewhat in the time period of interest. These recessions are also correlated with similar occurrences in nearby Lake Urmia:[26]

Apparently, further minor recessions occurred at 6300, 4800, and 3000 years B.P. The two oldest minima correlate with the playa phases 1 and 2 at Lake Urmia,[27] showing that these evaporative periods are not only local events. A high energy environment, that is, shallow conditions was recorded in Lake Urmia contemporary with the third minimum in Lake Van.

The B.P. (before present) terminology is standardized to a "present" of 1 January 1950. Thus, the recessions just mentioned are said to have occurred on the calendar dates of roughly 4350, 2850, and 1050 B.C. Greater interest here would be associated with information on a rapid increase in lake level (if any) between recessions and at a time that could be related to the biblical flood. Unfortunately, no suitable rapid rise in the level of Lake Van is discernible in the still very approximate data. The flood events that will be considered in Chapter 9 would seem to have occurred in the period between 2850 and 1050 B.C. The water level between those dates was as high as about 50 meters above its current level,[28] but studies thus far may not be adequate to show clearly an abrupt water level rise at any particular date in this range. Thus, if the date 2035 B.C. that has previously been suggested was actually the date of the flood,[29] then a sudden increase from the background level might, in principle, have been visible in Figure 7 of Landmann, et al. However, such an increase in water level may not have left any evidence that would be adequately resolvable today.

Another approximate interpretation of a lake level rise has been inferred by Kuzucuoglu, et al.:

7. Lake Van and the Flood

> Finally, the level rose up to today's, generating the filling of the present flood plains. During this later period, Bronze Age populations settled near the shores of a lake whose level was ~3 m below today's during the third millennium BC (EBA site in the Zilan delta), and ~+5 m during the second millennium BC (LBA site north of Van). Field evidence of these latest changes is scarce, however, and not confirmed by any section or excavation.[30]

In this quotation the letters EBA stand for the Early Bronze Age, which is understood to be from about 3300 to 2100 B.C. Similarly, the letters MBA would stand for the Middle Bronze Age from about 2100 to 1550 B.C., and the letters LBA stand for the Late Bronze Age from about 1550 to 1200 B.C. Thus, an EBA site of the third millennium B.C. would seem to have been occupied at a time between 3000 B.C. and 2100 B.C., while an LBA site of the second millennium B.C. would seem to have been occupied between 1550 and 1200 B.C. According to the quotation above, there was a net water level rise of about eight meters between some earlier time (in the period of 3000 to 2100 B.C.) and some later time (in the period of 1550 to 1200 B.C.).

A water-level rise of eight meters would probably be sufficient to float an ark, especially if that ark were constructed initially near the lake-shore. On the other hand, this calculation doesn't exclude the possibility of a larger rise (or multiple rises) in the lake level at some intermediate time, which level then had fallen to the second value reported above by a time between 1550 and 1200 B.C. Thus, the rise in water level responsible for the flood might, for example, have occurred after 2100 B.C., or at least after the EBA archeological records were created. Similarly, the time-dependent post-flood water level might have been inferred from a site that was created in about 1550 B.C., or at least after the water had reached its reported LBA level.

In principle, the time required for a water level of Lake Van that has been altered by a flood or other eventuality to return to its initial value can be calculated, at least approximately. A procedure for such a calculation has been outlined by Degens, et al. As an illustration, those authors considered the possibility that the lake started out empty but was subject to current evaporation rates and stream and rainfall inputs. The results are shown in Figure 7 of their report:

> Figure 7 shows the result of such a calculation, from which we conclude that the water rose at a dramatic rate as the lake started to be filled (5.2 m/yr at 400 m below present lake level), and slowed down considerably as it approached its present, more or less stable, level. The time required for the infilling of the lake to its present level may be estimated by integrating the curve of Figure 7. Only about 500 years appear to be necessary for the lake to acquire its current shape and level, almost an instantaneous event in terms of geological time and certainly far too rapid for geological processes like sedimentation or subsidence to play a significant role.[31]

The same sort of calculation could be carried out for any imagined circumstances associated with the flood. Thus, if the flood raised the lake level by a more modest ark-floating amount, the level might then drop to almost its initial value in a time period of at most a few decades. Estimates of historical water levels in Lake Van, as mentioned earlier in this section do not yet

7. Lake Van and the Flood

have the precision necessary for them to be useful in estimating the actual date of Noah's flood.

7.6 Causes of lake level changes

It is well-established that the water level of Lake Van is not constant, and the obvious example considered previously in this chapter is the likely occurrence of a rapid increase in the water level at the time of Noah. It is reasonable to wonder what might be the possible causes of up or down changes in the level of the surface of Lake Van (or other similar bodies of water). Several potential causes of lake level changes are suggested below:

A. Local weather

An obvious source for the water in a lake is the rainfall directly into the lake or into the streams of the catchment basin that feed water to the lake. On the other hand, the rate at which water leaves a lake to enter the atmosphere may depend on such factors as air and water temperature, air humidity, and wind. Some water may also flow into or out of the lake floor ("the fountains of the deep" Genesis 7:11).

B. Climate

Slower multiyear changes in water level may be associated with climate variations: "In any case, the high variability of Lake Van level, and hence the lake water volume, observed today indicates that the lake level is most probably highly sensitive to changes in the climatic components of the hydrological budget of the lake (mainly precipitation and evaporation)."[32]

The climate in some areas of the earth is subject to semi-periodic self-sustaining oscillations of winds and ocean currents.

Thus, the El Niño-Southern Oscillation effects vary with a period of a few years.[33] While these effects are of substantial interest, they seem not to be relevant to the behavior of Lake Van.

The number of sunspots usually exhibits a close correlation with the water level in Lake Van. Thus, rainfall rate and the lake level tend to rise and fall as the sunspot number rises and falls:

> The spring thaw as well as enhanced rains increase river discharge to as much as five times its monthly average, leading to a seasonal lake level rise that amounts to 50 ± 18 cm for the period 1944-1974. . . . The lake level record for this period may be cross-correlated to solar activity. . . . With a one-year delay the lake rises with increasing sunspot number, suggesting that there may be a causal relationship between them. At times of high solar activity, weather conditions tend to become more unstable with more rain, less evaporation and higher humidity, factors that promote a lake level rise.[34]

Data on sunspot numbers and water levels of Lake Van are too limited to be useful in interpreting lake levels of earlier times. However, the inferred modest changes are unlikely to have had any significant effect on the people or cultures that have existed in the vicinity of the lake.

C. Erosion and eruptions

The physical configuration of a lake such as Van can be altered by weathering and erosion of the shore materials. Ultimately these processes can lead to large-scale leakage of the lake water and a significant lowering of the water level. In contrast to erosion, a volcanic eruption near the lake can lead to an increase in the shore height and a reduction or elimination of

7. Lake Van and the Flood

direct leakage or seepage, together with a significant raising of the water level. Concerning such changes the following has been noted about specific previous events:

> These altitudes [of the lake] are higher than today's threshold (1736 m), which means that the threshold during C1' lake maximum was higher than the present one, suggesting that the lake basin has undergone significant morphological changes. Such changes can only be related either to volcanism (e.g. a volcanic flow damming water up to an altitude of 1755 m, followed by an erosion of this dam) or less probably, to tectonics (e.g. downfaulting of a block possibly forming an old threshold), or both. . . . These data suggest that the factor triggering C1' lake-level rise was possibly volcanic.[35]
>
> However, large quantities of pyroclastite flows and falls from the Nemrut volcano deeply affected the hydrographic drainage of Lake Van. Thus it is probable that volcanism also played an important role in the occurrence of these lake-level rises.[36]
>
> When, before 110 ka, high quantities of volcanic material (flows and falls) were emitted by the Nemrut volcano, these products dammed the outlet and caused the C1' lake-level rise. Since this period, the drainage area of the lake was subject to further volcanism-related changes such as damming (lake-level rises) and erosion of dams (lake-level falls).[37]

These quotations indicate that erosion and eruption have had substantial effects on the water level of Lake Van in the past. It is not unreasonable to imagine that other events might also have occurred in the past that have not previously been explicitly identified. One such possibility could account for recently discovered information concerning the kingdom of Urartu, and this interpretation will be considered in Section 7.7.

D. Asteroid impacts

Asteroids and comets sometimes carry with them large quantities of water or ice, and the impact of an asteroid on an ocean can also inject a large amount of water into the earth's atmosphere. This concept will be considered in Section 9.2 in connection with the flooding of Lake Van in the time of Noah.

E. Orbit variations

The rate at which rainfall lands in the catchment basin of Lake Van and is added to the water in the lake depends in part on the orientation of the earth with respect to the sun and its distance from the sun. Similarly, the temperature of the lake and the evaporation rate of its water also depend in part on the relative orientation of the sun and the earth. As it happens this orientation and distance vary slowly over a time scale of many thousands of years, and thus the water level of the lake isn't constant, and in fact it is a very complicated function of time. No theory explains all aspects of the earth's orbit, but the best contender at present is associated with what are called "Milankovitch cycles:"

> Milankovitch cycles describe the collective effects of changes in the Earth's movements on its climate over thousands of years. The term is named after Serbian geophysicist and astronomer Milutin Milankovic. In the 1920s, he hypothesized that variations in eccentricity, axial tilt, and precession of the Earth's orbit resulted in cyclical variation in the solar radiation reaching the Earth, and that this orbital forcing strongly influenced climatic patterns on Earth.[38]

7. Lake Van and the Flood

This theory is mathematically complicated, and the interested reader can learn more from the above-quoted web page. The inferred climate changes occur on the time scale of the ice ages, and thus they are not of particular importance for investigations relating to only the past few thousand years.

F. Global warming

For the first time in the long history of our planet, human activities are now causing increasingly dramatic changes in the earth's climate. As noted in Section 7.4 above, industrialization beginning in about the nineteenth century led to degraded visibility through the atmosphere from about the early twentieth century. By about 1965 atmospheric pollution had led to more substantial climate changes in what is referred to as global warming. Since that time there has been a rising water level in Lake Van that now amounts to about two meters:

> Global warming resulting from increasing greenhouse gases in the atmosphere and the local climate changes that follow affect local hydrospheric and biospheric environments. These include lakes that serve surrounding populations as a freshwater resource or provide regional navigation. Although there may well be steady water-quality alterations in the lakes with time, many of these are very much climate-change dependent. During cool and wet periods, there may be water-level rises that may cause economic losses to agriculture and human activities along the lake shores. Such rises become nuisances especially in the case of shoreline settlements and low-lying agricultural land. Lake Van, in eastern Turkey currently faces such problems due to water-level rises. The lake is unique for at least two reasons. First, it is a closed basin with no natural or artificial outlet and

233

second, its waters contain high concentrations of soda which prevent the use of its water as a drinking or agricultural water source.[39]

Efforts to control or reverse global warming have been inadequate, and it seems possible that destructive weather patterns will become more common, coastal properties will be inundated, and large areas of our planet may become increasingly uninhabitable. At the time of this writing, the U.S. government seems to have taken a leadership role in questioning even the existence of global warming.

7.7 Lake level change and the history of Urartu

It is probably clear from previous considerations here that changes in the water level of Lake Van could have had a substantial impact on the people living nearby. Thus, the lake could be a vital method for the rapid transportation of materials, food products, and armies. For economic reasons it would be advantageous for people to live and work close to the shore of the lake. For defensive purposes it could also be useful for fortresses and castles to be near the shore of the lake. However, it is also evident that rapid changes in water level would be very disruptive and dangerous. The water in the lake is toxic to plants and animals, so a sudden increase in water level would be extremely damaging to existing agricultural activities. On the other hand, a sudden drop in water level could encourage inhabitants to move their homes and livelihoods to locations nearer to the new shore, which might then be remote from agriculturally productive lands. An example of these effects seems to be discernible in recent discoveries relating to the kingdom of Urartu. This kingdom was

7. Lake Van and the Flood

always focused close to Lake Van and would have been forced to adapt to changes in lake level:

> Urartu, also known as Kingdom of Van, was an Iron Age kingdom centred on Lake Van in the Armenian Highlands. It corresponds to the biblical Kingdom of Ararat. Strictly speaking, Urartu is the Assyrian term for a geographical region, while "Kingdom of Urartu" or "Biainili lands" are terms used in modern historiography for the Urartian-speaking Iron Age state that arose in that region.[40]

Chronologies for the kingdom of Urartu are not all consistent. According to one model the kings of Urartu include Aramu (c. 860-843 B.C.), Sarduri I (c. 832-820 B.C.), Ishpuini (c. 820-800 B.C.), Menua (c. 800-785 B.C.), Argishti I (c. 785-760 B.C.), Sarduri II (753-735 B.C.), Rusa I (? B.C.), Argishti II (c. 714-685 B.C.), Rusa II (685-645 B.C.) Sarduri III (645-635 B.C.), Erimena (635-620 B.C.), Rusa III (620-609 B.C.), and Rusa IV (609-590 or 585 B.C.). A brief history of the later years of Urartu is the following:

> In 714 BC, the Urartu kingdom suffered heavily from Cimmerian raids and the campaigns of Sargon II. The main temple at Mushashir was sacked, and the Urartian king Rusa I was crushingly defeated by Sargon II at Lake Urmia. He subsequently committed suicide in shame. Rusa's son Argishti II (714–685 BC) restored Urartu's position against the Cimmerians, however it was no longer a threat to Assyria and peace was made with the new king of Assyria Sennacherib in 705 BC. This in turn helped Urartu enter a long period of development and prosperity, which continued through the reign of Argishti's son Rusa II (685–645 BC). After Rusa II,

235

however, the Urartu grew weaker under constant attacks from Cimmerian and Scythian invaders. As a result, it became dependent on Assyria, as evidenced by Rusa II's son Sarduri III (645–635 BC) referring to the Assyrian king Ashurbanipal as his "father."

According to Urartian epigraphy, Sarduri III was followed by three kings – Erimena (635–620 BC), his son Rusa III (620–609 BC), and the latter's son Rusa IV (609–590 or 585 BC). Late during the 7th century BC (during or after Sarduri III's reign), Urartu was invaded by Scythians and their allies – the Medes. In 612 BC, the Median king Cyaxares the Great together with Nabopolassar of Babylon and the Scythians conquered Assyria after it had been badly weakened by civil war. The Medes then took over the Urartian capital of Van towards 585 BC, effectively ending the sovereignty of Urartu.[41]

For purposes of these considerations, it is important to recall from Section 7.5 that the water level of Lake Van is said to have decreased in about the year 3000 B.P. In this calendar notation the letters B.P. stand for the words "before present," where "present" is defined conventionally as 1 January 1950 A.D. (= 0 B.P.). Thus, for example, 1 January 1949 would be exactly one year before present, which could be written 1 B.P. Also in this notation, 1 A.D. would be the same as 1949 B.P. On the other hand, 1 B.C. would be equivalent to 1950 B.P., because there is no year zero in the B.C./A.D, calendar system. The date of interest here is 3000 B.P. If 1950 B.P. is the same as 1 B.C., then 3000 B.P. could also be written as 1051 B.C.

The example just mentioned relates to the date at which Lake Van is said to have begun a substantial decrease in water level: "By 10,600 years B.P. the lake began to rise and reached,

7. Lake Van and the Flood

following another regression between 9000 and 8100 years B.P., the Holocene highstand by about 7500 years B.P., dropping to today's level at about 3000 years B.P."[42] This recession could not, of course, occur instantaneously; and a later restatement includes the possibility of a period of up to six hundred years for this event: "The water content curve suggests that lake level recessions may have occurred from 12,800 to 10,700 and from 3400 to 2800 years B.P. . . ."[43] Thus the recession at about 3000 B.P. may actually have taken much of the 600-year period from 3400 B.P. to 2800 B.P. During this time, it may be imagined that erosion and leakage occurred allowing water to drain from the lake. Thus, the lake could have reached its minimum level by about 851 B.C. rather than in 1051 B.C. as in the previous calculation.

One can assume that by about 851 B.C. the lake level would have stabilized, the soil would have largely recovered from its soda poisoning, and people would be reoccupying the previously flooded shore lands. It is of interest now to combine this information about the new lower lake level with events relating to the newly forming kingdom of Urartu. As indicated previously in this section, the first king of Urartu was Aramu (c. 860-843 B.C.). At his accession the Urartians would have been occupying their recently drained lands and working to establish their new monarchy.

The only uncertainty concerning this time is what the new lower water level would have been. In a quotation above in this section it has been assumed that the new level was essentially the same as the current water level. However, there is now significant reason to believe that the water level that was reached was actually much lower than the current water level. According to a

Causes And Effects Of The Flood

recent report, researchers have found the remains of an ancient Urartian castle that is currently far underwater:

> A team of Turkish researchers has discovered the remains of what is believed to be a 3000-year-old castle in the country's eastern Van province, the site of Lake Van, the largest lake in Turkey. . . . The researchers who went underwater believe that the ruins are supposedly from the Iron Age Urartu civilization, also known as the Kingdom of Van, thought to date back to the eighth to seventh centuries BC. . . . The current water level of the reservoir is about 150 meters higher than it was during the Urartu civilization."[44]

The earlier report quoted above suggested that in the recession of interest here the lake level dropped to its current level. The discovery of the underwater Urartian castle suggests that the lake level in this recession must actually have dropped to about 150 meters below the current level. This means, of course, that at some time between the beginning of the Urartian kingdom and the present the lake level must have increased by about 150 meters to its current level. As it happens, a possible explanation for a water-level increase near the end of the Urartu civilization may be contemplated.

As noted previously in Section 7.6C, one of the principal causes of water level increases in Lake Van is the occurrence of eruptions of the Nemrut volcano. One of the most striking biblical representations of a volcanic eruption has been discussed in Sections 6.6 and 6.7. That event was the major eruption of the Nemrut volcano that occurred in about 657 ± 25 B.C. The eruption could, like earlier eruptions, have raised the level of a lake outlet while still allowing a bridge to span the much-

7. Lake Van and the Flood

narrowed Ercis gulf. The circumstances of the eruption, including the submerging of castles near the lake shore, and the damage to agriculture and transportation would have been an enormous setback for the Urartian civilization. A later water level increase could have inundated the (possibly-legendary) ancient bridge mentioned in Section 7.4, and it would be of interest to inquire whether evidence of such an underwater bridge in the Ercis Gulf region might still exist.

From earlier discussions in this section, it may be seen that this eruption and many of its damaging consequences probably occurred during the reign of the Urartian king Rusa II (685-645 B.C.). The consequences of this eruption could even have included the end of the independent existence of the kingdom of Urartu. It may be worth recalling a few lines quoted previously concerning the reign following that of Rusa II: "After Rusa II, however, the Urartu grew weaker under constant attacks from Cimmerian and Scythian invaders. As a result, it became dependent on Assyria, as evidenced by Rusa II's son Sarduri III (645-635 BC) referring to the Assyrian king Ashurbanipal as his 'father.' "[45] Soon afterward, the sovereignty of Urartu came to an end.

7.8 Conclusion

One purpose of this chapter has been to consider the possibility that the biblical flood corresponded to a substantial increase in the water level of Lake Van. This interpretation is consistent with the ideas considered previously that the flood may have been associated with the land of Eden and that the ark came to rest in the mountains of Urartu (Ararat). Possible mechanisms for "the fountains of the great deep" were contemplated, and the reasonableness of the necessary rainfall rate was also considered. The water level of Lake Van has varied substantially in historic and prehistoric times, but it is difficult to demonstrate from evidence discoverable at the lake that the level changes necessary to create the biblical flood occurred at just the right time in history. The flood concept would be further supported if there were any historical evidence that major flooding occurred elsewhere in the world at about the same time as a reported rise in the water level of Lake Van. Some investigations into that possibility are included in Chapter 9. The relationship of changes in the level of Lake Van to the kingdom of Urartu have also been considered here.

---

1. J. Bell, *A System of Geography, Popular and Scientific; or a Physical, Political, and Statistical Account of the World and its Various Divisions*, Volume 4, Part 1 (A. Fullarton and Co., London, 1848), p. 262. An earlier and more extensive discussion of this event is included in "Mount Ararat. Official account of the late earthquake, drawn up by Major Voskoboinikof, of the Imperial Russian Engineers," (*St. Petersburgh Gazette*), *The Church of England Magazine*, Volume 10, Number 287, pp. 373-375 (5 June 1841).
2. "Ararat Volcano," Smithsonian Institution, *National Museum of Natural History*, Global Volcanism Program, Database Search, March 26, 2014.
3. "The ice cave phenomena of Eastern Washington," in D. W. Patten, *The Biblical Flood and the Ice Epoch, A Study in Scientific History* (Baker Book House, Grand Rapids, Michigan, 1976), pp. 120-124.
4. At this writing a convenient source for such information is "Global Weather and Climate Extremes," World Meteorological Organization; http://wmo.asu.edu. The

data for 5, 6, 7, 8, and 9 days are from H. Quetelard, P. Bessemoulin, R. S. Cerveny, T. C. Peterson, A. Burton, and Y. Boodhoo, "World-record rainfalls during tropical cyclone Gamede," *Bulletin of the American Meteorological Society*, Volume 90, Number 5, pp. 603-608 (May 2009).
5. G. Landmann, A. Reimer, and S. Kempe, "Climatically induced lake level changes at Lake Van, Turkey, during the Pleistocene/Holocene transition," *Global Biogeochemical Cycles*, Volume 10, Number 4, pp. 797-808 (December 1996); p. 798.
6. C. Kuzucuoglu, A. Christol, D. Mouralis, A.-F. Dogu, E. Akköprü, M. Fort, D. Brunstein, H. Zorer, M. Fontugne, M. Karabiyikoglu, S. Scaillet, J.-L. Reyss, and H. Guillou, "Formation of the upper pleistocene terraces of Lake Van (Turkey)," *Journal of Quaternary Science*, Volume 25, Number 7, pp. 1124-1137 (2010).
7. F. L. Whipple, "A comet model. I. The acceleration of comet Encke," *Astrophysical Journal*, Volume 111, pp. 375-394 (1950).
8. H. F. B. Lynch, *Armenia Travels and Studies*, Volume 2 (Longmans, Green, and Co., London, 1901), p. 39.
9. W. K. Loftus, "On the geology of portions of the Turko-Persian frontier, and of the districts adjoining," *The Quarterly Journal of the Geological Society of London*, Volume 11, pp. 247-344 (1855); pp. 317-320.
10. E.T. Degens, H. K. Wong, S. Kempe, F. Kurtman, "A geological study of Lake Van, Eastern Turkey," *Geologische Rundschau*, Volume 73, Issue 2, pp. 701-734 (1984); p. 705.
11. H. F. B. Lynch, op. cit., p. 42; considered also previously in Section 5.3.
12. W. K. Loftus, op. cit., pp. 319-320.
13. Ibid., p. 320.
14. H. F. B. Lynch, op. cit., pp. 29-30.

15. "Marco Polo, Yule's translation, London, 1874, vol. i. p. 47; and 'Merchant in Persia' in *Italian Travels in Persia*, Hakluyt Society, London, 1873, p. 160. The other six castles were Tadvan, Vostan, Van, Berkri, Adeljivas and Akhlat."
16. "Loftus, who visited Arjish in 1852, has collected the facts relative to the inundation (*Quarterly Journal of Geological Society*, London, 1855, vol. xi, p. 319)."
17. H. F. B. Lynch, op. cit., p. 52.
18. H. F. B. Lynch, op. cit., pp. 46-52.
19. E. T. Degens, et al., op. cit., pp. 716-717.
20. H. F. B. Lynch, op. cit., p. 51.
21. C. Kuzucuoglu, et al., op. cit., pp. 1125-1126, Figure 2.
22. L. W. Casperson, *Patterns of Biblical Chronology* (Westbow Press, Bloomington, IN, 2012), Chapter 22, p. 560; L. Oman, A. Robock, G. L. Stenchikov, and T. Thordarson, "High-latitude eruptions cast shadow over the African monsoon and the flow of the Nile," *Geophysical Research Letters*, Volume 33, L18711, pp. 1-5 (30 September 2006) caption to Figure 4.
23. L. W. Casperson, "The lunar dates of Thutmose III," *Journal of Near Eastern Studies*, Volume 45, Number 2, pp. 139-150 (April 1986); pp. 143-145 and Figure 1.
24. W. K. Loftus, op. cit., "Geological Sketch Map of the Turko-Persian Frontier," Plate IX.
25. E.T. Degens, et al., op. cit., p. 702.
26. G. Landmann, et al., op. cit., p. 807.
27. K. Kelts and M. Shahrabi "Holocene sedimentology of hypersaline Lake Urmia, northwestern Iran," *Palaeogeography, Palaeoclimatology, Palaeoecology*, Volume 54, Numbers 1-4, pp. 105-130, (15 May 1986).
28 G. Landmann, et al., op. cit., pp. 804-807, Figure 7.
29. L. W. Casperson, *Patterns of Biblical Chronology* (Westbow Press, Bloomington, IN, 2012), Chapter 21, p. 542, ref. 24; see also Section 9.2 of this study.

30. C. Kuzucuoglu, et al., op. cit., p. 1132.
31. E. T. Degens, et al., op. cit., p. 720-721, and Figure 7.
32. C. Kuzucuoglu, et al., op. cit., pp. 1124-1137 (2010); p. 1125.
33. See for example, "El Niño-Southern Oscillation," https://en.wikipedia.org/wiki/El_Niño-Southern_Oscillation, Web. 4 March 2018.
34. E. T. Degens, et al., op. cit., p. 716.
35. C. Kuzucuoglu, et al., op. cit., pp. 1133-1134.
36. Ibid., p. 1135.
37. Ibid., p. 1136.
38. "Milankovitch cycles," https://en.wikipedia.org/wiki/Milankovitch_cycles, Web. 4 March 2018.
39. M. Kadioglu, Z. Sen, and E. Batur, "The greatest soda-water lake in the world and how it is influenced by climatic change," *Annales Geophysicae*, Volume 15, pp. 1489-1497 (1997): p. 1489.
40. "Urartu," https://en.wikipedia.org/wiki/Urartu, Web. 4 March 2018.
41. Ibid.
42. G. Landmann, et al., op. cit., p. 797.
43. Ibid., p. 802.
44. "Ruins of 3,000-year-old Urartian castle reportedly found in Lake Van," www.panarmenian.net/eng/news/248744/, Web. 4 March 2018.
45. "Urartu," op. cit.

8. VOLCANIC ERUPTIONS AND EGYPT

"The gods [caused] the sky to come in a tempest of r[ain], with darkness in the western region and the sky being unleashed without [cessation, louder than] the cries of the masses" (Tempest Stela, lines 8-9).

8.1 Introduction

As discussed in Chapter 7, the Bible suggests two possible causes of the biblical flood. The first of these is the release of underground water, and the second is a period of heavy rainfall. In principle, both of these mechanisms could occur in the vicinity and at the time of a volcanic eruption. Earthquakes associated with an eruption can lead to the release of underground water, and an example considered in Chapter 7 involved an eruption of Mt. Ararat volcano in 1840 A.D. However, historical parallels may not seem likely to provide the amount of water implied in the biblical story.

On the other hand, a sustained period of heavy rainfall over a wide geographical area is often associated with major climate-changing eruptions. Thus, there is the possibility that flooding could also be an important consequence of such an eruption. In some cases of interest, the eruption could have been hundreds or

8. Volcanic Eruptions and Egypt

thousands of miles away from the area of flooding. Heavy rainfall for many days after an eruption would not be unusual.

Longer-term effects are also possible. A large eruption anywhere in the northern hemisphere, for example, could upset normal weather patterns, leading to a multi-year period of coolness with excess rainfall in some areas and severe droughts in others. It has been noted specifically that the Thera eruption near the center of the Mediterranean Sea seems to have led to unusually heavy rainfall in Anatolia for about seven years and drought in Egypt and China due to a diminished annual inundation for about the same period of time.[1] In the course of this chapter the effects of volcanic eruptions on the weather in Egypt will be considered.

General relationships between volcanic eruptions of the past and rainfall are considered in Section 8.2, including the famines at the times of the migrations of Abram and Jacob with their families to Egypt. Other records of Thera eruptions are considered in Section 8.3. The possibility that a volcanic eruption was responsible for the flooding conditions described in the Tempest Stela of Egypt, which was created during the Eighteenth Dynasty reign of Ahmose I, is discussed in Section 8.4. Possible volcanic effects just before the time of the exodus are considered in Section 8.5, and some events just after the exodus are reviewed in Section 8.6.

8.2 Biblical famines in Egypt

The references quoted in Section 7.3 show that most of the rainfall records indicated in Table 7.2 were set during tropical storms. This result is due in part to the high energy and humidity that is available to hurricanes and cyclones arising at tropical latitudes. Another factor, however, is the relative predictability of such storms and hence the widespread installation at present of appropriate weather-monitoring equipment. Extreme rainfall would tend to occur less often at more northern latitudes corresponding, for example, to Lake Van. Also, for such storms as did arise, it would be less likely that equipment would be in place to document features such as wind speed and rainfall rate.

One of the important causes of significant weather events at non-equatorial latitudes is the occurrence of major volcanic eruptions. As only recently fully appreciated, these eruptions can sometimes cause severe and relatively local rainfall events that last for many days:

> Volcanoes may release particles that can cause changes in local and regional weather at rates up to 100 million times higher than previously realized. . . . Volcanoes typically create two types of particles, big primary particles that quickly fall to the troposphere, the lowest portion of Earth's atmosphere, and smaller secondary particles, mostly composed of sulfuric acid, that react chemically with other molecules in the atmosphere and which are responsible for both local and global precipitation changes. These secondary particles can in turn both help form and seed clouds, changing precipitation levels over large areas. . . . It is possible that volcanic eruptions and other volcanic activities that release sulfur dioxide into the atmosphere may have a larger effect on climate than previously understood.[2]

The dimming of sunlight due to the injection of particles into the atmosphere can also affect the surface temperatures of the earth and alter rainfall patterns over periods of several years. This

8. Volcanic Eruptions and Egypt

long-term mechanism was explored previously as a possible cause of the famines that induced Abram and Jacob to travel to Egypt.[3] There it was suggested that atmospheric obscuration for several years following major volcanic eruptions caused unusual tree growth in Anatolia, presumably due to excess rainfall[4] or possibly also a volcanogenic fertilization effect.[5]

Among the most important volcanic eruptions of ancient times were those of the volcano on the island of Thera. Many dates have been proposed for the largest Thera eruption, and of those dates probably the most popular is about 1628 B.C.[6] Sometimes the season of the year is also inferred:

> Geography, oceanography and paleoclimatology tell us that in the autumn of 1628 B.C. the island of Thera was ripped open and blown apart and strewn about. Tidal waves spread outward from the rend in the earth, and archaeology tells us that the whole eastern Mediterranean was thrown into turmoil about this time.[7]

On the other hand, eruptions are not always distinct events having starting times that can be specified even to a single year. Thus, the Thera eruption, usually now dated to 1628 B.C., is said to have been preceded by one or more smaller eruptions that occurred earlier by many years. One effective means for dating eruptions on Thera has employed core samples of sediment from the bottom of the sea near the island. These samples show clearly the presence of layers of volcanic ash:

> More accurate examination of the layer of upper ash on one of the cores, however, has confirmed that the eruption occurred in three successive stages, since three alternating layers of coarse or finer ash have been distinguished. This

means that the volcano continued to be active for a long time and that it had three successive eruptions separated by some years. The second one caused the destruction of the settlements and the consequent devastation of the island, but the last was the one which caused the submersion of a portion of it. The three layers of ash were also distinguished on Thera.[8]

The first stage of the eruption was apparently relatively minor in its consequences. The second stage destroyed many buildings but didn't entirely displace the population:

> Exploration of a small area toward the southwest seems to confirm the theory that after the destruction of the principal buildings, squatters reoccupied parts of them and used them as workshops; for security they blocked some of the openings and the entrance of a staircase. This reoccupation shows that the definitive catastrophe which submerged the central part of the island had not yet happened. Two successive eruptions fifty to sixty years apart, the second [third stage] of unimaginable violence, would best explain the archaeological and geological evidence discovered on Thera and Crete.[9]

The preceding information indicates that a substantial eruption (second stage) caused widespread devastation and the destruction of many of the buildings on Thera and Crete but did not actually cause the evacuation of these sites. Many years later a further eruption of far greater magnitude (third stage) ended the occupation of Thera. This last eruption is the one often dated to 1628 B.C. The effects of that eruption led to the seven-year famine that brought Jacob and his family to Egypt. It may be of interest to also consider the possibility of written records from the

8. Volcanic Eruptions and Egypt

area of Egypt relating to the earlier but smaller (second stage) eruption.

In the above context it may be noted that an event in about 1728 B.C. has been suggested to have caused the briefer famine that led to the migration of Abram and his family to Egypt.[10] In view of the importance here of this smaller event, it would be of interest to identify its underlying cause. One possibility to consider is that the earlier and smaller eruption of Thera might have occurred in 1728 B.C. and might provide the explanation for the lesser famine-causing event. Evidence for this eruption could have been buried to such a degree by the later massive geological and archeological damage that the earlier occurrence did not, for the most part, achieve an independent historical identity.

If the two famine events were caused by similar (except for size) volcanic eruptions, then it would be of interest to compare them using the formalism described previously. It has been suggested that the duration Δt_f of an eruption-caused famine might be related in an approximate way to the volcanic explosivity index VEI of the eruption by the formula[11]

$$\Delta t_f = 3.2 VEI - 14. \qquad (8.1)$$

Solving for the VEI, one finds that according to this formula the explosivity index of the Thera eruption of 1628 B.C. having $\Delta t_f = 7$ years would be about $VEI = 6.6$. On the other hand, the famine that brought Abram to Egypt in about 1728 B.C. had a duration of about $\Delta t_f = 2$ years, so its volcanic explosivity index would be about $VEI = 5.0$.

As a next step, it may be recalled that the volcanic explosivity index for an eruption is related to the volume V of ejected tephra in km^3 by the following relationship:[12]

$$VEI = \log V + 5. \qquad (8.2)$$

With the value $VEI = 6.6$ for the eruption of 1628 B.C., the volume of ejected tephra from Equation (8.2) is given by

$$\log V = 6.6 - 5 = 1.6, \qquad (8.3)$$

or

$$V = 10^{1.6} \approx 40 \; km^3. \qquad (8.4)$$

On the other hand, with the value $VEI = 5.0$ for the eruption of 1728 B.C., the volume of ejected tephra from Equation (8.2) is given by

$$\log V = 5 - 5 = 0, \qquad (8.5)$$

or

$$V = 10^0 = 1 \; km^3. \qquad (8.6)$$

By this method of estimation, the eruption of 1628 B.C. produced about forty times the volume of tephra as the postulated eruption of 1728 B.C. If these two eruptions were both of the Thera volcano, then it would seem likely that much of the geological or archeological evidence for the earlier eruption might have been obscured by the later and much larger eruption.

8. Volcanic Eruptions and Egypt

Platon argued that there were two major eruptions of this same volcano separated in time by many years:

> One fact that remained curious was that, while the remains on Thera confirmed that the catastrophe occurred in the years around 1500 B.C. [perhaps to be replaced now by "many years before 1628 B.C."], in Crete in that period, though most of the Minoan centers suffered, the destruction seemed to have been less terrible than would be expected and to have had no irreparable effects. The Minoan centers were rebuilt, without substantial alteration, and life there continued for at least 50 years, until the final destruction [1628 B.C.], of incomparably greater extent, resulted in the complete ruin of the areas. It seemed plausible to assume two successive eruptions, the second of which was the one which caused the submersion of the greater part of Thera with the consequences already described.[13]

It should be noted at the outset that when Platon was conducting his research in the 1960s most dates estimated by archeological means were systematically low by more than a century. Thus, where Platon estimated that a major destruction occurred in "around 1500 B.C.," an earlier date is inserted in brackets. This earlier destruction could possibly be associated with the famine of Abraham's era, commencing with an eruption of Thera in about 1728 B.C. The actual identity of the famine-causing event at that time is not, however, essential for the present discussion.

Further comments on the date of the final destruction have also been given:

Causes And Effects Of The Flood

> The exact fixing of the chronology has particular significance in determining the nature of the catastrophe. There is definite proof that nearly all the later Minoan palaces suffered a similar general destruction at the same time. For a time specialists were doubtful as to whether the date of the final destruction of the palaces was about 1450 B.C. or about 1400 B.C. Lately, however, the problem of the exact chronology seems nearer solution. The palaces and other important centers in central and eastern Crete, Knossos included, were simultaneously destroyed about the middle of the fifteenth century B.C., apparently as a result of the same causes.[14]

Estimating that the final destruction might have occurred in about 1450 B.C. permitted the destruction to be associated with the biblical plagues of Egypt. This concept was advocated by Galanopoulos and others:

> Recently J. C. Bennett, Director of the Institute for the Comparative Study of History, Philosophy and the Sciences, has reached some further conclusions based on additional evidence. He has shown that the prodigious eruption of Santorin [Thera] was responsible not only for Deucalion's Flood and the destruction of Atlantis but also for the plagues of Egypt which made the Exodus possible.
> The Ten plagues of Egypt are referred to in the Bible as follows (Exodus 7-11):
> 1. The waters turned into blood
> 2. Frogs covering the land
> 3. Lice afflicting man and beast
> 4. Swarms of flies
> 4. Murrain attacking livestock
> 6. Boils and blains
> 7. Thunder and hail
> 8. Locusts

9. Darkness
10. Death of the first-born

If these calamities, which preceded the Exodus, are contemporary with the great eruption of Santorin [Thera], they may all be considered as direct or indirect consequences of it.[15]

Other researchers have also contributed extensively to the literature relating to the structure and eruption of Thera and the consequences of that eruption:

> The eruption and collapse of Thera is the greatest natural catastrophe that has occurred in historical times. Where Thera now lies, there was originally a limestone mountain whose three peaks of bright marble pierced the surface of the sea. The highest was 2,000 feet, the present Mt. Hagias Elias, the others Monolithos and Platinamos. They are connected by a stony submarine ridge to the island of Anaphi, 15 miles to the east.[16]

Many dates have been proposed for the major Thera eruption that may have led to the migration of Jacob and his family to Egypt. These dates vary considerably, and as seen above some of them are low compared to the chronologies being explored in this study. An example of an early and relatively satisfactory proposed date range is the following:

1750-1520 B.C.
RC 14 [radiocarbon] date from wood samples found in Phira quarry [on Thera] in 1967. Dating by E. Ralph, University of Pennsylvania.[17]

It may be recalled from the preceding paragraphs and their references that the eruption of Abram's time may have occurred in about 1728 B.C., while the eruption of Jacob's time may have been a century later in about 1628 B.C.

It has also been noted that there may be more references to volcanic eruptions and other volcanic effects in the Bible than one might have expected:

> The references to volcanic eruptions, floods, and earthquakes in the Bible are remarkable for their great number and influence upon the people of the time and the chroniclers of history. They are all the more remarkable because earthquakes and volcanic eruptions are not common in the lands usually associated with the Bible – Canaan and Egypt. Thera is the most likely candidate for vulcanism, and the lands of Greece and Turkey for earthquakes. Extensive references that could well refer to Thera are to be found in the Old Testament books of Exodus, Amos, Jeremiah, Zephaniah, and Ezekiel. The Book of Revelation (the Apocalypse), according to tradition written by Saint John the Divine on the island of Patmos, 60 miles from Thera, is dominated by vulcanism. In short, there was in Biblical times an unmistakable tradition of writing in terms of volcanic reference.[18]

8.3 Other records of Thera eruptions

Some of the preceding interpretations have suggested a close relationship between the major Thera eruption and the biblical seven-year famine in Egypt: "but after them there will arise seven years of famine, and all the plenty will be forgotten in the land of Egypt" (Genesis 41:30). Further support for these interpretations may be found in records of the corresponding climate incident in China. The Thera eruption of about 1628 B.C. is reported there to have caused very heavy rainfall initially, followed by seven years of famine:

> Plato's "story of Atlantis" tells of the island Atlantis submerged by the sea. Recent archaeological studies suggest that "Atlantis" may have been Minoan centers on Crete, destroyed by tidal waves from the Thera/Santorini eruption..., the most powerful historical eruption....
> Climatic and hydrologic anomalies recorded were:
> 1. Dry fog – "King Chieh (last of Xia) was unscrupulous the Earth emitted yellow fog." "At the time of King Chieh the sun was dimmed." "Three suns appeared..."
> 2. Cold seasons – "Winter and summer came irregularly." "Frosts in 6th month (July)." "Last year of King Chieh, ice formed in the morning" (very unusual since the climate of the Yellow River Valley was much warmer then).
> 3. Crop Failures – "The five cereals withered." "... Therefore famine occurred."
> 4. Hydrologic extremes – "There was heavy rainfall and communities were destroyed." The floods were followed by a severe drought that lasted seven years into the next (Shang) dynasty. The heavy rainfall is a characteristic atmospheric aftereffect of major volcanic eruptions....
>
> To date the Thera/Santorini eruption we use archaeologically verified predynastic and dynastic Shang, and Western Zhou royal genealogies, calibrated by absolute astronomical dates.... Fitting a line through the datum points we graphically solve for the Thera/Santorini eruption date.... The answer is 1600 ± 30 BC.[19]

The term "dry fog" was introduced by Benjamin Franklin, discoverer of the climatic effects resulting from volcanic eruptions.[20] The huge amount of dust and gas emitted into the atmosphere dimmed and colored the transmitted sunlight. The

Causes And Effects Of The Flood

diminished light transmission also resulted in a substantial cooling of the climate of much of the earth, as emphasized in item two of the above quotation.

The Chinese people, being in a normally warm climatic region, seem to have been startled to observe "three suns" (see item one in the above quotation). It may be worthwhile to comment briefly on this effect. On unusually cold days, it sometimes occurs that near sunrise or sunset the sun has two less-bright companion "suns" located at the same height above the horizon but at about twenty-two degrees to the right and left of the actual sun. These are frequently observable in the winter in the northern states of the U.S. The companion "suns" are often today called sun dogs, but they were known already in antiquity, and a description was given by Aristotle:

> Mock suns and rods [sun dogs] are always seen by the side of the sun, not above or below it nor in the opposite quarter of the sky. They are not seen at night but always in the neighbourhood of the sun, either as it is rising or setting but more commonly towards sunset.[21]

Sun dogs result from the refraction and reflection of light by oriented ice crystals in the atmosphere, even as rainbows result from the refraction and reflection of light by water droplets. It seems that Chinese observers, as quoted above, may have noticed sun dogs (for a total of three suns) more than twelve hundred years before Aristotle.

Items three and four in the above quotation refer to the more damaging climatic effects that can result from even a distant volcanic eruption. In the near term there can be localized heavy rainfall with disastrous flooding, and the drowning or washing away of vital food crops is also a possibility. As if such flooding weren't bad enough, it can be followed by multi-year drought, crop failure, and famine over a wide geographical area. It seems that in response to the Thera eruption all of these effects occurred in China, including a famine of seven years duration. In Egypt the same eruption was interpreted also to have caused the seven-year drought and famine of Joseph's era.[22]

Another hint of the eruption of 1628 B.C. may occur in a record associated with the island of Thera itself. Thus Herodotus relates a story of people seeking to migrate from Thera to new homes in Libya because of a seven year famine on Thera:[23]

> Then for seven years after this there was no rain in Thera; all their trees in the island save one were withered. The Theraeans inquired again at Delphi, and the priestess made mention of the colony they should send to Libya. So since there was no remedy for their ills, they sent messengers to Crete to seek out any Cretan or sojourner there who had travelled to Libya. These, in their journeys about the island, came to the town of Itanus, where they met a fisher of murex called Corobius, who told them that he had once been driven out of his course by winds to Libya, to an island there called Platea.[24] This man they hired to come with them to Thera; thence but a few men were first sent on shipboard to spy out the land, who, being guided by Corobius to the aforesaid island Platea, left him there with provision for some months, and themselves sailed back with all speed to Thera to bring news of the island. . . . As for the Theraeans, when they came to Thera after leaving Corobius on the island, they brought word that they had founded a settlemennt on an island off Libya.

It is mainly the mention of the seven-year famine that calls attention to this story. Except for the duration of the famine, the story would seem somewhat doubtful. As noted above, the seven-year famines in at least Canaan, Egypt, and China were probably caused by the devastating eruption of Thera in about 1628 B.C. It isn't likely that civilization continued on Thera with famine being the most significant eruption consequence for the inhabitants. Herodotus seems to acknowledge some scepticism concerning the records of Thera:[25] "Thus far in my story the Lacedaemonian and Theraean records agree; for the rest we have only the word of the Theraeans." One might have expected the Theraeans to have fled from Thera to Crete at the time of the eruption, and there they might have experienced the beginning of the seven year famine before migrating to Libya.

Similar effects have resulted from other eruptions of Thera:[26]

> No Thera eruptions in Classical and later periods approached the scale of the Bronze Age cataclysm. Nevertheless, accounts of these smaller eruptions provide a useful basis for extrapolation. . . . In 197 B.C., Thera exploded again, less violently, occasioning mention by Strabo, Seneca, Plutarch, Pausanias, Eusebius, and other Classical authors, who seem to have based their accounts on now-missing eyewitness documents or on local oral traditions.[27] We propose that the Babylonian astronomical diaries may afford the only extant, contemporaneous record of the aftereffects of this eruption. In February and March of 197 B.C., Babylon experienced exceptional weather: storms, heavy rains, flooding of the Euphrates, dense fog, and solar and lunar halos.[28]

It is evident that this later eruption of Thera caused substantial weather effects as far away as Babylon. The heavy rains and flooding near Babylon are of interest because this is somewhat close to the general area in which the biblical patriarchs lived at the time of the flood. The volcanic explosivity index for this eruption is reported to be $VEI = 3$,[29] and from calculations such as those in Section 8.2 above, this eruption magnitude would have been too small to cause multi-year climate changes.

8.4 Tempest in Egypt

A volcanic eruption of particular interest is the one that seems to have been responsible for a series of climatic disturbances in Egypt during the reign of the king Ahmose I, founder of the Eighteenth Dynasty of Egypt. The main evidence supporting the occurrence of this eruption is the description of a destructive country-wide storm documented on the so-called Tempest Stela of Ahmose. The following quotation includes the remaining text of this damaged stela as analyzed and translated by Ritner:[30]

> This translation follows the revised text edition in Helck,[31] line numbers following the parallel version on the recto [or verso[32]]. A damaged scene shows the king, accompanied by a female figure, facing a lost image of Amon-Ra, and offering "foodstuffs, fresh vegetables, and everything that the earth produces."
>
> (1) [Long live the Horus "Great of Manifestations," He of the] Two Ladies "Pleasing of Birth," the golden Horus "Who Binds the Two Lands," King of Upper and Lower Egypt, Neb-pehty-ra, son of Ra, Ahmose, living forever. Now then, His Majesty came [. . .]

Causes And Effects Of The Flood

(2) Ra himself had appointed him to be King of Upper Egypt. Then His Majesty dwelt at the town of Sedjefa-tawy

(3) [in the district just to] the south of Dendera, while A[mon-Ra, Lord of the Thrones of the Two Lands was] in Thebes. It was His Majesty

(4) who sailed south to offer [bread, beer, and everything good] and pure. Now after the offering [. . .]

(5) then attention was given in [. . .] this [district (?)]. Now then, the cult image [of this god. . .]

(6) as his body was installed in this temple while he was in joy.

(7) Now then, this great god desired [. . .] His Majesty [. . .] while the gods declared their

(8) discontent. The gods [caused] the sky to come in a tempest of r[ain], with darkness in the western region and the sky being

(9) unleashed without [cessation, louder than] the cries of the masses, more powerful than [. . .], [while the rain raged (?)] on the mountains louder than the noise of the

(10) cataract which is at Elephantine. Every house, every quarter that they reached [. . .]

(11) floating on the water like skiffs of papyrus opposite the royal residence for a period of [. . .] days,

(12) while a torch could not be lit in the Two Lands. Then His Majesty said: "How much greater this is than the wrath of the great god, than the plans of the gods!" Then His Majesty descended

(13) to his boat, with his council following him, while the crowds on the East and West had hidden faces, having no clothing on them

(14) after the manifestation of the god's wrath. Then His Majesty reached the interior of Thebes, with gold confronting(?) gold for this statue so that he (i.e., Amon-Ra) received that which he desired.

8. Volcanic Eruptions and Egypt

(15) Then His Majesty began to reestablish the Two Lands, to drain the flooded territories without his [. . .], to provide them with silver, with
(16) gold, with copper, with oil, and cloth of every bolt that could be desired. Then his Majesty made himself comfortable inside the palace (life! prosperity! health!).
(17) Then His Majesty was informed that the mortuary concessions had been entered (by water), with the tomb chambers collapsed, the funerary mansions undermined, and the pyramids fallen,
(18) having been made into that which was never made. Then His Majesty commanded to restore the temples which had fallen into ruin in this entire land: to refurbish
(19) the monuments of the gods, to erect their enclosure walls, to provide the sacred objects in the noble chamber, to mask the secret places, to introduce
(20) into their shrines the cult statues which were cast to the ground, to set up the braziers, to erect the offering tables, to establish their bread offerings,
(21) to double the income of the personnel, to put the land into its former state. Then it was done in accordance with everything that His Majesty had commanded.

From the stela one learns that the consequences of the storm included rain, darkness, thunder, and flooding over much of the populated land of Egypt. These occurrences would seem almost to be characteristic evidence for the eruption of a volcano. In an updated translation of the sentence in lines (10) to (12) the concept of human victims is also suggested:[33] "Then every house, every quarter that they (scil. the storm and rain) reached [. . . their corpses(?)] floating on the water like skiffs of papyrus outside the palace audience chamber for a period of [. . .] days [. . .] while no

Causes And Effects Of The Flood

torch could be lit in the Two Lands." This aspect of the flood described in the Tempest Stela will be recalled in Section 9.3 in connection with a flood in China and in Section 11.5B in a discussion of the flood of Noah's time.

Assuming that the tempest described on the stela may have been caused by a volcanic eruption event, it remains to determine the date and possibly the identity of that eruption. One proposal for the identity of the volcano is the volcanic island of Thera in the Aegean Sea that has been discussed in Section 8.2 above. The explosion of this island has often been thought to be the basis of early legends of the sinking of the land of Atlantis. In excavations of Thera in the 1960s by the discoverer of its port city Akrotiri, Marinatos estimated the date of the eruption to have been about 1520 B.C.[34] His conclusion was based mostly on archeology. An early radiocarbon dating placed the eruption in the range between about 1510 and 1310 B.C.[35] In more recent years, radiocarbon and ice-core dating methods have yielded earlier eruption estimates closer to 1600 B.C., while several applications of tree-ring dating (dendrochronology), have suggested the almost exact date of 1628 B.C. This date was adopted in our previous study.[36]

As long as the estimated date of the Thera eruption was believed to be close to popular dates for the exodus around 1440 B.C.,[37] it seemed plausible to consider an association of the effects of the eruption and the exodus plagues of Egypt.[38] With an earlier eruption date, such as 1628 B.C., a connection between the eruption and the plagues would seem to be difficult to support. However, by shifting conventional Egyptian chronology of the Eighteenth Dynasty earlier by a very large number of years (one hundred fifty years has been suggested), it has been possible

8. Volcanic Eruptions and Egypt

still to claim an association of the exodus plagues with the Thera eruption.[39]

A more recent idea is that the Thera eruption might instead be associated with the weather events described on the Tempest Stela of Ahmose I quoted above.[40] Several interpretations place the reign of Ahmose I between 1570 and 1546 B.C.[41] The Tempest Stela was created during the approximately twenty-five year reign of Ahmose I. However, in order to place the Thera eruption into this reign, some changes would still be needed in the dates suggested above for Ahmose I. Clearly, the tempest could not be associated with the Thera eruption (1628 B.C.) if the reign of Ahmose I occurred perhaps about seventy years after that eruption. In a chronology developed previously, the eruption of Thera was associated with the seven-year famine of Joseph's era rather than with the tempest in the reign of Ahmose I.[42]

If the Thera eruption was not related to the tempest, then it would seem reasonable to look for some other cause for the tempest including possibly the occurrence of a different volcanic eruption. In principle, one could look for evidence of such an eruption in many places, but it is natural here to consider the same Anatolian tree-ring database employed in our previous studies of eruption effects in Canaan and Egypt. Besides the eruption that caused the seven-year famine of Joseph's era mentioned above, that database also provided evidence for the apparently briefer famine experienced earlier by Abram.[43] There are, of course, risks in relying too much on a single data source, but these risks may be balanced in part by the advantages of format consistency and the nearby Middle-Eastern location of the Anatolian trees. As it happens, one probable eruption signature

Causes And Effects Of The Flood

discernible in the Anatolian tree-ring record does correspond closely in time to the tempest during the reign of Ahmose I. The data supporting this interpretation are listed in Table 8.1, and a plot of these data is shown in Figure 8.1. The relative tree-ring thickness data in the table are as reported by Kuniholm, et al.,[44] and the calendar-year calibration is as adopted previously for the eruptions in the eras of Joseph[45] and Abram.[46]

Table 8.1. Anatolian dendrochronology in relative tree-ring numbers and proposed calendar years B.C. for the time period around a major growth anomaly.

Relative Years	Calendar Years	Relative Thickness
925	1555 B.C.	89.7
926	1554	132.0
927	1553	148.0
928	1552	140.8
929	1551	117.8
930	1550	98.4
931	1549	109.5
932	1548	90.5
933	1547	96.7
934	1546	102.2
935	1545	108.8

8. Volcanic Eruptions and Egypt

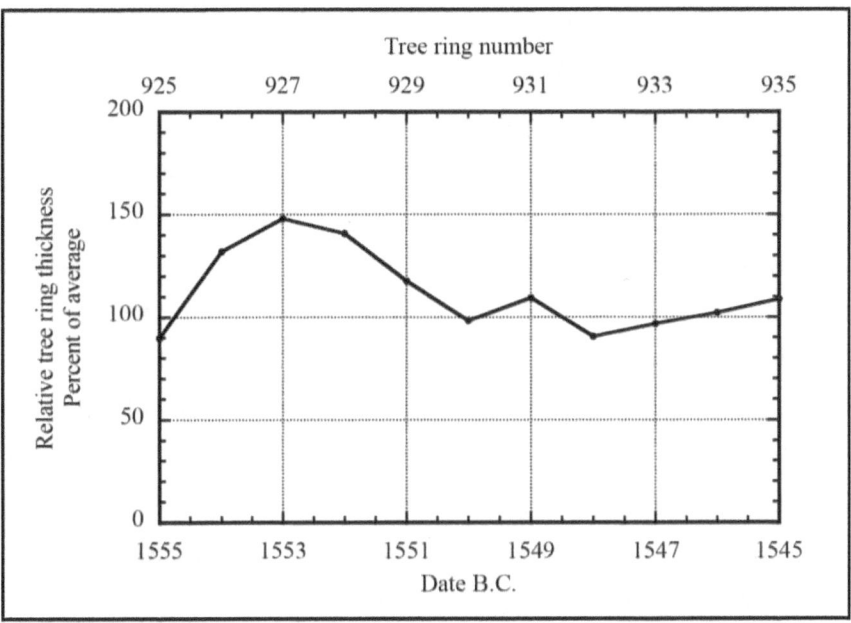

Figure 8.1. Relative tree-ring thickness for a growth anomaly in Anatolia that may be associated with the tempest during the reign of Ahmose I.

The substantial multi-year increase in the thickness of the growth rings of trees in Anatolia shown in Figure 8.1 is similar to the effects of two of the eruption events considered previously. Thus, one might infer that a climate-changing event such as a volcanic eruption occurred in about 1554 B.C. The atmospheric attenuation resulting from accompanying dust and gases created weather conditions that resulted in excess rainfall during the years from 1554 B.C. to about 1551 B.C. The enhanced growth resulting from this rainfall is visible in the tree rings for a period of about four years, and this period is longer than the two years that have been associated with the eruption-like event of about 1728 B.C. in Abram's time (see Section 8.2 above). However, the excess growth period shown in Figure 8.1 is shorter than the period of about seven years that has been associated with the famine of Joseph's era (Section 8.2).

It is of interest now to consider whether this postulated eruption could agree in time with the dating of the tempest stela. According to the data discussed previously, Ahmose I reigned from about 1570 to 1546 B.C. On the other hand, the tree-ring data just mentioned suggest that a major volcanic eruption may have occurred in about 1554 B.C., with widespread climatic effects probably lasting until about 1551 B.C. From the style of its text, the Tempest Stela was said to have been created before year twenty-two of Ahmose I.[47] By this argument, the stela documenting Ahmose I's planned reconstruction projects would likely have been created after the damaging eruption effects ended in about 1551 B.C. but before year twenty-two of Ahmose in about 1549 B.C.

Other evidence also suggests that the reconstruction project may have begun before about year twenty-two of Ahmose I:

> In year 22 of Ahmose, his treasurer Neferperet erected a stele to record the opening of a new quarry for extensive temple constructions throughout Egypt. So late in Ahmose's reign, these building projects seem unlikely to have been part of his post-Hyksos program, but rather, we propose, prime among the countrywide restoration efforts hailed in the Tempest Stele (ll. 18-19).[48]

An inquisitive person could wonder whether any now-identifiable volcanic eruption might have been responsible for the weather events recorded on the Tempest Stela. Table 8.2 is a listing of volcanic eruptions in the northern hemisphere that are

8. Volcanic Eruptions and Egypt

reported to have probably occurred between 1500 B.C. and 1600 B.C. and that have a volcanic explosivity Index (*VEI*) of four or more. This information has been obtained from the database of the Global Volcanism Program of the Smithsonian Institution:[49]

Table 8.2. Volcanic eruptions in the northern hemisphere between 1500 B.C. and 1600 B.C. that have an explosivity index of four or greater.

Name	Date	Uncertainty	*VEI*
Avachinsky	1500 B.C.	± ?	5
Etna	1500	± 50	5
Ibusuki	1500	± ?	4
Chikurachki	1500	± 250	4
Vesuvius	1550	± 75	4
Hayes	1550	± ?	5
Hekla	1550	± ?	4

Historical observations, geological interpretations, and radiocarbon dating aren't sufficiently comprehensive or precise to absolutely exclude the possibility of some other eruption being responsible for the tempest in Egypt. Nevertheless it will be assumed as an illustration that the actual cause of the tempest is an eruption included on the list.

If the climate disturbance resulting from the eruption could be said from Table 8.2 to have lasted for about four years, then from Equation (8.1) the volcanic explosivity index would have been roughly

$$VEI = (4+14)/3.2 \approx 5.6. \qquad (8.7)$$

Even a three-year climate disruption would require a volcanic explosivity index of about $17/3.2 = 5.3$. Thus, the eruptions having reported explosivities of four in Table 8.2 can probably be excluded as possible sources of the multi-year severe climate conditions reported on the stela. On the other hand, any of the three reported eruptions having explosivities of five could have been responsible for that event.

Eyewitness accounts may help to narrow the list even further. Line (8) of the tempest stela as quoted earlier in this section includes the following words: "The gods [caused] the sky to come in a tempest of r[ain], with darkness in the western region. . . ." Mt. Etna is located near the center of the Mediterranean Sea. Thus, clouds arising from an eruption of Mt. Etna, could readily be referred to by an Egyptian observer as "darkness in the western region." In the absence of further information, it seems reasonable to consider Mt. Etna, with a *VEI* of at least 5, as a viable candidate for the origin of the tempest in Egypt.

It should perhaps be noted here that the Mediterranean eruptions of interest are represented in the Anatolian tree ring records by excess growth for a period of time that is related to the volcanic explosivity index or *VEI*. The actual weather consequences at a location of interest may not be as simple as just excess growth. In other locations there sometimes seems to have been a period of severe weather conditions followed by a period of drought.

The biblical record of the Thera eruption event that brought Jacob and his family to Egypt mentions only the famine aspect.

8. Volcanic Eruptions and Egypt

On the other hand, the Chinese records of this event discussed in Section 8.3 mention explicitly the severe weather followed by the seven-year famine. The Tempest Stela event mentions the stormy weather, but no famine is indicated. As will be discussed in Sections 9.3 and 9.4, the Chinese and Sumerian records of the storm that may be associated with the flood of Noah's time mention both the immediate weather event and the subsequent drought.

8.5 Pre-exodus volcanic events

As indicated in previous sections, volcanic eruptions can have dramatic consequences. In the near term and at short range, the consequences could include flames, ashfalls, lava bombs, earthquakes, poisonous gases, acid rain, darkness, and severe storms of wind, rain, and hail. Major changes in climate can also occur over large distances and multi-year time periods. Thus, it has been natural to consider the possibility that some of the exodus plagues might have been associated with one or more volcanic eruptions.

As mentioned in Section 8.2, it was once thought that the exodus plagues might have been associated with an eruption of the Thera volcano. However, refined determinations of the date of the Thera eruption have placed that date much too early for it to have been related to the exodus events. As considered in the previous section, it has also been suggested that the Thera eruption could have been associated with the events described in the Tempest Stela. On the other hand, it is now usually considered that the Thera eruption was also too early for the tempest events, but instead that eruption may have been responsible for the seven-year famine in Egypt at the time of Joseph.[50]

With this background, it would seem natural to enquire whether some of the exodus "plagues" might have been caused by yet another volcano-like occurrence. As it happens, such climate events at about the time of the exodus may be discernible in the same Anatolian tree-ring records that have been employed here before. However, these events are not all so easily interpretable as those considered previously, and they seem to have occurred episodically over the years between about 1442 and perhaps 1425 B.C., while the exodus itself has previously been estimated to have occurred in about 1440 B.C.[51]

A listing of tree-ring thicknesses of trees in Anatolia between 1445 and 1435 B.C. is given in Table 8.3, and this same data is plotted in Figure 8.2. It appears from these results that there may have been a significant but somewhat complex series of climate events between about 1442 B.C. and 1436 B.C. These could perhaps be associated with a single eruption, but in most of the following discussions it will be imagined that there were two somewhat distinguishable events – one before and one after the exodus. From Figure 8.3 there appears to have been an isolated year of enhanced growth in about 1442 B.C. A second event discernible in this ten-year series includes an unusual decrease in growth in the year 1438 B.C., followed by two years of exceptional growth in the years 1437 and 1436 B.C. This second pattern is similar in form (though smaller in amplitude and duration) to the tree-ring growth effects associated with the earlier eruption of the Thera volcano in about 1628 B.C.[52] The growth sequence in Anatolia for the years from 1438 to 1436

8. *Volcanic Eruptions and Egypt*

B.C. is also somewhat similar in both magnitude and duration to the perhaps famine-related sequence of about 1728 to 1726 B.C. at the time of Abram as shown previously.[53]

Table 8.3. Anatolian dendrochronology in relative tree-ring numbers and proposed calendar years B.C. for the time period around two growth anomalies.

Relative Years	Calendar Years	Relative Thickness
1035	1445 B.C.	100.5
1036	1444	98.0
1037	1443	90.3
1038	1442	144.2
1039	1441	112.1
1040	1440	97.5
1041	1439	117.2
1042	1438	63.5
1043	1437	155.4
1044	1436	159.5
1045	1435	99.2

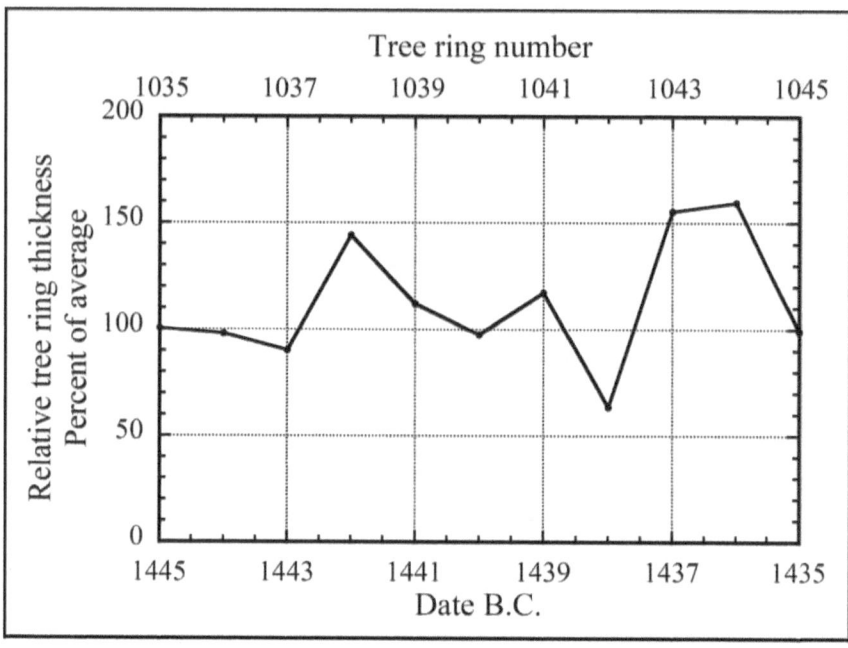

Figure 8.2. Relative tree-ring thickness for two growth anomalies in Anatolia that may be associated with events near the time of the exodus.

A possible candidate explanation for at least part of the exodus eruption sequence might be an eruption of Mt. Vesuvius in roughly 1430 B.C. having an estimated volcanic explosivity index *VEI* of between about 4 and 5.[54] It is perhaps reasonable to enquire whether any potentially volcano-related effects between 1442 and 1436 B.C. might be associatable with exodus events. As it happens, one of the later plagues before the exodus involved severe weather conditions such as those that are known sometimes to result from volcanic eruptions:

> And the Lord said to Moses, "Stretch forth your hand toward heaven, that there may be hail in all the land of

8. Volcanic Eruptions and Egypt

Egypt, upon man and beast and every plant of the field, throughout the land of Egypt." Then Moses stretched forth his rod toward heaven; and the Lord sent thunder and hail, and fire ran down to the earth. And the Lord rained hail upon the land of Egypt; there was hail, and fire flashing continually in the midst of the hail, very heavy hail, such as had never been in all the land of Egypt since it became a nation. The hail struck down everything that was in the field throughout all the land of Egypt, both man and beast; and the hail struck down every plant of the field, and shattered every tree of the field. Only in the land of Goshen, where the people of Israel were, there was no hail (Exodus 9:22-26).

It would seem from the description above that the hail and perhaps the thunder and lightning were the most unusual and fearful effects of this particular plague. In an eruption-caused storm, heavy rain would have been a possibility as well, depending on details of the temperature, humidity, wind, and cloud structure. The occurrence of such rain is indicated explicitly in the description of the ending of the storm:

Then Pharaoh sent, and called Moses and Aaron, and said to them, "I have sinned this time; the Lord is in the right, and I and my people are in the wrong. Entreat the Lord; for there has been enough of this thunder and hail; I will let you go, and you shall stay no longer." Moses said to him, "As soon as I have gone out of the city, I will stretch out my hands to the Lord; the thunder will cease, and there will be no more hail, that you may know that the earth is the Lord's. But as for you and your servants, I know that you do not yet fear the Lord God." (The flax and the barley were ruined, for the barley was in the ear and the flax was in bud. But the wheat and the spelt were not ruined, for they are late in coming up.) So Moses went

Causes And Effects Of The Flood

out of the city from Pharaoh, and stretched out his hands to the Lord; and the thunder and the hail ceased, and the rain no longer poured upon the earth. But when Pharaoh saw that the rain and the hail and the thunder had ceased, he sinned yet again, and hardened his heart, he and his servants. So the heart of Pharaoh was hardened, and he did not let the people of Israel go; as the Lord had spoken through Moses (Exodus 9:27-35).

As noted above in Section 8.3, severe weather conditions were also reported in China in association with the eruption of the Thera volcano in about 1628 B.C. Under the category of hydrologic extremes were the following observations: " 'There was heavy rainfall and communities were destroyed.' . . . The heavy rainfall is a characteristic atmospheric aftereffect of major volcanic eruptions. . . ."[55] The tempest stela discussed in Section 8.4 also reported rain and perhaps thunder: "The gods [caused] the sky to come in a tempest of r[ain], with darkness in the western region and the sky being unleashed without [cessation, louder than] the cries of the masses, more powerful than [. . .], [while the rain raged (?)] on the mountains louder than the noise of the cataract which is at Elephantine" (lines 8-10). Considering these possible eruption-related parallels, it would be tempting to associate the weather events of the plagues with the growth anomaly of 1442 B.C. shown in Figure 8.2

The penultimate plague, before the deaths of the first-born, was the plague of darkness:

> Then the Lord said to Moses, "Stretch out your hand toward heaven that there may be darkness over the land of Egypt, a darkness to be felt." So Moses stretched out his hand toward heaven, and there was thick darkness in all

8. Volcanic Eruptions and Egypt

the land of Egypt three days; they did not see one another, nor did any rise from his place for three days; but all the people of Israel had light where they dwelt. Then Pharaoh called Moses, and said, "Go, serve the Lord; your children also may go with you; only let your flocks and your herds remain behind." But Moses said, "You must also let us have sacrifices and burnt offerings, that we may sacrifice to the Lord our God. Our cattle also must go with us; not a hoof shall be left behind, for we must take of them to serve the Lord our God, and we do not know with what we must serve the Lord until we arrive there." But the Lord hardened Pharaoh's heart, and he would not let them go (Exodus 10:21-27).

This darkness incident just before the time of the exodus may have been known by Greek authors, and a possible mention will be considered in Section 15.5. That occurrence could also be associated with an eruption-related event beginning in about 1442 B.C. as suggested in Figure 8.2 above.

Some of the earlier exodus plagues might also have had some connection with volcanic activity. In fact it has sometimes been suggested that most or all of the exodus plagues were consequences of volcanic effects. This does not mean that a volcanic eruption occurred at or near the location of the exodus. Thus, as discussed in Section 8.4, the plague-like events associated with the tempest in Egypt may have been caused by the eruption of a volcano about a thousand miles to the west.

For later purposes, the first exodus plague will be of particular interest. In that plague the water in the Nile River is said to have turned to blood. Thus, the water had become undrinkable by the Egyptians and lethal to the fish:

Causes And Effects Of The Flood

Then the Lord said to Moses, "Pharaoh's heart is hardened, he refuses to let the people go. Go to Pharaoh in the morning, as he is going out to the water; wait for him by the river's brink, and take in your hand the rod which was turned into a serpent. And you shall say to him, 'The Lord, the God of the Hebrews, sent me to you, saying, "Let my people go, that they may serve me in the wilderness; and behold, you have not yet obeyed." Thus says the Lord, "By this you shall know that I am the Lord: behold, I will strike the water that is in the Nile with the rod that is in my hand, and it shall be turned to blood, and the fish in the Nile shall die, and the Nile shall become foul, and the Egyptians will loathe to drink water from the Nile." ' " And the Lord said to Moses, "Say to Aaron, 'Take your rod and stretch out your hand over the waters of Egypt, over their rivers, their canals, and their ponds, and all their pools of water, that they may become blood; and there shall be blood throughout all the land of Egypt, both in vessels of wood and in vessels of stone.' "

Moses and Aaron did as the Lord commanded; in the sight of Pharaoh and in the sight of his servants, he lifted up the rod and struck the water that was in the Nile, and all the water that was in the Nile turned to blood. And the fish in the Nile died; and the Nile became foul, so that the Egyptians could not drink water from the Nile; and there was blood throughout all the land of Egypt. But the magicians of Egypt did the same by their secret arts; so Pharaoh's heart remained hardened, and he would not listen to them; as the Lord had said. Pharaoh turned and went into his house, and he did not lay even this to heart. And all the Egyptians dug round about the Nile for water to drink, for they could not drink the water of the Nile (Exodus 7:14-24).

The coloration and poisoning of the water in the Nile may have been a result of the settling of volcanic ash from the atmosphere.

8. Volcanic Eruptions and Egypt

A different volcanic eruption that can be said to have turned water into blood will be mentioned in Section 19.6E.

One of the most important events associated with the exodus itself was the drowning of the Egyptian chariot troops and other soldiers by returning water in about 1440 B.C.:

> Then Moses stretched out his hand over the sea; and the Lord drove the sea back by a strong east wind all night, and made the sea dry land, and the waters were divided. And the people of Israel went into the midst of the sea on dry ground, the waters being a wall to them on their right hand and on their left. The Egyptians pursued, and went in after them into the midst of the sea, all Pharaoh's horses, his chariots, and his horsemen. And in the morning watch the Lord in the pillar of fire and of cloud looked down upon the host of the Egyptians, and discomfited the host of the Egyptians, clogging their chariot wheels so that they drove heavily; and the Egyptians said, "Let us flee from before Israel; for the Lord fights for them against the Egyptians."
> Then the Lord said to Moses, "Stretch out your hand over the sea, that the water may come back upon the Egyptians, upon their chariots, and upon their horsemen." So Moses stretched forth his hand over the sea, and the sea returned to its wonted flow when the morning appeared; and the Egyptians fled into it, and the Lord routed the Egyptians in the midst of the sea. The waters returned and covered the chariots and the horsemen and all the host of Pharaoh that had followed them into the sea; not so much as one of them remained. But the people of Israel walked on dry ground through the sea, the waters being a wall to them on their right hand and on their left (Exodus 14:21-29).

This text indicates that the waters were driven back by a strong east wind leaving the seabed as dry land. However, it is possible that earthquake activity or other wave-generation mechanisms associated with an eruption or a subsequent caldera collapse[56] may also have had a role in the initial retreat and subsequent return of the sea.

Other aspects of the general exodus story may have familiar interpretations. For example, the burning bush segment relates that a few years before the time of the exodus a bush-like object on Mount Horeb/Sinai surprised Moses as he was leading the sheep of his father-in-law. The bush appeared to be burning, but it wasn't consumed:

> Now Moses was keeping the flock of his father-in-law, Jethro, the priest of Mid'ian; and he led his flock to the west side of the wilderness, and came to Horeb, the mountain of God. And the angel of the Lord appeared to him in a flame of fire out of the midst of a bush; and he looked, and lo, the bush was burning, yet it was not consumed. And Moses said, "I will turn aside and see this great sight, why the bush is not burnt." When the Lord saw that he turned aside to see, God called to him out of the bush, "Moses, Moses!" And he said, "Here am I." Then he said, "Do not come near; put off your shoes from your feet, for the place on which you are standing is holy ground." And he said, "I am the God of your father, the God of Abraham, the God of Isaac, and the God of Jacob." And Moses hid his face for he was afraid to look at God.
>
> Then the Lord said, "I have seen the affliction of my people who are in Egypt, and have heard their cry because of their taskmasters; I know their sufferings, and I have come down to deliver them out of the hand of the

Egyptians, and to bring them up out of that land to a good and broad land, a land flowing with milk and honey, to the place of the Canaanites, the Hittites, the Amorites, the Per'izzites, the Hivites, and the Jeb'usites. And now, behold, the cry of the people of Israel has come to me, and I have seen the oppression with which the Egyptians oppress them. Come, I will send you to Pharaoh that you may bring forth my people, the sons of Israel, out of Egypt." But Moses said to God, "Who am I that I should go to Pharaoh, and bring the sons of Israel out of Egypt?" He said, "But I will be with you; and this shall be the sign for you, that I have sent you: when you have brought forth the people out of Egypt, you shall serve God upon this mountain" (Exodus 3:1-12).

The burning bush might seem to suggest the presence of extremely hot or combustible gases, and such gases could be present on an active volcano. Thus, this incident could be thought to correspond to modest volcanic activity. However, the Mount Sinai region is usually not considered to be a site of recent volcanism, and other interpretations should be considered.

Apart from the voice heard, the appearance of fire in a bush near the top of a mountain could have been an example of St. Elmo's fire:[57]

> Even the fact that the bush appeared to burn without being consumed can be explained by reference to a rare but hardly miraculous phenomenon. "The favourite explanation of the exegete has been a manifestation similar to St. Elmo's fire," writes the venerable Martin Noth,[58] "and in fact we must imagine something of this sort."

Causes And Effects Of The Flood

This phenomenon is basically an electrical discharge in an ionized gas resulting from unusually large electrostatic fields. Thus, a glowing and hissing "fire" (electrical discharge) has sometimes been observed at the tips of the masts on ships or at the tops of trees or towers, especially during electrical storms.

A more recent event that is now often associated with St. Elmo's fire was the disastrous crash of the Hindenburg airship on 3 May 1937 in Lakehurst, New Jersey. The Hindenburg arrived in stormy weather, which caused a three-hour delay in the landing. The actual landing disaster was captured on film, and a portion of the timeline of that event is the following:[59]

Timeline

7:00 p.m. The Hindenburg approaches the Lakehurst landing field.

7:07 p.m. It makes a sharp turn.

7:16 p.m. The airship is heavy in the stern, so some water ballast is released.

7:21 p.m. The first mooring rope is dropped to the ground.

7:25 p.m. The rear of the Hindenburg bursts into flames. Within 35 seconds, the whole airship is a blazing wreck.

A possible interpretation of this disaster is the following. The final sharp turn was required by the erratic wind conditions. It has been suggested that the abrupt turn may have caused the snapping of a rudder control cable, which resulted in a massive leak in one or more of the hydrogen cells near the tail of the ship. The loss of

8. Volcanic Eruptions and Egypt

hydrogen caused the stern to sink rapidly toward the ground, so the crew dropped 1100 kilograms of ballast water from the rear of the ship in a sequence of three releases. This partially stabilized the ship but probably left the tail region with a dangerous amount of highly combustible free hydrogen. The dropping of the rain-wet mooring ropes to the ground would have quickly brought the ship to the same electrical potential as the earth, which was very different from the potential of the atmosphere at the height of the airship. The result may have been St. Elmo's fire near the angular tail structure (like the mast structure on a sailing ship), which could have ignited the hydrogen from the gas leak in the same area.

The present author had a brief encounter with one aspect of St. Elmo's fire while mountain climbing in the Sierra Nevada in California. While descending from a climb of Mt. Whitney many years ago, the author and two friends decided to also climb to the top of nearby Mt. Muir. As we climbed, we noticed that the sky was looking threatening, thunder could be heard in the distance, and the rocks around us were hissing (a characteristic feature of St. Elmos's fire). We didn't wish to be the highest points on the mountain in a thunderstorm, and at about 50 feet below the summit we abandoned our climb and headed down. On a different climb a friend of the author was actually struck by a lightning-caused current flowing along a wet rock face. (He made a full recovery.) Another physical consequence of large electrostatic fields during stormy weather near the top of a mountain is that a climber's hair may stand on end. This effect caused the early termination of a climb by the author's daughter-in-law. A non-hazardous demonstration of a related hair-raising

Causes And Effects Of The Flood

phenomenon is shown in Fig. 8.3, where the electrostatic field generated as the author's grandson descended a playground slide caused his hair to stand out.

Fig. 8.3 Electrostatic fields and one of their non-hazardous consequences.

8.6 Post-exodus volcanic events

A few years after the burning bush incident and the exodus, seemingly similar events occurred at the same location. Moses led the Israelites to the mountain as he had been instructed previously (Exodus 3:12, quoted above):

> On the morning of the third day there were thunders and lightnings, and a thick cloud upon the mountain, and a very loud trumpet blast, so that all the people who were

8. Volcanic Eruptions and Egypt

in the camp trembled. Then Moses brought the people out of the camp to meet God; and they took their stand at the foot of the mountain. And Mount Sinai was wrapped in smoke, because the Lord descended upon it in fire; and the smoke of it went up like the smoke of a kiln, and the whole mountain quaked greatly. And as the sound of the trumpet grew louder and louder, Moses spoke, and God answered him in thunder (Exodus 19:16-19).

Now when all the people perceived the thunderings and the lightnings and the sound of the trumpet and the mountain smoking, the people were afraid and trembled; and they stood afar off, and said to Moses, "You speak to us, and we will hear; but let not God speak to us, lest we die." And Moses said to the people, "Do not fear; for God has come to prove you, and that the fear of him may be before your eyes, that you may not sin." (Exodus 20:18-20).

The glory of the Lord settled on Mount Sinai, and the cloud covered it six days; and on the seventh day he called to Moses out of the midst of the cloud. Now the appearance of the glory of the Lord was like a devouring fire on the top of the mountain in the sight of the people of Israel. And Moses entered the cloud, and went up on the mountain. And Moses was on the mountain forty days and forty nights (Exodus 24:15-18).

The weather events reported above may have been related to the biblical stories of other activities on Mount Horeb/Sinai after the exodus. The stories do not seem to suggest in any way the presence in this region of characteristic eruption products such as ash clouds, lava flows, or tephra accumulations; and in the absence of such products local volcanism would be difficult to suggest. On the other hand, the biblically reported thunder and

Causes And Effects Of The Flood

lightning atop the mountain are prominent, and actual fires would not seem to be unlikely. The major natural cause of wildfires is lightning, and lightning strikes would be especially common on mountaintops. Wildfires can spread rapidly in many types of terrain such as forests, brush, or grasslands. If such fires occurred, it would not be unreasonable for the mountain to be topped with a "devouring fire" and to be "wrapped in smoke."

Geological activity is also mentioned in the exodus account, and earthquakes may be of particular interest. The earth's crust is a semi-rigid spherical structure that is broken up into several segments referred to as tectonic plates. These plates move about very slowly with typical velocities of a few millimeters per year. In this movement process they routinely run into or slide past each other. Earthquakes occur most frequently at boundaries between plates. The boundary between the African Plate and the Arabian Plate passes beside the Sinai Peninsula, and more specifically Mount Sinai is close to the intersection of the "Red Sea spreading axis" and the "Aqaba-Dead Sea transform" sections of the plate boundary. Earthquakes in the Sinai region are not uncommon, and a recent (at this writing) issue of *The Times of Israel* mentioned three that had occurred in the previous few months:

> An earthquake rumbled in the Red Sea early Monday morning, shaking buildings in the Sinai peninsula and as far north as Israel's southern tip. There were no immediate reports of injuries or damage in Israel from the temblor, which struck at about 4:45 a.m. Monday morning. According to the US Geological Survey (USGS), the tremor measured 5.0 on the Richter scale with an epicenter in the Red Sea, 25 kilometers northeast

8. Volcanic Eruptions and Egypt

of Dhahab, Egypt, and 100 kilometers southwest of Eilat in Israel.

Last month, a very small earthquake hit parts of southern Israel as local residents of the city of Arad and the Dead Sea area reported experiencing minor tremors.

Prior to that incident, the previous quake in Israel occurred in July 2015, when a tremor measuring 4.4 on the Richter scale was felt throughout the country. That, too had its epicenter in the Dead Sea area.

The last major earthquake to hit the region was in 1927 – a 6.2 magnitude tremor that killed 500 people and injured another 700.[60]

Thus, large earthquakes are possible along the plate boundaries of interest, and smaller earthquakes may occur there several times a year.

In the context of earthquakes, one event included in the exodus accounts may be of special interest. In this story Dathan, Abiram, and their colleagues were punished for rebelling against Moses.

> Then Moses rose and went to Dathan and Abi'ram; and the elders of Israel followed him. And he said to the congregation, "Depart, I pray you, from the tents of these wicked men, and touch nothing of theirs, lest you be swept away with all their sins." So they got away from about the dwelling of Korah, Dathan, and Abi'ram; and Dathan and Abi'ram came out and stood at the door of their tents, together with their wives, their sons, and their little ones. And Moses said, "Hereby you shall know that the Lord has sent me to do all these works, and that it has not been of my own accord. If these men die the common

death of all men, or if they are visited by the fate of all men, then the Lord has not sent me. But if the Lord creates something new, and the ground opens its mouth, and swallows them up, with all that belongs to them, and they go down alive into Sheol, then you shall know that these men have despised the Lord."

And as he finished speaking all these words, the ground under them split asunder; and the earth opened its mouth and swallowed them up, with their households and all the men that belonged to Korah and all their goods. So they and all that belonged to them went down alive into Sheol; and the earth closed over them, and they perished from the midst of the assembly. And all Israel that were round about them fled at their cry; for they said, "Lest the earth swallow us up!" (Numbers 16:25-34).

The possibility of the earth burying people or things is well known, and typical mechanisms include landslides and collapsing sinkholes. Such events are often triggered by earthquakes, and more rarely earthquake-associated cracks and fissures in the surface of the earth have swallowed people alive.[61]

8.7 Hunger and thirst

During the many years that the Israelites lived in Egypt, they seem initially to have been free people under the sympathetic rule of the Hyksos conquerors. There is no evidence that in those circumstances they ever experienced exceptional periods of starvation or thirst. There was, of course, an eruption-caused famine at the time of the entry of Jacob and his family into Egypt, but Joseph was vizier of Egypt at that time, and he was in charge of managing food resources for the entire country. Thus, it is

8. Volcanic Eruptions and Egypt

doubtful that food and drink were in short supply for the Israelites.

After the Egyptians defeated the Hyksos, conditions for the Hebrews worsened until most of them were serving as slaves. Of course, slaves are useless if they are hungry and thirsty, and it seems that they were generally well supplied by the Egyptians. Concerning water, it may be noted that Egyptian civilization tended to be concentrated along the Nile River, together with its various branches, tributaries, and canals. Thus, these waters were usually an abundant resource for drinking, washing, irrigation, and transportation. Besides direct consumption in drinking, the same waters could also have been the basis for other consumable liquids, and one of the favorites seems to have been beer.

One recollection of beer consumption may be recognized in a descripton of the brickmaking and construction activities supervised by Rekhmire. He was vizier of Egypt under the king Thutmose III, who is often considered to have been Egypt's greatest ruler. A description of these activities is included on the walls of Rekhmire's tomb. Some of the brickmakers have been identified as Semitic foreigners, and these could correspond to the captive Hebrews.[62] In this tomb an inscription near the bricklayers has them say of their supervisor "He supplies us with bread, beer, and every good sort."

After their exodus from Egypt, the Israelites began their long and difficult journey to the promised land of Canaan (see also Section 16.3). The terrain through which they had to travel was harsh, but actual famine-like conditions seemed especially prevalent near the beginning of their march:

They set out from Elim, and all the congregation of the people of Israel came to the wilderness of Sin, which is between Elim and Sinai, on the fifteenth day of the second month after they had departed from the land of Egypt. And the whole congregation of the people of Israel murmured against Moses and Aaron in the wilderness, and said to them, "Would that we had died by the hand of the Lord in the land of Egypt, when we sat by the fleshpots and ate bread to the full; for you have brought us out into this wilderness to kill this whole assembly with hunger" (Exodus 16:1-3).

And the Lord said to Moses, "I have heard the murmurings of the people of Israel; say to them 'At twilight you shall eat flesh, and in the morning you shall be filled with bread; then you shall know that I am the Lord your God.' "
In the evening quails came up and covered the camp; and in the morning dew lay round about the camp. And when the dew had gone up, there was on the face of the wilderness a fine, flake-like thing, fine as hoarfrost on the ground. When the people of Israel saw it, they said to one another, "What is it?" [or "It is manna."] For they did not know what it was. And Moses said to them, "It is the bread which the Lord has given you to eat" (Exodus 16:11-15).

On the sixth day they gathered twice as much bread, two omers apiece; and when all the leaders of the congregation came and told Moses, he said to them, "This is what the Lord has commanded: 'Tomorrow is a day of solemn rest, a holy sabbath to the Lord; bake what you will bake and boil what you will boil, and all that is left over lay by to be kept till the morning' " (Exodus 16:22-23).

8. Volcanic Eruptions and Egypt

The above description of the food that the Lord provided may also be compatible with the stated means of preparation. Thus the words "bake what you will bake" could refer to the preparation of bread from the manna," while the words "boil what you will boil" could refer to the cooking of the quail. Thus the term "fleshpots" would refer to the implements used in the cooking of the quail and perhaps other food.

The bread and quail that the Hebrews are said to have eaten during their escape from Egypt also has parallels in the diet of the Egyptians of a later era:

> They eat bread, making loaves which they call "cylestis" of coarse grain. For wine, they use a drink made of barley; for they have no vines in their country. They eat fish uncooked, either dried in the sun or preserved with brine. Quails and ducks and small birds are salted and eaten raw; all other kinds of birds, as well as fish (except those that the Egyptians hold sacred) are eaten roast and boiled.[63]

In place of wine, the Egyptians at the time of Herodotus were apparently fond of beer made from barley, perhaps similar to the beer mentioned above and provided to the Hebrews at the time of Rekhmire. In the story of a plague before the exodus, the Bible informs us that an important crop of the Egyptians was indeed barley (Exodus 9:31).

The Hebrews soon moved on from the wilderness of Sin to Rephidim. There they encountered an unanticipated shortage of water:

> All the congregation of the people of Israel moved on from the wilderness of Sin by stages, according to the

commandment of the Lord, and camped at Reph'idim; but there was no water for the people to drink. Therefore the people found fault with Moses, and said, "Give us water to drink." And Moses said to them, "Why do you find fault with me? Why do you put the Lord to the proof?" But the people thirsted there for water, and the people murmured against Moses, and said, "Why did you bring us up out of Egypt, to kill us and our children and our cattle with thirst?" So Moses cried to the Lord, "What shall I do with this people? They are almost ready to stone me." And the Lord said to Moses, "Pass on before the people, taking with you some of the elders of Israel; and take in your hand the rod with which you struck the Nile, and go. Behold, I will stand before you there on the rock at Horeb: and you shall strike the rock, and water shall come out of it, that the people may drink." And Moses did so, in the sight of the elders of Israel. And he called the name of the place Massah and Mer'ibah, because of the faultfinding of the children of Israel, and because they put the Lord to the proof by saying, "Is the Lord among us or not?" (Exodus 17:1-7).

Following his flight from Egypt in about 1505 B.C.,[64] Moses is said by *Jasher* to have reigned over Ethiopia for forty years.[65] This possibility seems not to be endorsed by the Bible and other texts. However, Moses may in any case have spent some of this time period or the following pre-exodus time elsewhere away from Egypt, as will be considered in Chapter 16. After the exile period, Moses arrived in Midian in about 1462 B.C.[66] In Midian he married Zipporah, daughter of Reuel (Jethro). Moses helped Jethro care for his flock of sheep, and in that process he inevitably became familiar with the geography of the lands near Mt. Sinai/Horeb: "Now Moses was keeping the flock of his father-in-law, Jethro, the priest of Mid'ian; and he led his flock to

8. Volcanic Eruptions and Egypt

the west side of the wilderness, and came to Horeb, the mountain of God" (Exodus 3:1).

It is striking that, in the paragraphs quoted above, Moses seems to have led his "flock" of Israelites into the same region near Mt. Sinai where he had formerly led Jethro's flock of sheep. This would seem logical in the sense that Moses would already have been very familiar with the means of survival in this land for both the Israelites and their sheep and cattle. The surprise then is that they encountered a very severe shortage of water, as noted above. A possible interpretation is that the variations in tree-ring growth rate in Anatolia following the time of the exodus and represented in Figure 8.2 may be a consequence of widespread weather disruptions associated with a volcanic event. Those disruptions may have caused the unanticipated period of famine near Egypt reflected in the biblical reports of hunger and thirst. This would be similar to the way in which a major climate-related famine brought Abraham and his family to Egypt,[67] and then a later famine brought Jacob and his family to the same destination.[68] Based on the existing biblical texts this exodus famine may have lasted only a few years, corresponding perhaps to the weather disruptions in Anatolia from about 1437 B.C. to 1436 B.C. as suggested by Figure 8.2. Other water-from-rock incidents will be considered in Section 16.3.

8.8 Conclusion

A major volcanic eruption at a distant location can have a dramatic effect on climate. Thus, the Thera eruption of about 1628 B.C. led to seven years of abundant rainfall in Anatolia, as evident in the tree-ring record, and at the same time there were seven years of famine as recorded in Egypt and China. Among other topics, this chapter has explored the possibility that the heavy rain at the time of the tempest in Egypt or a famine after the exodus may also have been caused by such eruptions.

Heavy rainfall downwind of a major eruption is a well-known and almost immediate effect. In the longer term of many years the changes in weather patterns can be more complex, resulting mostly from the reduction of sunlight hitting the surface of the earth. Thus, these changes can lead either to increases in precipitation, often beneficial to plant growth, or decreases in precipitation often leading to famine. In Bible-related stories distinct volcanic effects have been associated with the famines that brought first Abram and his family and later Jacob and his family to visit Egypt, with the tempest in Egypt while the Hebrews were there, and with the exodus.

---

1. L. W. Casperson, *Patterns of Biblical Chronology* (Westbow Press, Bloomington, IN, 2012), Section 21.3, "The Thera eruption," pp. 526-532; Section 21.4, "Contemplated chronology," pp. 532-540.
2. E. Weise, "Volcanoes may cause more rain than realized," *USA Today*, (Tuesday, 12 July, 2011); see also J. Boulon, K. Sellegri, M. Hervo, and P. Laj, "Observations of nucleation of new particles in a volcanic plume," *Proceedings of the National Academy of Sciences*, Volume 108, Number 30, pp. 12223-12226, (26 July 2011).
3. L. W. Casperson, op. cit., Section 21.3, "The Thera eruption," pp. 526-532; Section 24.4, "Abram and the famine," pp. 605-609.

4. C. L. Pearson, D. S. Dale, P. W. Brewer, P. I. Kuniholm, J. Lipton, S. W. Manning, "Dendrochemical analysis of a tree-ring growth anomaly associated with the Late Bronze Age eruption of Thera," *Journal of Archaeological Science*, Volume 36, pp. 1206-1214 (2009); p. 1207.
5. Ibid., p. 1211.
6. L. W. Casperson, op. cit., Section 21.3, "The Thera eruption," pp. 526-532.
7. C. Pellegrino, *Unearthing Atlantis, An Archaeological Odyssey* (Random House, New York, 1991), p. 22.
8. N. Platon, *Zakros, The discovery of a lost palace of ancient Crete* (Charles Scribner's Sons, New York, 1971, pp. 278-279.
9. Ibid., pp. 283-284.
10. L. W. Casperson, op. cit., Sec. 24.4, "Abram and the famine," pp. 605-609.
11. Ibid., p. 560, Equation (22.18).
12. Ibid., p. 559, Equation (22.15).
13. N. Platon, op. cit., pp. 276-277.
14. Ibid., p. 286.
15. A. G. Galanopoulos and E. Bacon, *Atlantis, The truth behind the legend* (The Bobbs-Merrill Company, Indianapolis/New York, 1969), p. 197.
16. J. W. Mavor, Jr., *Voyage to Atlantis* (G. P. Putnam's Sons, New York, 1969), pp. 56-58.
17. Ibid., p. 271.
18. Ibid., p. 286.
19. K. D. Pang, R. Keston, S. K. Srivastava, and H. H. Chou, "Climatic and hydrologic extremes in early Chinese history: possible causes and dates," *EOS (Transactions of the American Geophysical Union)*, Volume 70, Number 43, p. 1095 (1989); also considered in L. W. Casperson, op. cit., Section 21.4, "Contemplated chronology," pp. 532-540; pp. 538-539.

20. M. R. Rampino, S. Self, and R. B. Stothers, "Volcanic winters," *The Annual Review of Earth and Planetary Sciences*, Volume 16, pp. 73-99 (1988); p. 74.
21. *The Works of Aristotle*, Translated into English under the editorship of W. D. Ross, Volume 3 (Oxford at the Clarendon Press, London, 1931), *Meteorologica*, Translated by E. W. Webster, Book 3, Part 2, paragraph 4, written c. 350 B.C.
22. L. W. Casperson, op. cit., Section 22.5, "Famine duration," pp. 557-560.
23. *Herodotus*, With an English translation by A. D. Godley, In Four Volumes (Loeb Classical Library, Harvard University Press, Cambridge, Massachusetts, 1981), Volume II, Book IV, paragraphs 151-153, pp. 353, 355.
24. "The island now called Bomba, east of Cyrene."
25. *Herodotus*, op. cit., Volume II, Book IV, paragraph 150, p. 351.
26. K. P. Foster, R. K. Ritner, and B. R. Foster, "Texts, storms, and the Thera eruption," *The Journal of Near Eastern Studies*, Volume 55, Number 1, pp. 1-14 (January 1996); pp. 4, 8.
27. "F. A. Fouqué, *Santorin et ses éruptions* (Paris, 1879), pp. 1-6."
28. "For translations of the relevant entries, see A. J. Sachs and H. Hunger, *Astronomical Diaries and Related Texts from Babylonia* (Vienna, 1989), pp. 247-253."
29. "Santorini Volcano [Thera]" Smithsonian Institution, *National Museum of Natural History*, Global *Volcanism Program*, Database Search, August 3, 2016.
30. R. K. Ritner, "A new translation of the Tempest Stele of Ahmose," Appendix A of K. P. Foster, R. K. Ritner, and B. R. Foster, op. cit.
31. W. Helck, *Historisch-biographische Texte der 2. Zwischenzeit und neue Texte der 18. Dynastie*, 2nd ed. rev. (Otto Harrassowitz, Wiesbaden, 1983), pp. 104-110.

32. R. K. Ritner and N. Moeller, "The Ahmose 'Tempest Stela', Thera and comparative chronology," *Journal of Near Eastern Studies*, Volume 73, Number 1, pp. 1-19, (April 2014); Note 27.
33. Ibid., pp. 6-7.
34. A. G. Galanopoulos and E. Bacon, op. cit., pp. 120-121.
35. Ibid., p. 121.
36. L. W. Casperson, op. cit., Section 21.3, "The Thera eruption," pp. 526-532.
37. See for example, ibid., Chapter 6, "Astronomy and the exodus date," pp. 111-132.
38. A. G. Galanopoulos and E. Bacon, op. cit., Appendix B, pp. 192-199; J. V. Luce, 1969, *Lost Atlantis, New Light on an Old Legend* (McGraw-Hill Book Company, New York, 1969), p. 174; J. W. Mavor, Jr., *Voyage to Atlantis* (G. P. Putnam's Sons, New York, 1969), Appendices A and B, pp. 267-290.
39. See for example, C. Pelligrino, op. cit., pp. 239-246.
40. K. P. Foster, R. K. Ritner, and B. R. Foster, op. cit.; R. K. Ritner and N. Moeller, op. cit.
41. W. C. Hayes, "Chronological Table," *The Scepter of Egypt, A Background for the Study of the Egyptian Antiquities in The Metropolitan Museum of Art, Part II: The Hyksos Period and the New Kingdom* (1675-1080 B.C.) (The Metropolitan Museum of Art, New York, 1959), pp. xiv-xv; "Chronological Tables: (A) Egypt, Kings from the Thirteenth to the Eighteenth Dynasties," in *The Cambridge Ancient History*, Third Edition, Volume II, Part 1, *History of the Middle East and the Aegean* Region c. 1800-1380 B.C., Edited by I. E. S. Edwards, C. J. Gadd, N. G. L. Hammond, and E. Sollberger (Cambridge at the University Press, 1973), pp. 818-819; G. Steindorff and K. C. Seele, *When Egypt Ruled the East* (The University of Chicago Press,

Chicago, 1957), pp. 32-33; N. Grimal, *A History of Ancient Egypt* (Barnes and Noble Books, New York, 1997), p. 193); E. F. Wente and C. C. Van Siclen III, "A chronology of the New Kingdom," *Studies in Honor of George R. Hughes*, Studies in Ancient Oriental Civilization, Number 39 (The Oriental Institute of the University of Chicago, January 12, 1977), pp. 217-261, Table 1; L. W. Casperson, op. cit., Section 13.2, "Chronology of the early eighteenth dynasty in Egypt," pp. 303-304, Table 13.1, p. 303.
42. L. W. Casperson, op. cit., Chapter 21, "Cause of the famine," pp. 523-542.
43. Ibid., Section 24.4, "Abram and the famine," pp. 605-609.
44. P. I. Kuniholm, M. W. Newton, C. B. Griggs, and P. J. Sullivan, "Dendrochronological dating in Anatolia: The second millennium BC," *Der Anschnitt*, Anatolian Metal III, Beiheft 18, pp. 41-47 (2005).
45. L. W. Casperson, op. cit., Section 21.4, "Contemplated chronology," pp. 532-540; 532-533.
46. Ibid., Section 24.4, "Abram and the famine," pp. 605-609; pp. 605-606.
47. C. Vandersleyen, "Deux nouveaux fragments de la stèle d'Amosis relatant une tempête," *Revue d'Egyptologie*, Volume 20, pp. 127-134 (1968), p. 132; K. P. Foster, R. K. Ritner, and B. R. Foster, op. cit., p. 9.
48. K. P. Foster, R. K. Ritner, and B. R. Foster, op. cit., p. 10.
49. Smithsonian Institution, *National Museum of Natural History, Global Volcanism Program*, Database Search, November 29, 2017.
50. L. W. Casperson, op. cit., Section 22.5, "Famine duration," pp. 557-560.
51. Ibid., Chapter 6, "Astronomy and the exodus date," pp. 111-132.

52. Ibid., Section 21.4, "Contemplated chronology," pp. 532-540.
53. Ibid., Figure 24.1, p. 607.
54. Smithsonian Institution, op. cit., Database Search, November 29, 2017.
55. K. D. Pang, et al., op. cit.
56. See for example, J. P. Gray and J. J. Monaghan, "Caldera collapse and the generation of waves," *Geochemistry, Geophysics, Geosystems*, Volume 4, Issue 2, pp. 1-28, 15 February 2003.
57. J. Kirsch, *Moses, A Life* (Ballantine Books, New York, 1998), p. 121.
58. M. Noth, *Exodus: A Commentary*, Translated by J. S. Bowden, Old Testament Library (Westminster Press, Philadelphia, 1962), p. 39.
59. J. Bingham, *The Hindenburg 1937: A Huge Airship is Destroyed by Fire* (Raintree, Chicago, Illinois, 2006), p. 50.
60. "5.0-magnitude earthquake shakes southern Israel," *The Times of Israel*, May 16, 2016.
61. A. G. Galanopoulos and E. Bacon, op. cit., p. 90.
62. L. W. Casperson, op. cit, Section 18.8, pp. 461-466; J. H. Breasted, *Ancient Records of Egypt*, Volume II (Russell and Russell, New York, 1962), pp. 292-294.
63. *Herodotus*, op. cit., Volume I, Book II, paragraph 77, p. 365.
64. L. W. Casperson, op. cit., Section 16.5, "Chronology of the exile," pp. 403-409, Table 16.1.
65. *The Book of Yashar*, Translated from the Hebrew and published by Manuel Mordecai Noah (Hermon Press, New York, 1972), (cited here in the form *Jasher*), *Jasher* 73:2.
66. L. W. Casperson, op. cit., Section 16.5, "Chronology of the exile," pp. 403-409, Table 16.1.

67. Ibid., Section 24.4, "Abram and the famine," pp. 605-609.
68. Ibid., Section 21.3, "The Thera eruption," pp. 526-532.

9. EVIDENCE OF THE BIBLICAL FLOOD

"In the six hundredth year of Noah's life, in the second month, on the seventeenth day of the month, on that day all the fountains of the great deep burst forth, and the windows of the heavens were opened. And rain fell upon the earth forty days and forty nights" (Genesis 7:11-12).

9.1 Introduction
Determining the date of the biblical flood could be an important step in establishing the chronology of some of the earliest events reported in the Bible. As already discussed, unusually heavy rainfall was probably a contributor to the flood. Thus, one should seek for any evidence that might reveal an ancient datable period of heavy rainfall.

It has been suggested several times in preceding chapters that plant growth can be a sensitive indicator of rainfall events. Tree ring thickness in particular has proven to be extremely useful in studying the ancient chronologies of climates. Many samples of wood have been found in which the tree rings have recorded clear information about rainfall rates for time periods thousands of years in the past. Thus, dendrochronology will be considered in Section 9.2 as a possible means of determining the date of the flood. This approach suggests 2035 B.C. as the date of a major climate event, which may correspond to the biblical flood. Section 9.3 includes a consideration of historical records from China relating to a flood in about 2035 B.C. that was followed by a severe drought.

A popular chronology for a major climate event in Sumer is considered in Section 9.4. The sequence of heavy rain in Sumer followed by drought and famine seems also to be compatible with the idea of a major climate-altering occurrence having happened in about 2035 B.C. Various efforts to survive the flood are considered in Section 9.5 A much later biblical weather event is considered in Section 9.6. In this incident a cloud rising from the Mediterranean area west of Israel was followed by heavy rainfall in the era of King Ahab and the prophet Elijah. The cause might have been a moderate eruption $(VEI = 4)$ of the volcano Vesuvius known to have occurred at about the time of the reign of King Ahab. A much larger eruption $(VEI = 5)$ of Vesuvius in 79 A.D. is discussed in Sections 19.5 and 19.6.

9.2 An occurrence in 2035 B.C.

The examples in the previous chapter have been intended to show that large eruptions of the Thera volcano and others may have caused major floods, rainstorms, darkness, and other effects, at least in Egypt and presumably over large areas of the Middle East and Asia. It is reasonable to suppose that large eruptions of volcanoes in other places had similar effects, and more examples could be cited. One purpose here is to see whether there might be additional specific support for the idea that the flood associated in the Bible with Noah and his family could have been caused by a volcanic eruption.

A climate event considered previously in the context of Noah's flood was tree-ring-dated to 2035 B.C.[1] That event has been associated with the most severe frost-ring occurrence in a continuous tree-ring sequence from 3435 B.C. to 1971 A.D. As with the Thera eruption in about 1628 B.C., the climate event of 2035 B.C. was also responsible for a major multi-year growth anomaly in the Anatolian tree-ring record. The cause of the earlier incident had been ascribed to an eruption of Mt. St. Helens in Washington State. The date 2035 B.C. could be conformable to the biblical flood chronology, but other possibilities may also be considered. To begin, one may enquire into the significance of frost rings in an ancient tree-ring record:

> Frost damage zones (frost rings) provide a time-equivalent internal marker that can also be used for relative dating of tree ring sequences and, in some circumstances, for absolute dating of growth layers. Frost damage is caused by the occurrence of temperatures well below freezing at some time during the growing season. Under such conditions, extracellular ice formation causes dehydration and physical disruption of immature xylem cells, leaving a permanent record in the wood.[2]

In principle, severe freezing could lead to the loss of one or more growth rings, causing an error in the implied chronology:[3] "[A] frost event in 2035 B.C. is the most severe in the entire record, as it occurs in all the trees in the sample and caused severe anatomical damage." Only this frost event caused an interuption of the ring count in some of the trees that were studied. Besides the frost, this incident is also, of course, a promising candidate for having caused the largest flood of that same period. One of the lesser frost-damage events corresponds to the dramatic eruption of the Thera volcano. This eruption caused an unusual July frost in China (Section 8.3). These results suggest that the event of 2035 B.C. may have led in some areas to even more severe weather conditions than were caused by the Thera eruption. A comment concerning this frost follows:

> This unique event, which is dated at 2035 B.C., caused the complete disruption of the physical continuity of the wood in the radial direction in nearly all the specimens. As a result, specimens fall apart along this surface, which has also been an avenue for weathering and decay. Most measured ring width series thus end a few years prior to 2035 B.C. and are continued a few years afterward. The fact that this traumatic frost damage is not repeated anywhere in the record strongly supports the conclusion that no gross duplications have been made that would result in large chronology errors and that might yield erroneous tree ring ages substantially greater than the true age of the wood.[4]

It should also be noted that the date 2035 B.C. has been verified by comparison with the independently constructed Methuselah chronology for bristlecone pine.[5]

The fact that this dramatic frost damage is not matched at any other date implies that climate-related effects of the underlying event could also be the most significant in the frost ring record. It has been suggested that the incident of 2035 B.C. corresponds to a volcanic eruption, possibly of the Mt. St. Helens volcano, as mentioned above. This would seem to be a natural suggestion in

9. Evidence of the Biblical Flood

the present context, but it might be appropriate to consider other possiblities as well. One concern with the Mt. St. Helens interpretation is that the estimate of the date of that eruption is about 1900 B.C. with a $VEI = 5$.[6] A more recent representation of what was probably the same eruption gives it a date of about 1860 B.C. with a $VEI = 6$.[7] These date estimates are not as close as one might wish to the tree-ring date of 2035 B.C. for the flood incident, and the estimates of the volcanic explosivity index are smaller than $VEI = 6.6$ inferred in Section 8.2 for the 1628 B.C. eruption of Thera. Thus, besides the difficulty with the eruption dates, the explosivity inferred for the supposed eruption of 2035 B.C. is much less than the explosivity of the Thera eruption of 1628 B.C. In disagreement with this interpretation, the frost event of 2035 B.C. was more severe than that for the Thera eruption of 1628 B.C.[8]

Considering the difficulties with the eruption interpretation, it may be worthwhile to consider the possibility that the frost damage of 2035 B.C. was caused by some incident other than a conventional eruption of the Mt. St. Helens volcano. Instead there could have been some other eruption event in 2035 B.C. That eruption, if real, remains unidentified and could have occurred elsewhere on the surface of the earth (or underwater).

A different possibility is that the climate event of 2035 may have been unrelated to any volcanic eruption. As an example, it might be considered that this incident involved the impact of a small comet or asteroid. Thus, the idea that comets consist mostly of water (the "dirty snowball model") was mentioned prevously in Section 7.3. This model has been extended in recent years including its relevance to the formation of our planet. In fact water-carrying comets and asteroids have been recognized as major sources of the water that now exists on earth:[9]

> [T]he prevailing opinion is that the water presently on Earth came from the outer solar system, carried by comets colliding with the freshly-formed Earth (see Owen and Bar-Nun, 1995).[10] This is usually called the late veneer scenario.
>
> According to the nebular model by Drouart et al. (1999),[11] the Jupiter-Saturn region and the outer asteroid belt should be the only regions where planetesimals could have included water with essentially the isotopic composition of that of the Earth.
>
> Asteroids and the comets from the Jupiter-Saturn region were the first water deliverers, when the Earth was less than half its present mass. The bulk of the water presently on Earth was carried by a few planetary embryos, originally formed in the outer asteroid belt and accreted by the Earth at the final stage of its formation. Finally, a late veneer, accounting for at most 10% of the present water mass, occurred due to comets from the Uranus-Neptune region and from the Kuiper Belt.

While such water-rich impacts would not be as common now as in the earlier development of the solar system, they must still occur from time to time. One might wonder how the collision of a comet or asteroid with earth could have been reflected in the weather or perceived by earth's inhabitants. One early description of the onset of the flood events associated with Noah describes circumstances that could bring to mind such an impact:

9. Evidence of the Biblical Flood

> And on that day, the Lord caused the whole earth to shake, and the sun darkened, and the foundations of the world raged, and whole earth was moved violently, and the lightning flashed, and the thunder roared, and all the fountains in the earth were broken up, such as was not known to the inhabitants before; and God did this mighty act, in order to terrify the sons of men, that there might be no more evil upon earth.
>
> And still the sons of men would not return from their evil ways, and they increased the anger of the Lord at that time, and did not even direct their hearts to all this. And at the end of seven days, in the six hundredth year of the life of Noah, the waters of the flood were upon the earth.[12]

Thus, *Jasher* reports that the actual flood events were preceded by violent global disturbances. These disturbances, if real, could be imagined as consequences of a comet or asteroid impacting the earth.

Significant comet or asteroid impacts with earth are not common events in human history, and it is natural to be cautious about assigning the cause of the biblical flood to such an unfamiliar occurrence. It should be noted, however, that smaller asteroid interactions are certainly not unknown even in very recent times. Several recent impacts are listed in Table 9.1.[13] The nature of these events range from near-horizontal trails in the atmosphere to localized craters in the earth's surface, depending on such characteristics as angle of incidence, mass, velocity, and composition. Much larger impacts than those listed in the table have occurred in the more distant past, and a few comments on those major incidents are included in Section 20.5.

Table 9.1. Recent asteroid impacts.

Name	Date	Location	Diameter (m)
Chelyabinsk	15 February 2013	Chelyabinsk, Russia	17-20
Indonesia	8 October 2009	Indonesia	5-10
Sudan	7 October 2008	Sudan	2-3
Peru	15 September 2007	Carancas, Peru	3
Sikhote-Alin	12 February 1947	Vladivostok, Russia	2.5
Tunguska	30 June 1908	Tunguska, Siberia	50

Asteroid encounters such as those listed in Table 9.1 are usually not accompanied by helpful warnings, and during this writing a close encounter occurred unexpectedly:

> While most of us were sleeping last Sunday morning [15 April 2018, 2:41 A.M. E.D.T], Earth had a close call with an asteroid that had been detected a mere 21 hours before it zipped past . . . at 66,174 miles per hour. . . . [T]he asteroid, officially known as 2018 GE3, was about the size of a football field, measuring between 157 and 361 feet in diameter. At its closest point to Earth, it passed by some 119,500 miles away – about half the distance between the Earth and the moon. . . . The asteroid is much larger than many of the other space rocks that have passed by or exploded over Earth, causing some curiosity about how it went undetected for so long. After all, asteroids do have the potential to wreak havoc on Earth.
>
> A meteor that blew up over Chelyabinsk, Russia, in 2013, for example, resulted in nearly 1,500 injuries and thousands of buildings damaged. . . . Asteroid 2018 GE3 is actually three to six times the size of the Chelyabinsk

9. Evidence of the Biblical Flood

meteor and roughly equal to the size of the space rock that exploded over Tunguska in 1908.[14]

It remains now to consider whether an asteroid-like impact on the earth could have had any flood-like consequences. An interagency program in the U.S. is working to discover other near-earth objects (NEOs) that have the potential to collide with the earth. The program also seeks to find methods of deflecting such objects and preventing collisions. Aspects of those studies include the observation that most collisions will occur over oceans, even though most historical observations have been of impacts over land:

> Chelyabinsk is about 1600 km from the Kara Sea and 3200 km from the Arabian Sea. Although the meteor that crashed there opened eyes to the threat, an asteroid strike would be most likely to occur somewhere over the three-quarters of earth that is ocean.[15]

The importance here of this fact is that an ocean impact is likely to have particularly wide-ranging consequences for climate:

> Unless the NEO crashed near a coastal city, where its effects would be catastrophic, the most significant impact of an ocean strike would be the injection of huge quantities of water vapor into the stratosphere, where it would not quickly rain out and could affect climate. . . . A Chelyabinsk-like airburst over the ocean would dissipate and have a much weaker impact than a marine strike.[16]

This supports the possibility that the substantial frost effects of 2035 B.C. could well have been associated with Noah's flood, but the frost and flood may have been the result of an asteroid or comet impact on the ocean rather than a volcanic eruption.

To gain a better sense of the possible consequences of a large asteroid or comet impact, it could be helpful to consider an ancient and more dramatic example. Thus, the Chicxulub Crater in the Yucatan provides a physical record of the climate catastrophe that followed an asteroid impact on earth 66 million years ago: "The event is blamed for the demise of three-quarters of plant and animal species, including the dinosaurs. The researchers' investigations suggest the impact threw more than 300 billion tonnes of sulphur into the atmosphere. This would have dropped temperatures globally below freezing for several years."[17] The asteroid is thought to have been about 12 km in diameter in comparison to the Tunguska asteroid, which as indicated in Table 9.1 was about 50 m in diameter.

As it happens, an asteroid impact that may have caused Noah's flood has already been partially identified. Studies of the unusual lava flows and ice caves in areas of Washington, Oregon, Idaho, Nevada, and California suggest that a substantial impact by a large ice-carrying object occurred there only a few thousand years ago.[18] The impact resulted in a lava plateau covering approximately 150,000 square miles, and it is likely to have caused the extinction of the last mammoths on earth. Those mammoths were already restricted in their habitation to small Wrangel Island in the Arctic Ocean. Based on radiocarbon dating methods, it has been concluded that no mammoths survived after about 2000 B.C.[19] With the advantage of precise tree-ring data, this date would seem to be refinable to 2035 B.C., as discussed here previously. The idea that the asteroid carried large amounts of water ice is compatible with the uniquely extensive frost damage associated with that date and the worldwide heavy rainfall.

It should be noted that the idea of a comet or asteroid interaction being responsible for the flood is not a new concept. William Whiston wrote about this possibility several centuries ago in 1696 A.D.: "[A] Comet is capable of passing so close by the Body of the Earth as to involve it in its Atmosphere and Tail a considerable time, and leave prodigious quantities of the same Condensed and Expanded Vapours upon its Surface; we shall easily see that a Deluge of Waters is by no means an impossible thing."[20]

Given the magnitude of the flood event, it would seem likely that weather was modified over a large area. As noted previously, the effects of the Thera eruption of 1628 B.C. were observed over much of the northern hemisphere. The occurrence of flooding in about 2035 B.C. would, of course, be of particular interest. It would be especially helpful to find actual historical records of such flooding. Except in the Middle East and Asia, there are few written records for the third millennium B.C. Therefore, it is to the Middle East and Asia that one might first look.

9.3 Effects in China

One of the best places to consider for information on early floods is China. Many ancient records have been found in China, and those records often tend to be more objective in style than the sometimes myth-encumbered legends that have come down to us from Mesopotamia and the Middle East. Within China, the Yellow River region has been an important source for early historical information: "The Yellow River Valley is known as the Cradle of Chinese Civilisation,"[21] and major floods have occurred in that region:

> The earliest extraordinary floods recorded in Chinese historical texts occurred shortly before the beginning of Xia, the first hereditary dynasty in China. Yu, the founder of Xia, is credited with having successfully controlled these floods. . . .
> With the invention of writing, the occurrence of floods and celestial phenomena began to be recorded. The earliest extant textual record of extraordinary floods and the practice of astronomy both appear in the Canon of Yao in the *Book of Documents* edited by Confucius in the 6th century B.C. The events happened over a millennium earlier, i.e., during the reign of King Yao, one of the so-called sage kings who ruled China before Xia . . .[22]

Thus, the extraordinary floods in China occurred during the reigns of the sage kings Yao and Shun, shortly before the beginning of the Xia dynasty established by Yu.

"The passage referring to the extraordinary floods appears later in the Canon, where King Yao commissioned K'wan to control the floods:"[23]

> . . . 'See! The floods assail the heavens', said the king. 'Oh! Chiefs, destructive in their overflow are the waters of inundation. In their vast extent they embrace the mountains and overtop the hills, threatening the heavens with their floods, so that the people groan and murmur'
> . . .

Other writings convey similar remarks: "The flood water spread, inundating all under heaven."[24] "The people climbed up to the

9. Evidence of the Biblical Flood

hilltops and treetops"[25] in an effort to escape the fate of drowning, as seems to have occurred during the later tempest in Egypt (Section 8.4).

A reasonable question to ask is whether the date of the "extraordinary floods" in China can be established. One approach has involved royal genealogies together with records of astronomical observations. Using this approach, the following chronology has been obtained: King Yao was followed by King Shun, whose successor King "Yu ['the founder of Xia'] thus ruled in about 2000 B.C."[26] This approximate date is compatible with the idea that the extraordinary floods during the reign of King Yao may have commenced in about 2035 B.C., the date of the extraordinary frost ring event in the White Mountains and perhaps also the date of the biblical flood.[27]

Spectroscopic and archeological evidence from the Yellow River area may also be compatible with this date for the extraordinary floods:[28] "The flood events were OSL [Optically Stimulated Luminescence] dated to between 4100 and 4000 a BP [approximately 2150 and 2050 B.C. respectively] and checked by archaeological dating of the anthropogenic remains retrieved from the sequences." This approximate date range for the occurrence of flooding in China is roughly compatible with the specific date 2035 B.C. Thus, this date will continue to be considered as at least a reasonable place-holder for the time of the flood recorded in the Bible, in China, in the *Gilgamesh Epic* (to be considered in Chapter 10), and elsewhere.

Another interesting question concerns the drought and famine that sometimes follow a major volcanic eruption. As noted in Section 8.3, the heavy rainfall and flooding in China that was associated with the Thera eruption of 1628 B.C. was followed by a seven-year famine. A seven-year famine following the same eruption may have been the event that brought Jacob and his family to Egypt in about 1625 B.C. Since the Thera eruption of 1628 B.C. and the event of 2035 B.C. have both been documented in the bristlecone pine frost-ring records of the White Mountains, it seems reasonable to enquire whether the earlier event might have led to similar flooding and famine. The biblical flood and the extraordinary floods in China have already been tentatively associated with the frost rings of 2035 B.C. Concerning the possibility of famine following this flood period, the following should be noted: "As the extraordinary floods (and a severe drought) occurred immediately prior to (and at the end of) Xia, determining the dynastic chronology would give the dates of the said climatic anomalies and vice versa."[29] The extraordinary floods (and a severe drought) prior to Xia are seemingly similar to events in China after the Thera eruption in 1628 B.C. That occasion was also associated with unusually cold weather (Section 8.3).

9.4 Effects in Sumer

Sumer is considered to have been the first urban civilization in the region now referred to as southern Mesopotamia. The earliest written texts date from about 3300 B.C., while cuneiform writing appeared in about 3000 B.C. The first Sumerian historical writings are from about the 27th century B.C. Another possible source of volcano-related (or asteroid/comet-related) history or at least imagery for the period around 2035 B.C. is available in Mesopotamian documents:

9. Evidence of the Biblical Flood

Sumerian laments, describing utter destruction of human settlement, refer to "(flaming) 'potsherds' " raining from the sky, "deep shadow(?) of a fiery glow(?)," dust, abnormal darkness and light, and other phenomena suggestive of volcanic activity or its aftermath. In these sources, "(flaming) 'potsherds' raining down" is usually interpreted as a hailstorm, but even if this proposal could be contextually correct, the origin of the topos may well be a description of volcanic activity that survived in Sumerian literature of the outgoing third millennium B.C. Although the Sumerian passages are far too early for the Thera eruption, they raise the possibility that the aftereffects of some earlier eruption left their imprint on Mesopotamian literature.[30]

Weather anomalies of about 2035 B.C. could, of course, be considered to be "of the outgoing third millennium B.C."

As a perhaps clearer parallel, a Sumerian record of a fearful storm near the Euphrates River includes evidence of widespread disaster:

80β. The storm was a harrow coming from above, the city was struck (as) by a hoe.
81. On that day, heaven rumbled, the earth trembled, the storm worked without respite.
82. The heavens were darkened, they were covered by a shadow, the mountains roared,
83. The sun lay down at the horizon, dust passed over the mountains,
84. The moon lay at the zenith, the people were afraid.
85. The city ... stepped outside.
86. The foreigners in the city (even) chase away the dead.
87. Large trees were being uprooted, the forest growth was ripped out,

Causes And Effects Of The Flood

88. The orchards were being stripped of their fruit, they were being cleaned of their offshoots,
89. The crop was drowning while it was still on the stalk, the (yield) of the grain was being diminished.
90. . . .
91. . . .
92. . . .
93. [. . .] they piled up in heaps [. . .] they spread out like sheaves.
94. There were corpses (floating) in the Euphrates, brigands roamed [the roads].[31]

It may be noted here that this description of the flood in Sumer has some similarities to *Jasher*'s description of the biblical flood, as quoted above in Section 9.2:

Jasher
And on that day, the Lord caused the whole earth to shake, and the sun darkened, and the foundations of the world raged, and whole earth was moved violently, and the lightning flashed, and the thunder roared. . . .

Sumer:
81. On that day, heaven rumbled, the earth trembled, the storm worked without respite.
82. The heavens were darkened, they were covered by a shadow, the mountains roared,

Both of these representations include the shaking of the earth, the darkening of the sun, and the roaring of thunder. The shaking of the earth in Mesopotamia and elsewhere may cause one to contemplate a significant astrophysical interaction as an alternative to the previously suggested volcanic eruption in Washington state. The reference above to "(flaming) 'potsherds'

9. Evidence of the Biblical Flood

raining down" could of course relate to asteroid fragments burning in the atmosphere rather than volcanic activity.

The storm excerpts quoted above are also similar in concept but more detailed than the storm description in lines (8) to (12) of the tempest stela quoted in Section 8.4. In both accounts there was a rainstorm with darkness, thunder, and flooding. In the Sumerian text "There were corpses (floating) in the Euphrates," while in a suggested restoration of the much later tempest stela (mentioned in Section 8.4) human corpses were "floating on the water [of the Nile] like skiffs of papyrus."[32]

As in major volcano-caused disruptions of the atmosphere that have been considered here, the storm described above (whatever its interpretation) may have been followed by an extended famine. A portion of a plea to An, the god of heaven, for an ending to such a famine follows:

498. That the Tigris and Euphrates (again) carry water – may An not change it,
499. That there (again) be rain in the skies and good crops on the ground – may An not change it,
500 That there be water courses with water and fields with grain – may An not change it,
501 That the marshes support fish and fowl – may An not change it,
502 That fresh reeds and new shoots grow in the canebrake – may An not change it,[33]

While political turmoil could have been responsible for some famine-like consequences, it would seem that disruptions of the normal weather patterns may also have been important. The weakened Ur III dynasty was eventually conquered by invaders:[34] "During the twenty-fourth year of the reign of King Ibbi-Sin

Causes And Effects Of The Flood

(2028-2004 B.C.) the city of Ur fell to an army from the east." "[T]he final blow to Ur was dealt by an army from the east, from the lands of Elam and Simaski." These events were described in the words of a lamentation:

> 32. That its people no longer dwell in their quarters, that they be given over (to live) in an inimical place,
> 33. That (the soldiers of) Simaski and Elam, the enemy, dwell in their place,
> 34. That its shepherd be captured by the enemy, all alone,
> 35. That Ibbi-Sin be taken to the land of Elam in fetters,[35]

"[T]he decline of the Ur III kingdom . . . had already begun during the reign of Ibbi-Sin's predecessor, Su-Sin. . . ."[36] With the middle chronology of Mesopotamia, Ibbi-Sin's reign lasted from about 2028 to 2004 B.C., as quoted above; and then Su-Sin's reign was from about 2037 to 2029 B.C. But the frost ring records suggest that a major climate event occurred in 2035 B.C., early in the reign of Su-Sin. Thus, in this chronology the frost event, perhaps followed by heavy rainstorms and/or drought, might have occurred near the beginning of the reign of Su-Sin.

The mechanism of drought and famine is one of the important ways in which a climate event can have long-term negative consequences for a civilization. If flooding had occurred near the beginning of Su-Sin's reign, then a following drought and famine may have contributed to the weakening of the reigns of Su-Sin and Ibbi-Sin. Widespread famine also followed the Thera eruption, and as suggested in Section 9.3 flooding followed by drought in China may have been associated with the 2035 B.C. frost-ring event.

It should be noted that the Sumerian chronology employed by Michalowski and quoted above is what is known as the middle chronology for the Middle East. On the other hand, when considering the later era of Hammurabi in a previous study, the low (or short) chronology was suggested.[37] In fact several different absolute chronologies have been proposed for the ancient kingdoms of Mesopotamia including at least the following: ultra-low, low (or short), middle, high (or long). None of these seems individually able to accomodate all of the chronological data and interpretations that are available, and none of them has gained near-unanimous acceptance:[38] "At present, there is no conclusive evidence to favor any of the competing approaches to absolute chronology for the early to mid-second millennium B.C. Scientific methods and historical sources all involve problems and uncertainties." Fortunately for these studies, dating uncertainties relating to Mesopotamia of the second millenium B.C. have little effect on most of the results relating to the floods and famines that have been considered here.

9.5 Surviving the flood

One of the most dramatic aspects of the flood story, as recorded in our Bibles, is that the flood was to result in the deaths of all human beings on earth except for Noah and his family. This same flood was also to destroy most animals and birds and creeping things:

> The Lord saw that the wickedness of man was great in the earth, and that every imagination of the thoughts of his heart was only evil continually. And the Lord was sorry that he had made man on the earth, and it grieved

him to his heart. So the Lord said, "I will blot out man whom I have created from the face of the ground, man and beast and creeping things and birds of the air, for I am sorry that I have made them." But Noah found favor in the eyes of the Lord (Genesis 6:5-8).

For behold, I will bring a flood of waters upon the earth, to destroy all flesh in which is the breath of life from under heaven; everything that is on the earth shall die. But I will establish my covenant with you; and you shall come into the ark, you, your sons, your wife, and your sons' wives with you. And of every living thing of all flesh, you shall bring two of every sort into the ark, to keep them alive with you; they shall be male and female. Of the birds according to their kinds, and of the animals according to their kinds, of every creeping thing of the ground according to its kind, two of every sort shall come in to you, to keep them alive. Also take with you every sort of food that is eaten, and store it up; and it shall serve as food for you and for them." Noah did this; he did all that God commanded him (Genesis 6:17-22).

Previous content of this chapter and other sources suggest that there may have been flood survivors besides Noah and his family. As discussed in Section 9.3, the earliest recorded extraordinary flooding in China seems to have coincided in time with the flood date identified here with Noah's flood. In the Chinese accounts people climbed up to the hilltops and treetops to escape the flood. In the Sumerian accounts considered in Section 9.4, there is said to have been widespread drowning, but there is no suggestion that this drowning was universal. Also, of course, if everyone in China or Sumer had drowned, there would have been no one to write the historical records that have survived.

9. Evidence of the Biblical Flood

Records of the flooding from the Middle East exist beyond those mentioned previously. According to Josephus, the Armenians point out the remains in their country of the ark, and this is said to have been mentioned also by Berosus the Chaldean, Hieronymus the Egyptian, Mnaseas, and many others.[39] According to the historian Nicolaus of Damascus, many people were saved by climbing to the top of a mountain in Armenia, the same mountain upon which the ark later landed:

> Nicolaus of Damascus, in his ninety-sixth book, hath a particular relation about them; where he speaks thus: "There is a great mountain in Armenia, over Minyas, called Baris, upon which it is reported that many who fled at the time of the Deluge were saved; and that one who was carried in an ark came on shore upon the top of it; and that the remains of the timber were a great while preserved. This might be the man [Noah] about whom Moses the legislator of the Jews wrote.[40]

Thus, many who fled from the rising waters of Noah's flood survived by climbing a "great mountain in Armenia," and by extension others may have survived by finding high ground elsewhere.

The fear of future floods made the survivors hesitant to come down from their mountain homes:

> Now the sons of Noah were three, – Shem, Japhet, and Ham, born one hundred years before the Deluge. These first of all descended from the mountains into the plains, and fixed their habitation there; and persuaded others who were greatly afraid of the lower grounds on account of the flood, and so were very loth to come down

Causes And Effects Of The Flood

from the higher places, to venture to follow their examples.[41]

The fear of similar floods in the future also led to a different survival strategy. This strategy involved direct disobedience to God and the building of their own "mountain" for survival in the event of future floods. Josephus reports this circumstance as follows:

> Now it was Nimrod who excited them to such an affront and contempt of God. He was the grandson of Ham, the son of Noah, a bold man, and of great strength of hand. He persuaded them not to ascribe it to God, as if it was through his means they were happy, but to believe that it was their own courage which procured that happiness. He also gradually changed the government into tyranny, seeing no other way of turning men from the fear of God, but to bring them into a constant dependence on his power. He also said he would be revenged on God, if he should have a mind to drown the world again; for that he would build a tower too high for the waters to be able to reach! and that he would avenge himself on God for destroying their forefathers!
>
> Now the multitude were very ready to follow the determination of Nimrod, and to esteem it a piece of cowardice to submit to God; and they built a tower, neither sparing any pains, nor being in any degree negligent about the work: and, by reason of the multitude of hands employed in it, it grew very high, sooner than any one could expect; but the thickness of it was so great, and it was so strongly built, that thereby its great height seemed, upon the view, to be less than it really was. It was built of burnt brick, cemented together with mortar, made of bitumen, that it might not be liable to admit water.[42]

9. Evidence of the Biblical Flood

The story of this tower, built to save lives in the event of another flood, leads directly (in the writings of Josephus) to the story of the tower of Babel. This well-known story is also recorded in the Bible (Genesis 11:1-9), but Josephus provides further details:

> When God saw that they acted so madly, he did not resolve to destroy them utterly, since they were not grown wiser by the destruction of the former sinners [by the flood]; but he caused a tumult among them, by producing in them divers languages, and causing that, through the multitude of those languages, they should not be able to understand one another. The place wherein they built the tower is now called Babylon, because of the confusion of that language which they readily understood before; for the Hebrews mean by the word Babel confusion.[43]

The main purpose here is, of course, to emphasize that there are many records that confirm the occurrence of a major flood event. This flood seems to have destroyed large numbers of people, while many others survived. Besides the Bible, approximately datable flood stories have survived in the written histories and climate records of Mesopotamia and China. Less detailed flood myths are found in many other cultures. The closest parallels to the biblical flood story seem to be those that are traceable to Mesopotamia. Among these are flood stories from Nippur, the Atrahasis Epic, the deluge account of Berossus, and others.[44] One of the most important flood stories is included in the *Gilgamesh Epic*, and that work will be considered in Chapters 10 and 11.

Causes And Effects Of The Flood

9.6 A later biblical eruption

The emphasis in this chapter has been on the possibility that a major climate-altering event, such as a volcanic eruption or asteroid/comet impact, might have caused the heavy rainfall associated with the flood of Noah's time. Of special interest have been chronological implications based on tree-ring records and other evidence from China and Mesopotamia. Indications of volcanic eruptions during the presence of the Hebrews in Egypt have been considered in Chapter 8 and elsewhere. As a final example, we consider here a possible eruption-related rainfall event that occurred several centuries after the Hebrews departed from Egypt.

The event of interest occurs in a well-known story involving the prophet Elijah and the king Ahab (1 Kings 17-18). In a confrontation between Elijah and the prophets of Baal, the false god Baal was proven to be powerless and his hundreds of prophets were executed. Following the defeat of the prophets, Elijah was able to announce to the king the approaching end of a famine that had been of about three years duration:[45]

> And Eli'jah said to Ahab, "Go up, eat and drink; for there is a sound of the rushing of rain." So Ahab went up to eat and to drink. And Eli'jah went up to the top of Carmel; and he bowed himself down upon the earth, and put his face between his knees. And he said to his servant, "Go up now, look toward the sea." And he went up and looked, and said, "There is nothing." And he said, "Go again seven times." And at the seventh time he said, "Behold a little cloud like a man's hand is rising out of the sea." And he said, "Go up, say to Ahab, 'Prepare your chariot and go down, lest the rain stop you.' " And in a little while the heavens grew black with clouds and wind,

9. Evidence of the Biblical Flood

and there was a great rain. And Ahab rode and went to Jezreel. And the hand of the Lord was on Eli'jah; and he girded up his loins and ran before Ahab to the entrance of Jezreel (1 Kings 18:41-46).

This incident was only part of a longer story about the activities of Elijah, and it is perhaps not inappropriate to mention an aspect that connects this story to a New Testament sermon by Jesus. Before Elijah's confrontation with the prophets of Baal, he had been a low-profile refugee in territories outside of Israel. In the event of interest, Elijah had been sent to assist a widow who was not an Israelite, and while there he helped to miraculously provide food for the widow and her son. Later when her son had become ill and died, "The Lord hearkened to the voice of Eli'jah; and the soul of the child came into him again, and he revived" (1 Kings 17:8-24).

Many years later, Jesus referred to the story just mentioned as he preached to an increasingly hostile audience in the synagogue of his hometown of Nazareth:

And he said, "Truly, I say to you, no prophet is acceptable in his own country. But in truth, I tell you, there were many widows in Israel in the days of Eli'jah, when the heaven was shut up three years and six months, when there came a great famine over all the land; and Eli'jah was sent to none of them but only to Zar'ephath, in the land of Sidon, to a woman who was a widow. And there were many lepers in Israel in the time of the prophet Eli'sha; and none of them was cleansed, but only Na'aman the Syrian." When they heard this, all in the synagogue were filled with wrath. And they rose up and put him out of the city, and led him to the brow of the hill on which their city was built, that they might throw him down headlong. But passing through the midst of them he went away (Luke 4:24-30).

The New Testament book of James used this same story as an example of the power of prayer:

> Therefore confess your sins to one another, and pray for one another, that you may be healed. The prayer of a righteous man has great power in its effects. Eli'jah was a man of like nature with ourselves and he prayed fervently that it might not rain, and for three years and six months it did not rain on the earth. Then he prayed again and the heaven gave rain, and the earth brought forth its fruit (James 5:16-18).

One comment here could be related especially to Elijah's short-range weather forecast. He told king Ahab that he knew rain was coming, because he had heard the sound of the rain storm ("rushing rain"). Falling rain doesn't usually make much sound, and one thinks immediately that what Elijah heard may have been the rumbling of thunder or perhaps a distant volcanic eruption. The absence of any clouds in the sky makes thunder seem unlikely, and in any case thunder can usually only be heard for a distance of at most a few miles from the lightning bolt which caused it.

On the other hand, a volcanic eruption can often be heard for a distance of many hundreds of miles. The eruption explosions from a distant volcano could be heard long before the cloud of dust and condensation droplets might be visible, if they ever became visible at all. The most powerful eruption explosions of modern times are considered to be those of the volcano Krakatoa in August 1883. Those explosions were heard for a distance of

9. Evidence of the Biblical Flood

about 3000 miles, and the lower-frequency pressure waves were documented on recording barometers around the world. Depending on the details of an eruption event, a wide range of sounds can be created, and thunder-like rumbling at a distance is not uncommon. Thus it is worth considering that the sound Elijah heard may have been the sound of a distant volcanic eruption. As discussed in previous chapters, such eruptions can lead in the short term to heavy rainfall.

If the sound and the actual rain in the Elijah story could have been a consequence of a volcanic eruption, it would seem reasonable to see if there might have been any suitable volcanic eruptions at the right time and place. The right time would have been during the reign of King Ahab, and his reign is said to have been between about 874/873 B.C. and 853 B.C.[46] The right place would have been in the Mediterranean area far to the west of Israel. In a quotation above, Elijah is said to have been at the top of Mount Carmel. This mountain or mountain range in Israel has an excellent view out over the Mediterranean Sea to the west. Its setting makes the actions and words of Elijah (quoted also previously) seem very reasonable: "Eli'jah went up to the top of Carmel; and he bowed himself down upon the earth, and put his face between his knees. And he said to his servant, 'Go up now, look toward the sea' " (1 Kings 18:43). After checking seven times the servant was able to report as follows: "Behold a little cloud like a man's hand is rising out of the sea" (1 Kings 18:44). In this interpretation the "little cloud" would represent the first visibility on the western horizon of a growing cloud of volcanic dust and gas. "And in a little while the heavens grew black with clouds and wind, and there was a great rain" (1 Kings 18:45).

There is only one somewhat promising identity for a substantial volcanic eruption during the reign of King Ahab and located in the Mediterranean area well to the west of Israel. Thus, there is said to have been an eruption of Mt. Vesuvius at the time 880 B.C. ± 50 years.[47] That eruption is said to have had an explosivity of $VEI = 4$. As shown in previous considerations here, this explosivity is sufficient to cause a significant localized rainfall event but probably not a multi-year global climate change.

As indicated above and according to Luke, Jesus employed the example of Elijah in his sermon to the people of Nazareth. He also may have quoted from the well-known story of Elijah's successful weather forecasting:

> He also said to the multitudes, "When you see a cloud rising in the west, you say at once, 'A shower is coming'; and so it happens. And when you see the south wind blowing, you say, 'There will be scorching heat'; and it happens. You hypocrites! You know how to interpret the appearance of earth and sky; but why do you not know how to interpret the present time? (Luke 12:54-56).

9.7 Conclusion

Of special interest here has been a major climate event in about 2035 B.C. Heavy rainfall on a wide scale may have resulted from this event, and the rainfall may have been associated with the biblical flood. Based on the appearance of frost rings in bristlecone pine trees at the White Mountains of California in about that year, the climate event may have been the largest in a period of several millennia. Besides the chronologically precise tree-ring evidence related to this date, there are other less specific chronological suggestions in flood records, historical texts, and

9. Evidence of the Biblical Flood

archeological data from China and Mesopotamia. Of particular relevance to the biblical flood is the *Gilgamesh Epic*, which will be considered in following chapters.

As a final example here of a weather-changing volcanic eruption, a probable event of this type during the reign of King Ahab of Israel has been considered. In this example the rainfall caused by the eruption brought an end to a multi-year famine in Israel. Jesus quoted from this story involving the prophet Elijah.

---

1. L. W. Casperson, *Patterns of Biblical Chronology* (Westbow Press, Bloomington, IN, 2012), Section 21.4, "Contemplated Chronology," pp. 532-540; pp. 537, 542, Reference 24; V. C. LaMarche, Jr., and K. K. Hirschboeck, "Frost rings in trees as records of major volcanic eruptions," *Nature*, Volume 307, pp. 121-126 (12 January 1984); pp. 125-126.
2. V. C. LaMarche, Jr., and T. P. Harlan, "Accuracy of tree ring dating of bristlecone pine for calibration of the radiocarbon time scale," *Journal of Geophysical Research*, Volume 78, Number 36, pp. 8849-8858 (20 December 1973); pp. 8851, 8853.
3. V. C. LaMarche, Jr., and K. K. Hirschboeck, op. cit., p. 125.
4. V. C. LaMarche, Jr., and T. P. Harlan, op. cit., p. 8855.
5. V. C. LaMarche, Jr., and T. P. Harlan, op. cit., pp. 8855-8856.
6. V. C. LaMarche and K. K. Hirschboeck, op. cit., "Table 3, Notable frost-ring events in the White Mountains, California, and dates of possible associated eruptions," p. 125.
7. Smithsonian Institution, *National Museum of Natural History, Global Volcanism Program*, database search, 5 August 2017. The *VEI* values are reported with a precision of one significant figure.

8. V. C. Lamarche and K. K. Hirschboek, op. cit., p. 125.
9. A. Morbidelli, J. Chambers, J. I. Lunine, J. M. Petit, F. Robert, G. B. Valsecchi, and K. E. Cyr, "Source regions and timescales for the delivery of water to the Earth," *Meteoritics and Planetary Science*, Volume 35, pp. 1309-1320 (2000); p. 1309.
10. T. Owen and A. Bar-Nun, "Comets, impacts and atmospheres," *Icarus*, Volume 116, pp. 215-226 (1995).
11. A. Drouart, B. Dubrulle, D. Gautier, and F. Robert, "Structure and transport in the solar nebula from constraints on deuterium enrichment and giant planets formation," *Icarus*, Volume 140, pp. 129-155 (1999).
12. *The Book of Yashar*, Translated from the Hebrew and Published by Mordecai Manuel Noah (Hermon Press, New York, 1972), (cited here and below in the form *Jasher*); *Jasher* 6:11-13.
13. "ATLAS - Historical Impacts," http://www.fallingstar.com/historical.php. Web. 11 August 2017.
14. J. Trevino, "An Unexpected Asteroid Buzzed by Earth Last Sunday," 18 April 2018, https://www.smithsonianmag.com/smart-news/asteroid-flew-earth-unexpectedly-sunday-180968818/. Web. 28 April 2018.
15. D. Kramer, "Effort in asteroid defense under way despite funding constraints," *Physics Today*, Volume 70, Issue 2, pp. 31-33, February 2017.
16. Ibid.
17. J. Amos, "Asteroid impact plunged dinosaurs into catastrophic 'winter'," *Science & Environment*, www.bbc.com/news/science-environment-41825471. Web. 31 October 2017.
18. D. W. Patten, *The Biblical Flood and the Ice Epoch* (Pacific Meridian Publishing Company, Seattle,

Washington, 1976), Section 6.7 "The Ice Cave Phenomena of Eastern Washington," pp. 120-124.
19. S. L. Vartanyan, Kh. A. Arslanov, T. V. Tertychnaya, and S. B. Chernov, "Radiocarbon dating evidence for mammoths on Wrangel Island, Artic Ocean, until 2000 B.C.," *Radiocarbon,* Volume 37, Number 1, pp. 1-6 (1995).
20. W. Whiston, *A New Theory of the Earth* (R. Roberts for B. Tooke at the Middle-Temple-Gate in Fleet-Street, London, 1696), Book IV, Chapter IV, p. 301.
21. K. D. Pang, "Extraordinary floods in early Chinese history and their absolute dates," *Journal of Hydrology,* Volume 96, pp. 139-155 (1987); p. 139.
22. Ibid., pp. 139-140.
23. Ibid., p. 140.
24. Ibid.
25. Ibid., p. 141.
26. Ibid., p. 146.
27. L. W. Casperson, op. cit., p. 537, p. 542 reference 24.
28. C. C. Huang, J. Pang, X. Zha, Y. Zhou, H. Su, and Y. Li, "Extraordinary floods of 4100-4000 a BP recorded at the late neolithic ruins in the Jinghe River Gorges, middle reach of the Yellow River, China," *Palaeogeography, Palaeoclimatology, Palaeoecology,* Volume 289, pp. 1-9, (2010); p. 1.
29. K. D. Pang, op. cit., p. 145.
30. B. R. Foster, "Volcanic phenomena in Mesopotamian sources?," Appendix B of K. P. Foster, R. K. Ritner, and B. R. Foster, "Texts, storms and the Thera eruption," *The Journal of Near Eastern Studies,* Volume 55, Number 1, pp. 1-14 (January 1996); p. 12.
31. P. Michalowski, *The Lamentation over the Destruction of Sumer and Ur* (Eisenbrauns, Winona Lake, 1989), pp. 41, 43, lines 80β-94.

32. R. K. Ritner and N. Moeller, "The Ahmose 'Tempest Stela', Thera and comparative chronology," *Journal of Near Eastern Studies*, Volume 73, Number 1, pp. 1-19, (April 2014); pp. 6-7.
33. P. Michalowski, op. cit., p. 69, lines 498-502.
34. Ibid., pp. 1, 2.
35. Ibid., p. 39, lines 32-35.
36. Ibid., p. 1.
37. L. W. Casperson, op. cit., p. 618.
38. G. Schwartz, "Problems of Chronology: Mesopotamia, Anatolia, and the Syro-Levantine Region," in *Beyond Babylon: Art, Trade, and Diplomacy in the Second Millennium B.C.,* Edited by Joan Aruz, Kim Benzel, and Jean M. Evans (The Metropolitan Museum of Art, New York; Yale University Press, New Haven, 2008), pp. 450–452; p. 452.
39. F. Josephus, *The Life and Works of Flavius Josephus*, Translated by William Whiston (Holt, Rinehart and Winston, New York), *The Antiquities of the Jews*, Book 1, 3:5-6, p. 38.
40. Ibid., Book 1, 3:6, p. 38.
41. Ibid., Book 1, 4:1, p. 39.
42. Ibid., Book 1, 4:2-3, p. 39.
43. Ibid., Book I, 4:3, p. 39.
44. A. Heidel, *The Gilgamesh Epic and Old Testament Parallels* (The University of Chicago Press, Chicago, 1963), Chapter 2, pp. 102-136.
45. L. W. Casperson, op. cit., Section 20.6B, "Biblical famines," Table 20.1, pp. 513-514, 1 Kings 18:1, Luke 4:25, James 5:17.
46. Ibid., p. 50, Table 3.2.
47. Smithsonian Institution, *National Museum of Natural History, Global Volcanism Program,* Database Search, August 3, 2016. The *VEI* values are reported with a precision of one significant figure.

10. THE GILGAMESH EPOCH

"Though it be in sorrow and pain, in cold and heat, in sighing and weeping, I will go! Open now the gate of the mountains."[1]

10.1 Introduction

As indicated previously, the development of a chronology for the stories in the Bible is dependent on finding connections between the biblical genealogies and datable historical events. For the early Genesis accounts a particular event that could lend itself to geophysical dating methods is the flood of Noah's era. The possibility that the flood might be associated with a volcanic eruption (or asteroid/comet impact) was considered in Sections 9.2–9.4, and the implied effects of such an event provided several approaches to determining the date of the flood. Records of possible eruption-related flooding events from China and Sumer were also considered. This same flood might be documented directly in other non-Hebrew cultures without reference to volcanism, and if so other approaches to dating might be possible. As it happens, evidence of what could be the same flood event is indeed found in other cultures, and this chapter focuses on records from Mesopotamia. These records are of particular

Causes And Effects Of The Flood

interest because the flood event recorded in the Bible must have occurred in the general vicinity of Mesopotamia.

The Mesopotamian flood records that are most important for this purpose are those associated with a young ruler named Gilgamesh. This king was a member of the first dynasty of Uruk, and many stories are associated with his name. The best known collection of these stories is now usually known as the *Gilgamesh Epic*. Following the death in this epic of his closest friend Enkidu, Gilgamesh became obsessed with the idea of gaining immortality. He had heard of Utnapishtim (the Babylonian flood hero corresponding to Noah in the Bible) who had been granted immortality, and he decided to seek out Utnapishtim to learn from him the secret of eternal life. The latter part of the epic is concerned with Gilgamesh's adventures and disappointments in this ultimately unsuccessful effort.

The *Gilgamesh Epic* has various similarities with both the Genesis garden of Eden and flood stories. Parallels include an evil serpent, a tree of life, a massive flood, an ark, and much more. The date at which the Gilgamesh flood story may have been written is considered in Section 10.2. The story itself is summarized in Section 10.3 including quotations of some of the texts reporting Gilgamesh's search for Utnapishtim. The path to Utnapishtim was guarded by scorpion-people, and these and other supernatural creatures are considered in Section 10.4. After being granted entrance by the scorpion-people, Gilgamesh travelled many hours on a path that seems to have included a tunnel through a mountain, and a modern identification of this path is suggested in Section 10.5. Ancient representations of the underground path and neighboring caves are available today, and

10. The Gilgamish Epoch

some of these are discussed in Section 10.6. After emerging from the tunnel, Gilgamesh found himself in the presence of two kinds of fruit-bearing shrubs or trees, and these are discussed in Section 10.7. The next part of this story will be considered in Chapter 11. There, Gilgamesh crosses a "sea" to arrive at the home of Utnapishtim/Noah shortly after the flood. Utnapishtim then relates his version of the flood story and addresses Gilgamesh's questions about eternal life.

10.2 The age of the epic

In the discussions of Section 4.4, it was found that the biblical creation account gives some possible information on the location of the garden of Eden (near the headwaters of the Tigris and Euphrates Rivers). On the other hand, the story of the flood of Noah's time provides possible information about the location of the landing of the ark (in the mountains of Ararat). The descriptions of these locations are similar enough and imprecise enough that Noah's landing may be considered to have been in the same general area as the garden of Eden. Further insight into the geography of the garden and the flood may possibly be obtained from a consideration of the *Gilgamesh Epic*. This story gives us a review of the amazing adventures of the hero-king Gilgamesh as he seems first to be seeking new challenges and excitement and later as he desperately attempts to gain a longer life.

How an important ancient story originated is always a subject of interest. As a starting point it may be helpful to mention the *Sumerian King List*. This document provides a useful but incomplete summary of the first known dynasties in the land of Sumer in southern Mesopotamia:

> Besides the impossibly long reigns of the mythical or legendary pre- and early post-Flood rulers, historians have other good reasons for questioning the accuracy of some aspects of the *Sumerian King List*. The list omits the names of monarchs such as Mesilim (or Mesalim) who titled himself "King of Kish" on several inscribed objects found at Lagash and Adab. . . . Kish was probably the first city-state to achieve some preeminence in southern Mesopotamia by forcing several of the other cities to become its vassals. According to the *Sumerian King List*, Kish was the first city on which "kingship was lowered from heaven" after the Flood (the event that seems to separate mythological events from legendary ones in Mesopotamian lore). . . .
>
> The earliest king known from actual inscriptions is Enmebaragesi, a ruler of Kish. According to the *King List*, he conquered Elam and had a reign of 900 years. Though the *King List* fancifully gives Enmebaragesi an impossibly long reign, two Early Dynastic II inscriptions show that he was a real person. . . . Enmebaragesi's son Agga (or Aka) was the last ruler of the First Dynasty of Kish according to the *King List*. A later epic poem described a conflict between this King Agga and Gilgamesh, King of Uruk. Gilgamesh, who was a semi-divine hero of several Sumerian epics, is also named in the *King List* where he is the fifth king of the First Dynasty of Uruk. However, because the *King List* claims that kingship over Sumer passed to Uruk only after the end of Kish's First Dynasty, and because all of these early kings are given fantastically long reigns, it dates Gilgamesh more than a thousand years after Agga. Because Enmebaragesi was a real person, some scholars are convinced that Gilgamesh and Agga were real as well

10. The Gilgamish Epoch

and that their respective dynasties overlapped considerably rather than being successive. . . . Unfortunately, even if Sumerologists find an inscription of the historical figure behind the Gilgamesh legends, they may not recognize it. Gilgamesh is possibly an epithet rather than a real name; it seems to mean "heroic ancestor." . . . Thus, some historians refer to the Early Dynastic II Period as Mesopotamia's "Heroic Age," an era whose rulers (real or imagined) later became larger than life and the subjects of myths, legends, and epic poems.[2]

In spite of possible uncertainty concerning the historical reality of Gilgamesh, there is considerable interest in the stories that came to be associated with his name. Fortunately, the beginning of the epic itself provides a summary of how Gilgamesh's story came to be recorded. Less fortunately, the summary is damaged and very brief:

> Tablet I
> Column i
> 1. [He who] saw everything [within the confi]nes(?) of the land;
> 2. [He who] knew [all things and was versed(?) in] everything;
> 3. [. . .] together [. . .];
> 4. [. . .] wisdom, who everything [. . .].
> 5. He saw [se]cret thing(s) and [revealed] hidden things(s);
> 6. He brought intelligence of (the days) before the flo[od];
> 7. He went on a long journey, became weary and [worn];
> 8. [He engra]ved on a table of stone all the travail.[3]

Thus, Gilgamesh is said to have lived in, or otherwise obtained information concerning, the time before the flood. He engraved his story on stone, including information about the era of the flood and times shortly thereafter.

Assuming that the biblical flood was an actual historical event, an important part of these chronological studies is to at least attempt an estimation of the date of that event. A previous preliminary estimate of the time of the biblical flood was approximately 2035 B.C.[4] With the clear parallels between the biblical and Babylonian floods that will be noted more fully in Section 11.4, it is not unreasonable to imagine that the stories from these two sources refer to the same event.

As a starting point, it is worthwhile to consider how and when the composite document we call the *Gilgamesh Epic* came to be written:

> But the question as to the origin of the material of the various episodes cannot as yet be answered with any certainty. To judge from the Sumerian fragments of the epic which have so far come to light and from the fact that the Semitic Babylonians became in general the heirs of Sumerian culture and civilization, it appears reasonable to assume that also the other episodes in the Gilgamesh Epic were current in Sumerian literary form before they were embodied in the composition of this Semitic Babylonian poem. From this, however, it does not necessarily follow that all this material had its origin with the Sumerians, either in their former home or after they had occupied the plains of the Tigro-Euphrates Valley. Instead, the material itself may have originated, at least in part, with the Semitic Babylonians, from whom the Sumerians may have taken it over, adapting it to their own views and

10. The Gilgamish Epoch

beliefs and giving it expression in their own script and language. . . .

The prominence given to the old Sumerian ruler deities Anu and Enlil in our epic and the complete absence of the name of Marduk, in sharp contrast with the main Babylonian creation story, indicate that our epic was composed before Anu and Enlil, in the days of Hammurabi, "committed the sovereignty over all the people to Marduk," and before Hammurabi "brought about the triumph of Marduk." The date of the composition of the Gilgamesh Epic can therefore be fixed at about 2000 B.C.[5]

As is already evident, the greatest interest here will be in the flood story as described in the epic:

Of the various episodes in the Gilgamesh Epic, the one which enjoyed perhaps the most popular favor among the Babylonians and Assyrians, and which certainly is the most important from the viewpoint of the Old Testament scholar, is the story of the great flood and of Utnapishtim's enviable attainment of blessed immortality. This episode originally formed an independent account and has come down to us in a number of different recensions. The oldest of them undoubtedly is the Sumerian version inscribed on a tablet excavated at Nippur. While the tablet itself dates only to about the time of Hammurabi, the story it relates is unquestionably older.[6]

Thus, the oldest written copy of the flood story is the Sumerian version as written on a tablet found at Nippur dating from "about the time of Hammurabi." As indicated above the overall epic was

Causes And Effects Of The Flood

composed before the time of Hammurabi and probably in about 2000 B.C.

By most estimates Hammurabi's reign would have been within the eighteenth century B.C. or a little later. In the "low chronology" adopted previously, his reign was estimated to be from about 1728 B.C. to 1686 B.C.[7] Much of the latter part of the *Epic* (Tablets VIII to XI) relates the story of Gilgamesh seeking out Utnapishtim/Noah shortly after the flood. Gilgamesh's objective was to learn from him how to obtain eternal life. Gilgamesh's visit must have been soon enough after the flood that the water level had not completely returned to its pre-flood level, since he had to dive underwater to reach a still-living tree-of-life plant of the sort familiar from the garden of Eden story. This misadventure will be considered in more detail in Section 11.6. A flood date of about 2035 B.C., as considered in Chapter 9, would not be incompatible with Gilgamesh visiting Utnapishtim after the flood and with the composition date of the epic being in about 2000 B.C., as quoted above.

10.3 The Gilgamesh story

A convenient starting point here is a condensed version of the part of the Gilgamesh story in which he seeks out then-immortal Utnapishtim. The following paragraphs are from an introduction by Heidel. Braces have been added around two phrases beginning with the word "probably" to indicate that the geography suggested there by Heidel doesn't agree with the geography suggested in this chapter. In this excerpt Gilgamesh has just lost his closest friend Enkidu, and his purpose in seeking Utnapishtim is to learn from him the secret of immortality so that he can avoid the death that he so greatly dreads:

> Steeped in sorrow at the death of his friend who has turned to clay [i.e., died], Gilgamesh leaves Uruk and roams over the desert, lamenting: "When I die, shall I not be like unto Enkidu?" His grief-sticken spirit is obsessed with the fear of death and finds no comfort in the glory of his past accomplishments. His sole interest now lies in finding ways and means to escape the fate of mankind; he is willing to go through the greatest perils and the most extraordinary hardships to gain immortal life! He thinks of faraway Utnapishtim, the Babylonian Noah, who, Gilgamesh has heard, has received blessed immortality, and decides to hasten to him with all possible speed to obtain from him the secret of eternal life.
>
> But to reach the dwelling place of Utnapishtim, Gilgamesh must go on a long and arduous journey fraught with many dangers. He arrives at the towering mountain range of Mâshu, {probably the Lebanon and Antilebanon range}. Here is the gate through which the sun passes on his daily journey. The gate is guarded by a terrifying pair of scorpion-people, "whose look is death" and "whose frightful splendor overwhelms mountains." At the sight of them the face of even a demigod like Gilgamesh becomes gloomy with fear and dismay, and he falls prostrate before them. But the scorpion-people, recognizing the partly divine nature of Gilgamesh, receive him kindly and permit him to enter the gate and to traverse the mountain range. After a journey of twelve double-hours of utter darkness, which does not permit him to see what lies ahead of him or what lies behind him, he comes out on the other side and stands before a beautiful garden of precious stones, with trees and shrubs, fruit and vines, all of glittering stone. [Or perhaps it was a garden of ordinary plants.]

> And there in the distance, at the edge of the sea, {probably the Mediterranean Sea on the Phoenician coast}, dwells Siduri, the divine barmaid! Gilgamesh hastens thither and inquires of her how he can get to Utnapishtim, to obtain from him the secret of immortality. The barmaid at first tries to persuade him that his quest is vain, for there is no escape from death. She therefore advises him to enjoy life in full measure and to abandon his hazardous, yet hopeless, undertaking. Nevertheless, Gilgamesh persists in his plan, and at last the barmaid directs him to Utnapishtim's boatman, who has come across from the other side of the sea, where Utnapishtim dwells, and is now in the woods, in search of something. "Him let thy face behold," she tells Gilgamesh. "[If it is possi]ble, cross over with him; if it is not possible, turn back (home)." Gilgamesh leaves the goddess and goes to the boatman, who at length agrees to take him along. With much difficulty the two cross the sea and the waters of death and finally arrive at the shores of the land of blessed Utnapishtim.[8]

Sadly, from Gilgamesh's point of view, he was never granted immortality. Instead, he eventually decided to make the best of his mortal life, as barmaid Siduri had suggested.

The quotation above introduces the "double hour" as a unit of time. Because of its importance here and also our general indebtedness for the mathematical innovations of the Sumerians, a few sentences on these subjects may be of interest:

> The Sumerians developed a mathematical system based on 60 (the sexagesimal system), but which used the factor 10 as well. Because of this system, Mesopotamians tended to order things in units of 6, 12 or 60, or in numbers evenly divisible by 60. Thus, they divided circles into 360 degrees, each of which could be broken into 60

minutes. The day, which began at sunset, was partitioned into 12 "double hours" of 60 "double minutes" each. We still use these divisions of circles and time. . . .

The flexibility of the sexagesimal system of numeration and the positional notation system developed to write it allowed later Mesopotamian mathematicians to create tabulations of reciprocals, squares, square roots, cubes, cube roots, and the sums of squares and cubes. These tables were used to help solve various types of equations and calculations of area and volume.[9]

Several features of the Gilgamesh story bring to mind aspects of not just the biblical flood story, but also of the biblical story of the garden of Eden. In considering the Genesis and Gilgamesh garden stories, it is well to remember that they would have occurred at two different times. Adam and Eve were driven from the biblical garden of Eden a few centuries before the flood of Noah's time. Utnapishtim's flood (probably the same as Noah's flood) occurred prior to Gilgamesh's visit to the garden reported in the epic. Thus, the two garden stories probably differ in both time and location.

There may, however, be common features between the descriptions of the entryways into these gardens. The biblical description is as follows: "He [the Lord God] drove out the man; and at the east of the garden of Eden he placed the cherubim, and a flaming sword which turned every way, to guard the way to the tree of life" (Genesis 3:24). Similarly, when Gilgamesh sought to enter the region including both the garden and also the home of Utnapishtim, he had to be granted passage by the scorpion-people who guarded the gate to the entrance path. The role of the scorpion-people was quoted above in paraphrase, as an introduction to Gilgamesh's quest for eternal life. To clarify the similarities and differences between the two gardens and their respective guardians, this portion of the story is quoted directly from Tablet IX of the *Gilgamesh Epic*, beginning with Gilgamesh's arrival at a mountain or mountain range:[10]

Tablet IX
Column ii

1. The name of the mountain is Mâshu.
2. As he arri[ves] at the mountain of Mâshu,
3. Which every day keeps watch over the rising [and setting of the sun],
4. Whose peak(s) r[each] (as high as) the "banks of heaven,"
5. (And) whose breast reaches down to the underworld,
6. The scorpion-people keep watch at its gate,
7. Those whose radiance is terrifying and whose look is death,
8. Whose frightful splendor overwhelms mountains,
9. Who at the rising and setting of the sun keep watch over the sun.
10. When Gilgamesh saw them,
11. His face became gloomy with fear and dismay,
12. (But) he collected his thoughts and bowed down before them,
13. The scorpion-man calls to his wife:
14. "He who has come to us, his body is the flesh of gods!"
15. The wife of the scorpion-man answers him:
16. "Two-(thirds) of him is god, one-third of him is man."
17. The sco[rpion-m]an calls the man,
18. Speaking (these) words [to the offspri]ng of the gods:
19. "[Why hast thou traveled such] a long journey?
20. [Why hast thou come all the way] to me,

10. The Gilgamish Epoch

21. [Crossing seas] {or crossing mountains?} whose crossings are difficult?
22. [The purpose of] thy [comi]ng I should like to learn."
    (Remainder broken away)

Column iii

1-2. (Destroyed)
3. "[For the sake of] Utnapishtim, my father, [have I come],
4. Who entered into the assembly of [the gods . . . .].
5. Concerning life and death [I would ask him]."
6. The scorpion-man opened his mouth [and said],
7. Speaking to [Gilgamesh]:
8. "There has not (yet) been anyone, Gilgamesh, [who has been able to do that].
9. No one has (yet) [traveled] the paths of the mountains.
10. At twelve double-hours the heart [. . . .].
11. Dense is the darkness and [there is] no [light].
12. To the rising of the sun [. . .].
13. To the setting of the sun [. . .].
14. To the setting of [the sun . . .]."
15-20. (Too badly damaged)
    (Remainder broken away)

Column iv
(Top broken off)

33. "[Though it be] in sorrow [and pain],
34. In cold and [heat],
35. In sighing [and weeping, I will go]!
36. [Open] now [the gate of the mountains]."
37. The scorpion-man [opened his mouth and said]
38. To Gilgamesh [. . .]:
39. "Go, Gilga[mesh, . . .].
40. The mountains of Mâshu [I permit thee to cross];
41. The mountains (and) mounta[in ranges thou mayest traverse].

Causes And Effects Of The Flood

42. Safely may [thy feet carry thee back].
43. The gate of the mountain(s) [is open to thee]."
44. [When] Gilga[mesh heard this],
45. [He followed(?)] the word of [the scorpion-man].
46. Along the road of the sun [he went(?)].
47. One double-hour [he traveled];
48. Dense is the dark[ness and there is no light];
49. Neither [what lies ahead of him nor what lies behind him] does it per[mit him to see].
50. Two double-hours [he traveled].

Column v
(Top broken off)

23-44. {repetitive text on increasing numbers of double-hours traveled}
45. [After he has traveled eleven double-hours he co]mes out before sun(rise).
46. [After he has traveled twelve double-hours], it is light.

Lines 33 to 36 of Tablet IX, Column iv, quoted above, show clearly Gilgamesh's desperation to reach Utnapishtim. This was his only hope for achieving eternal life. A simplified version of these lines was also quoted at the top of this chapter. The scorpion-man finally agreed to let Gilgamesh take the underground path through the mountain.

In the biblical story of the garden of Eden, it was the responsibility of the cherubim to see that no one ever be allowed to enter onto the garden path. On the other hand, it seems from the above quotation that Gilgamesh was the first person granted entry by the scorpion people onto a pitch-black path through the mountains on the way to reaching Utnapishtim. Given their somewhat parallel roles, it seems appropriate to consider at least

10. The Gilgamish Epoch

briefly what sort of creatures were intended to be represented by cherubim and scorpion-people.

10.4 Cherubim and scorpion-people

As noted by Heidel, "*Enûma elish* is the principal source of our knowledge of Mesopotamian cosmology."[11] This work is usually thought to have been written at some time during or shortly after the reign of King Hammurabi of the First Babylonian Dynasty.[12] This date is also after the writing of the *Gilgamesh Epic*. Nevertheless, *Enûma elish* provides perspective on an early time in Bablyonian history including possible background on the scorpion people and other supernatural creatures said to have been created by Tiamat:[13]

132. Mother Hubur [Tiamat], who fashions all things,
133. Added (thereto) irresistible weapons, bearing monster serpents
134. [Sharp] of tooth (and) not sparing the fang(?).
135. [With poison] instead of blood she filled [their] bodies.
136. Ferocious [dra]gons she cl[othed] with terror,
137. She crowned (them) with fear-inspiring glory (and) made (them) like gods,
138. So that he who would look upon them should pe[rish] from terror,
139. So that their bodies might leap forward and none turn back [their breasts].
140. She set up the viper, the dragon, and the lahâmu,
141. The great lion, the mad dog, and the scorpion-man,
142. Driving storm demons, the dragonfly, and the bis[on],
143. Bearing unsparing weapons, unafraid of ba[ttle].
144. Powerful were her decrees, irresistible were they.

> 145. Altogether (?) eleven (kinds of monsters) of this sort she brought [into being].

These descriptions are not all easy to interpret. The serpents with sharp teeth and poisonous blood are reminiscent of the serpent in *Adam and Eve* as discussed in Section 6.2. For comparison a description of the serpent in the garden of Eden before and after its having been cursed may be of interest here:

> And whereas aforetime the serpent was the most exalted of all beasts, now it was changed and become slippery, and the meanest of them all, and it crept on its breast and went on its belly. And whereas it was the fairest of all beasts, it had been changed, and was become the ugliest of them all. Instead of feeding on the best food, now it turned to eat the dust. Instead of dwelling, as before, in the best places, now it lived in the dust. And, whereas it had been the most beautiful of all beasts, all of which stood dumb at its beauty, it was now abhorred of them. And, again, whereas it dwelt in one beautiful abode, to which all other animals came from elsewhere; and where it drank, they drank also of the same; now, after it had become venomous, by reason of God's curse, all beasts fled from its abode, and would not drink of the water it drank; but fled from it.[14]

The potential intimidation value of the serpent is suggested by the inclusion of a large serpent on the helmets of the Urartian soldiers, as considered also in Section 5.2.

If one of the supernatural creatures attributed to Tiamat might somehow correspond to the biblical cherubim, then their "unsparing weapons" could in this case be a reference to the "flaming sword which turned every way, to guard the way to the tree of life" (Genesis 3:24). That these various creatures or monsters are an important part of the Gilgamesh story is suggested also by the fact that the same text appears at multiple places in the epic.[15]

Actual arthropod scorpions are common in modern Turkey, but one might wonder about the imagined appearance of a creature referred to as a scorpion-man (line 141 in the above quotation) or the nature of any weapons with which he might be armed. Fortunately for the curious, there are ancient representations of scorpion-men. A Babylonian boundary stone of the twelfth century B.C. includes a carving of a scorpion-man in Register 5 armed with a bow and arrow, and a large serpent; and a picture of this stone is reproduced in Figure 10.1.[16] Unlike many supernatural creatures, the scorpion-man in this picture is seen to stand on two legs, while still having the standard poisonous tail of a more conventional scorpion. The representation of a scorpion-man armed with a bow and arrow on a boundary stone is consistent with the idea of the scorpion-man in the *Gilgamesh Epic* protecting the entrance to a path or territory that is off-limits to human beings. The biblical cherubim with their "flaming sword" had a similar responsibility in protecting the entrance to the garden of Eden (Genesis 3:24). Discussions of creatures such as scorpion-people and cherubim and their appearances are available.[17]

Causes And Effects Of The Flood

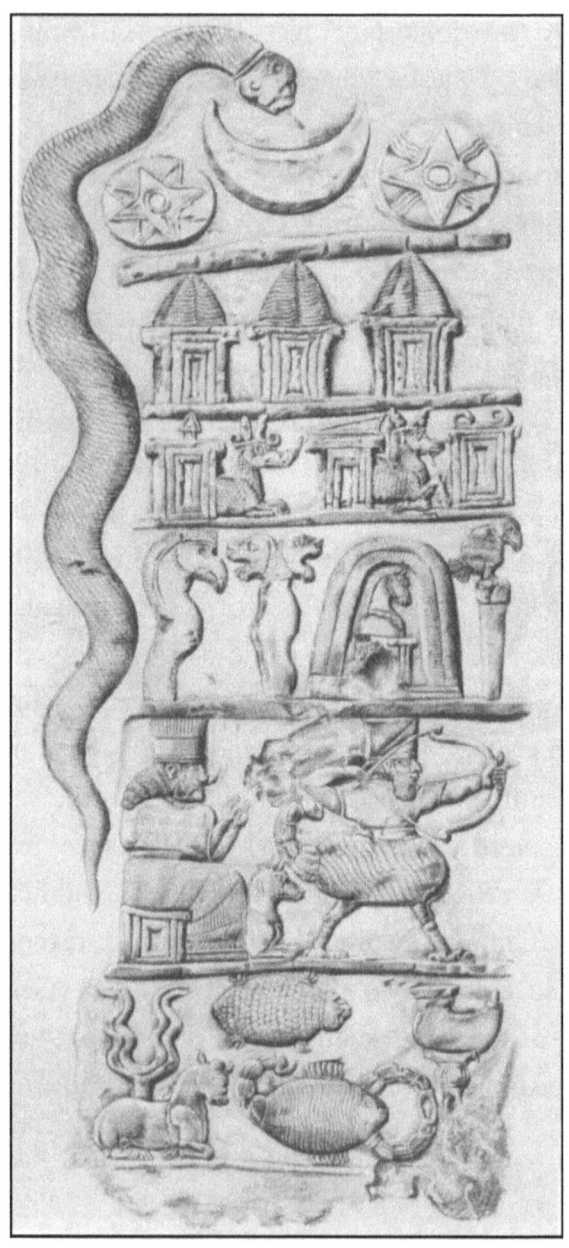

Figure 10.1. A Babylonian boundary stone of the twelfth century B.C., showing a scorpion man in Register 5, counting scenes downward from the top. (Reproduced with permission)

10.5 The Tigris Tunnel

In a quotation from the *Gilgamesh Epic* in Section 10.3, Gilgamesh is said to have obtained permission from the scorpion-man to travel the paths of the mountains: "[When] Gilga[mesh heard this], [He followed(?)] the word of [the scorpion-man]. Along the road of the sun [he went(?)]."[18] It might be tempting to imagine that "the road of the sun" was a path over the top of the mountains. If that were the case, then a modern parallel would seem to be the Going-to-the-Sun Road across the Rocky Mountains of Glacier National Park, Montana. However, there are two fundamentally different ways to cross steep hills or mountains. The first, as just suggested, is to go over the top of the mountains by any means necessary. The second is to go through the mountains via natural or artificial tunnels. Crossing mountains by automobile in real situations often includes a combination of steep roads and artificial tunnels. Thus, the Going-to-the-Sun Road over the Rocky Mountains also includes two automobile tunnels.

In the *Gilgamesh Epic* it is emphasized that at least an important part of Gilgamesh's travel across the mountains was in total darkness: "Dense is the dark[ness and there is no light]; Neither [what lies ahead of him nor what lies behind him] does it per[mit him to see]."[19] This strongly suggests that his crossing may have involved a tunnel, and given the early date that tunnel would probably have been natural rather than man-made. Support for such an unusual sounding possibility is provided also by the following statement: "[After he has traveled eleven double-hours he co]mes out before sun(rise). [After he has traveled twelve double-hours] it is light."[20]

Causes And Effects Of The Flood

Thus, the travelling involved eleven double-hours in total darkness. Depending on the terrain of his path in darkness, the total distance traveled might have been modest. This presupposes also that the information provided in the epic is a reasonable representation of some actual historical event. It would be especially helpful to identify some possible underground path on which also there is no light. While this may not seem a particularly likely possibility, as it happens there is even today a famous natural tunnel, referred to as the Tigris Tunnel, through which one might approach Lake Van from the southwest. This tunnel goes beneath what is called Korha Mountain.[21]

The Tigris Tunnel was explored by Taylor in 1862, and a few excerpts from his desciption are included here:

> About three miles below the sources the river [Dibeneh Su] enters a high cave, 80 feet high and two miles long, running north-east and south-west, and emerges from it near the village of Korkhar, at a point where the rocks are smooth and hard. Here, just outside the cave, on the right bank, and some twenty feet up the face of the rock, is the figure of an Assyrian king, with ten lines of a cuneiform inscription, in excellent preservation. Further inside the cave, but on an uneven and misshapen part of the rock, is another figure and inscription, but unfortunately, owing to the irregularities of the surface and other causes, in a nearly illegible state. During the spring floods, the river, confined in a narrow gorge with high perpendicular cliffs, comes down with immense force; the north-east end of the cave is naturally, therefore, a mass of fallen rock and smaller fragments; so, if at any time another inscription existed there, it must from these causes have disappeared long ago. I am inclined to believe that from the numerous *débris* which now choke the stream, and the appearance

10. The Gilgamish Epoch

through which it runs, this subterranean channel of the Tigris, or Dibeneh Su, extended close up to its sources, and thus gave some countenance to the fabulous length of its underground course as mentioned by Strabo....

The early Arab geographers seem to have been perfectly aware of this source of the Tigris, which they regarded as the main branch, and described it as being north of Miafarkeyn, and close to the castle of Dhul Karneyn. Near it is an immense stalactite cave, called Bakireyn, with innumerable passages branching off in every direction: the natives say it extends to Erzerum. Accompanied by a large party of Kurds, with torches and candles, I followed the main passage for about one hour, and was then forced to return without finding an opening at the other end.

The fanciful imagination of the Kurds had induced them to believe that the fantastic shapes of the stalactites [stalagmites] were representations of men and beasts, idols of an earlier age; and they thought my visit to a spot which they all avoid was a pilgrimage to the Pantheon of heathen ancestors.[22]

As quoted by Taylor, Rawlinson has commented as follows on the inscriptions referred to above:

Sir H. Rawlinson says one inscription and figure is that of Tiglath Pileser, B.C. 1110, and the other that of Ashur Izir Pal, B.C. 880, the King of the Nimroud Monolith. See Professor Rawlinson's 'Ancient Monarchies.' The inscription, as translated by Sir H. Rawlinson, is "By the grace of Asshur, Shamas and Iva the great gods, I Tiglath Pileser, King of Assyria, son of Asshur-ris-illim, King of Assyria, who was the son of Mutaggil-Nebo, King of Assyria, marching from the great sea of Akhiri (the Mediterranean) to the sea of Nairi (Lake Van), for the third time have invaded the country of

Nairi." This monument exhibits the earliest Assyrian sculpture known to exist, and is mentioned by Asshur-idanni-pal, the father of the black obelisk king, in his great inscription. – Professor Rawlinson's 'Monarchies,' vol. ii. p. 331.[23]

While the young ruler Gilgamesh may have visited the Tigris Tunnel as suggested above, his visit would have been many centuries before the visit by Tiglath Pileser I (1115-1076 B.C.), considered to be the first king to have had his image carved at the Tigris Tunnel.[24] Such image carvings by the ancient kings are sometimes said to have had the purpose of "marking territory,"[25] a terminology more commonly associated today with canines. Another image of Tiglath Pileser I was found in Lebanon near the Mediterranean Sea.[26] Interestingly, Tiglath Pileser I encountered the tunnel as he and his army journeyed from the Mediterranean Sea to Lake Van. He was the first Assyrian king to extend his kingdom as far as the Mediterranean.[27] The king mentioned above should not be confused with Tiglath Pileser III (745-727 B.C.), who is mentioned by name in 2 Kings 15:29, 16:7, and 16:10 – and also by the name Pul in 2 Kings 15:19 and 1 Chronicles 5:26.

10.6 Representations of the tunnel

The Tigris Tunnel is one of the ancient places for which we still today have pictorial representations. Shalmaneser III visited the Tunnel area in at least his seventh and fifteenth regnal years, or 852 and 844 B.C.[28] He left records of his visits in various places, and these include bronze reliefs from large timber gates at ancient Imgur-Enlil (Tell Balawat). Some general comments on

Relief Panel 10 are given below,[29] and the righthand portion of that panel is reproduced in Figure 10.2.

> The episode on Relief Panel 10, which narrates the king's seventh-year campaign, culminates with a ceremonial scene, understood as Shalmaneser's visit to the Source of the Tigris, where he received the submission of local kings, made offerings to the gods, celebrated a banquet and had his craftsmen carve his images and inscriptions on the rock faces (figure 11). In this outstanding scene, the cultic and ceremonial activity is depicted in great detail, while the rockfaces, the river and the caves from which the Tigris emerges are represented in articulate spatial specificity. In the upper register a sacrifice takes place, while within a large cave-like space the image of the king is carved attended by a high Assyrian official standing on a raised platform. In the lower register, while a series of sacrificial animals are being led to the scene from the left, the stele-shaped image of the king is carved by the artisan on the rock face. Given our knowledge of the actual topography of the Birkleyn caves, the interpretation of this scene as a depiction of those performative acts at the Source of the Tigris is compelling. The cave on the upper register represents a large and clear cave space, just as it is in the upper cave at the Birkleyn site, whereas the topography of the lower register reminds us of the multiple breaks in the bed-rock that covers the river. The spatial representation on the Balawat bronze bands is remarkably commensurate with the topography of Birklincay caves.[30]

Causes And Effects Of The Flood

Figure 10.2. Right-hand side of Relief Panel 10 of the Bronze door reliefs of Shalmaneser III at Tell Balawat (Imgur-Enlil). This relief includes scenes representing the visit of Shalmaneser to the Tigris Tunnel area in the king's seventh year 852 B.C. (Reproduced with permission)

More detailed comments by King relating to this same panel are also quoted here:

> A bull and a ram are being led forward for sacrifice before the image of Shalmaneser, which is being carved on the rock-face of the grotto, in front of the tunnel's mouth, by a sculptor standing on a block in the stream (Pl. LIX). The subterranean course of the river is conventionally shown by means of rectangular openings, through which men are seen wading waist-deep and carrying plants or torches. The trees, which appear to be growing in the stream and protruding from the openings, explain the convention: at a point near its mouth the roof

10. The Gilgamish Epoch

of the tunnel has fallen in, and one can still look down on to the stream from above through a wide opening, on the steep sides of which brushwood and small trees have found a footing. A sentry on the hill above the natural tunnel closes the register. The sacrificial scene in the Upper Register (Pl. LIX) is taking place at the head of a neighbouring valley. The objects which are usually explained as four rows of posts across the valley, may perhaps be altars of incense, the rising smoke from which is represented conventionally by disks.[31]

Concerning the last sentence in the above quotation, another possibility might also be suggested. Instead of altars of incense, the four objects in the "large cave-like space" could be stalagmites with the spots above them representing falling drops of water. One of the major caves in the Tigris Tunnel area has been identified by modern cave explorers as the Stalagmite Cave.[32] The striking appearance of the stalagmites in this cave is suggested by Taylor's comment quoted above: "The fanciful imagination of the Kurds had induced them to believe that the fantastic shapes of the stalactites [stalagmites] were representations of men and beasts, idols of an earlier age." The noteworthy appearance of these stalagmites at the time of Shalmaneser's visit may have prompted him to include their representation in the bronze reliefs documenting his visit. A photograph of these stalagmites is shown in Figure 10.3.[33]

Causes And Effects Of The Flood

Figure 10.3. A cave with stalactites and stalagmites in the Tigris Tunnel region.

It may be recalled that, in an early episode of the *Gilgamesh Epic*, Enkidu and Gilgamesh had encountered and cut off the head of the terrible ogre Humbaba, who was the guardian of a vast cedar forest (Tablets III-V). The description of this cedar forest would seem to imply that the ogre had lived in what is now Lebanon near the Mediterranean Sea. In a later episode Gilgamesh is found to be bemoaning the death of Enkidu and undertaking a quest to obtain eternal life (Tablet IX). It isn't entirely clear how any historical events underlying these episodes of the epic might have followed after each other, so the direction by which Gilgamesh began his approach to the home of Utnapishtim isn't known with certainty. If he had been coming from the direction of Lebanon, where Humbaba had previously been killed, then his path would have been almost the same as that recorded by Tiglath Pileser I many centuries later as he marched from the vicinity of the Mediterranean Sea to Lake Van. Tiglath Pileser I and Shalmaneser III had intentionally visited the Tigris Tunnel, and Gilgamesh may at an earlier date have done so as well.

As seen in Taylor's report quoted above, he suggests that the Tigris Tunnel could be the underground water course of the Tigris River that was imagined by the ancient geographers. Concerns with this concept include the possibility that the Tigris Tunnel may have been known by them to be traversable, and at least the immediate source of its water could have been known not to be the lake Thospites (Lake Van). On the other hand, the association of this tunnel with the underground path taken by Gilgamesh in total darkness may be quite plausible.

In the Gilgamesh story as quoted earlier in this section, the path in darkness was referred to as the "road of the sun," emphasizing perhaps the welcome return of daylight upon emergence from the tunnel. In some Hittite texts the Tigris Tunnel is sometimes referred to as the "Divine Road of the Earth," perhaps indicating more explicitly that the path was underground. The tunnel is also said today to be nine hundred meters (0.559 miles) in length.[34] Taylor, quoted above, estimated the length at two miles and believed that it could have been much longer in the distant past.

10.7 The fruit trees

The quotation from the *Gilgamesh Epic* in Section 10.3 ended with the emergence of Gilgamesh back into daylight. The trans-

Causes And Effects Of The Flood

lation and interpolations of the text on this subject have been reported as follows:

> 45. [After he has traveled eleven double-hours he co]mes out before sun(rise).
> 46. [After he has traveled twelve double-hours], it is light.
> 47. Be[fore him stand] shrubs of (precious) [stone]s; as he sees (them) he draws nigh.
> 48. The carnelian bears its fruit;
> 49. Vines hang from it, good to look at.
> 50. The lapis lazuli bears . . .
> 51. Also fruit it bears, pleasant to behold.[35]

It would seem that immediately after emerging Gilgamesh encountered an unusual garden. The above brief text is all the information we have about the fruit trees in this garden. As indicated, there were two kinds of trees called carnelian and lapis lazuli, and both kinds bore fruit. It is difficult to estimate the original author's intention, but a "restoration" in the text above suggests a strange garden in which all of the plants and their fruit are made of stone.

It might be mentioned as a starting point that the mysterious garden may not have been in this part of the original story at all. It can be recalled from Section 10.5 that when Tiglath Pileser I left the vicinity of the Tigris Tunnel he moved on to Lake Van. One could imagine that if a supernatural garden had been encountered in his journey it would have been mentioned somewhere in his annals. Nevertheless, a few comments on a possible non-supernatural interpretation will be suggested here. It may be noted that Heidel has carefully defined the use of special

10. The Gilgamish Epoch

symbols in his translation of the *Gilgamesh Epic*, and two of these symbols are the following:[36]

> Parentheses "enclose elements not in the original but desirable or necessary for a better understanding in English."

> Brackets "enclose restorations in the cuneiform text."

With these definitions it seems less than certain that in the original text shown above the author intended anything more than rather conventional fruit-bearing shrubs. Only two types are indicated, carnelian and lapis lazuli, and these color names may refer to the colors of the fruit of ordinary plants.

The biblical story of the garden of Eden also emphasizes the beauty of the plants in that garden: "And out of the ground the Lord God made to grow every tree that is pleasant to the sight and good for food . . ." (Genesis 2:9). Thus, the biblical trees are "pleasant to the sight," while in the Gilgamesh story carnelian plants are "good to look at" and the lapis lazuli fruit is "pleasant to behold." An imaginative reader could consider the possibility that an author contributing to the Gilgamesh story wanted explicit names for attractive plants bearing fruit of red (carnelian) or blue (lapis lazuli) color, so he simply used the color names.

If the missing word in the epic as quoted above could have represented wood instead of stone, for example, the meaning of the text would have been changed entirely. In that case, the plants could have been actual fruit-bearing trees of wood instead of mysterious tree-like structures made of stone. In this connection it is of interest to consider a few lines from the preceding Tablet VIII:

> Table VIII:
> Column v
> 45. As soon as the first shimmer of morning beamed forth, Gilgamesh fashioned [. . . .].
> 46. He brought out a large table of elammaqu-wood.
> 47. A bowl of carnelian(?) he filled with honey.
> 48. A bowl of lapis lazuli he filled with butter.[37]

This text in the epic follows upon the death of Gilgamesh's best friend Enkidu, and it precedes the beginning of Gilgamesh's quest for immortality. Unfortunately, the column following this text is mostly destroyed. However, it seems that Gilgamesh had (or fashioned) a table of wood, a bowl of carnelian, and a bowl of lapis lazuli. He filled the carnelian bowl with honey, and he filled the lapis lazuli bowl with butter. Perhaps he was setting out these foods in some form of worship of the gods or as a memorial to his friend Enkidu. If the words carnelian and lapis lazuli could refer to actual fruit bearing trees, then one could easily imagine further that the bowls were made of the wood of those trees rather than somehow being shaped from large gemstones. It is striking that in this earlier part of the story only the names carnelian and lapis lazuli are mentioned, and these are the same two names used later in identifying the fruit trees growing in the garden.

The connection between the carnelian fruit tree of the Gilgamesh story and a likely fruit tree of Eden also finds more direct support, including a possible identification of the tree species. Thus, the carnelian or cornelian cherry (Cornus mas) is well-known today in the west as an attractive ornamental plant. It is a species of dogwood and bears carnelian-colored cherry-like fruit. Several explanations have been given for the origin of the names:

10. The Gilgamish Epoch

Botanically, cornelian cherry is a species of dogwood, unrelated to grocers' cherries. The word "cornelian" refers to the similarity in color of the fruit to cornelian (or carnelian) quartz, which has a waxy lustre and a deep red, reddish-white, or flesh red color. (Carnis is Latin for flesh.)[38]

The common name, Cornelian Cherry, or shortened to "Cornel," was given to the plant because its cherry-like fruits are about the color of the gemstone carnelian.[39]

Thus, the carnelian cherry may have been named for the resemblance of the color of its fruit to the carnelian gemstone. This resemblance was likely well known in ancient times, and the same naming convention may have been employed. The gemstone called carnelian was known in the kingdom of Urartu, and small beads of this stone resembling carnelian cherries were, for example, used in necklaces.[40] The tart cherry-like fruit of carnelian cherry trees has also had many culinary and medical applications.

It is natural to wonder whether the modern carnelian cherry could have any other connections to the carnelian fruit tree of the *Gilgamesh Epic*. The description quoted above seems to suggest that the fruit was to be found on relatively narrow branches (vines) of the carnelian cherry plants,[41] and this is also true of the fruit of the modern carnelian cherry. It is significant that the carnelian cherry is considered to be native to southern Europe, Armenia, Azerbaijan, Georgia, Iran, and southwest Asia. It is also striking that the carnelian cherry is said to be the national fruit of Turkey, because as noted previously a likely location for the garden of Eden and also the landing place of the ark is within the area known today as Turkey. The carnelian cherry thrives there at present, and it may well have done so also four thousand years ago as a beautiful and useful fruit tree in the garden encountered by Gilgamesh.

As suggested above it might be worth considering whether a different restoration of the text could allow the trees in the garden to have been of "(precious) [wood]s" rather than of "(precious) [stone]s." Having tentatively identified one of these trees as the carnelian cherry, one might enquire whether the "precious" wood of that tree ever actually had any practical applications. This consideration also reveals another explanation for the name of this dogwood:

> The best known of the Cornelian Cherries, and the one to which the name is most properly applied, Cornus mas, has been valued in Europe as a utilitarian plant since classical Greek and Roman times, and was mentioned in the writings of Homer and Virgil. The Latin name now given the plant is derived from the names applied to it in ancient times, and the common name also has been long established. Both are derived from its utilitarian attributes.
> The wood of Cornus mas has been valued over the centuries for its hardiness, durability, and flexibility. Although put to more mundane use in recent times, such as for the manufacture of wheel spokes, ladder rungs, and tool handles, it was favored by the Romans to make the shafts of javelins. The modern generic name Cornus is the name they used for the plant, its derivation being from the Latin cornu, meaning "horn," because of the hardness of the wood."[42]

That the Latin word cornu is related to "horn" is familiar from its use in cornucopia or "horn of plenty." The numerous uses to which the unique wood of the carnelian cherry has been put over the millennia could be consistent with the interpretation "precious wood" suggested above. The trunks of carnelian cherry trees can be over a foot in diameter, and the idea of a bowl being made of the durable wood of this tree may not be implausible.

One can also wonder if some other unknown fruit tree may long ago have been identified with lapis lazuli and might also have had similar useful applications. Lapis lazuli is known for its blue color, so one might imagine that a fruit tree or shrub with this name would have blue-colored fruit. There are, of course, many plants with blue-colored fruit, including for example blueberries and some grapes. After the flood Noah is said to have used grapes in wine-making: "Noah was the first tiller of the soil. He planted a vineyard; and he drank of the wine, and became drunk. . . ." (Genesis 9:20-21). Grapes have long flourished in Armenia, and the oldest known winery in the world was found in the land that came to be called Urartu.[43] Other blue-fruited plants could be considered as well, and a particular one will be mentioned in Section 11.7.

The idea of Gilgamesh encountering fruit trees as he travelled from the Tigris Tunnel toward a sea (as summarized in Section 10.3) also is compatible with the geography in the book *Adam and Eve*. As suggested in previous chapters, the sea in Noah's flood may correspond to Lake Van. If so, the fruit trees encountered by Gilgamesh would have been located part way between the Tigris Tunnel and Utnapishtim's post-flood home beside Lake Van.

In the interpretation considered in Section 5.5, after Adam and Eve left the garden of Eden they lived in a cave home near or on Mt. Nemrut. While they lived there, they learned of a fertile land downhill and to the west. Later Cain and his wife moved away from their parents Adam and Eve to the same rich land:

> Then the Word of God came and said unto him, "O Adam, go down to the westward of the cave, as far as a land of dark soil, and there thou shalt find food." And Adam hearkened unto the Word of God, took Eve, and went down to a land of dark soil, and found there wheat growing, in the ear and ripe, and figs to eat; and Adam rejoiced over it.[44]

> He [Cain] then went down to the bottom of the mountain, away from the garden [of Eden], near to the place where he had killed his brother. And in that place were many fruit trees and forest trees.[45]

This place of dark soil and fruit trees, if factual, could be related to the garden of fruit trees encountered by Gilgamesh as he approached Lake Van from the southwest a few centuries later.

10.8 Conclusion

After the death of his young friend Enkidu, Gilgamesh began his seemingly desperate quest to find the secret of eternal life. That quest led him to meet the scorpion-people who were guardians of a path through the mountains. With their permission, he travelled that path in complete darkness and arrived at a mysterious garden in which were two (or possibly more) types of fruit tree. Gilgamesh's quest will continue in Chapter 11 as he meets Siduri by the side of a sea, and she directs him onward across the sea to find Utnapishtim/Noah.

10. The Gilgamish Epoch

1. A. Heidel, *The Gilgamesh Epic and Old Testament Parallels* (The University of Chicago Press, Chicago, 1963), p. 66, Tablet IX, Column iv, lines 33-36 (translator indications deleted, full text in Section 10.3).
2. W. H. Stiebing, Jr., *Ancient Near Eastern History and Culture*, Second Edition (Pearson Longman, New York, 2009), pp. 44-46.
3. A. Heidel, op. cit., p. 16, Tablet I, Column i, lines 1-8.
4. L. W. Casperson, *Patterns of Biblical Chronology* (Westbow Press, Bloomington, Indiana, 2012), Chapter 21, p. 542, reference 24; see also Chapter 9 of this study.
5. A. Heidel, op. cit., pp. 14, 15.
6. Ibid., p. 102.
7. L. W. Casperson, op. cit., Sec. 24.6, "Abram in Canaan and Egypt," pp. 610-618; p. 618.
8. A. Heidel, op. cit., pp. 8-9.
9. W. H. Stiebing, Jr., op. cit., p. 59.
10. A. Heidel, op. cit., pp. 65-68, Tablet IX, Column ii, line 1 – Column v, line 46.
11. A. Heidel, *The Babylonian Genesis, The Story of Creation,* Second Edition (The University of Chicago Press, Chicago, 1951), p. 10.
12. Ibid., p. 14.
13. Ibid., pp. 23-24, Tablet I, lines 132-145.
14. *Adam and Eve*, Translated by S. C. Malan from the Ethiopic edition edited by E. Trumpp, included in *The Forgotten Books of Eden*, Edited by Rutherford H. Platt, Jr. (Alpha House, Inc., 1927); Book 1, 17:3-6.
15. A. Heidel, *The Babylonian Genesis*, op. cit., including pp. 23-24, Tablet I, lines 132-145; p. 26, Tablet II, lines 19-32; pp. 31-32, Tablet III, lines 23-36; and pp. 33-34, Tablet III, lines 81-94.
16. Ibid., Figure 4 at the end of the book.
17. See for example, A. Wood, *Of Wings and Wheels, A Synthetic Study of the Biblical Cherubim* (Walter de

Gruyter, Berlin, 2008), Part III: The Archaeological Evidence, pp. 157-204.
18. A. Heidel, *The Gilgamesh Epic*, op. cit., p. 67, Tablet IX, Column iv, lines 44-46.
19. Ibid., p. 67, Tablet IX, Column iv, lines 48-49.
20. Ibid., p. 68, Tablet IX, Column v, lines 45-46.
21. Ö. Harmansah, " 'Source of the Tigris'. Event, place and performance in the Assyrian landscapes of the Early Iron Age," *Archaeological Dialogues*, Volume 14, Number 2, pp. 179-204 (2007); p. 185.
22. J. G. Taylor, "Travels in Kurdistan, with notices of the sources of the eastern and western Tigris, and ancient ruins in their neighbourhood," *Journal of the Royal Geographical Society of London*, Volume 35, pp. 21-58 (1865); pp. 41-43.
23. Ibid., note ‡, p. 41.
24. Ö. Harmansah, op. cit., p. 190.
25. Ibid., p. 196.
26. Ibid., p. 192; S. Yamada, *The Construction of the Assyrian Empire: A Historical Study of the Inscriptions of Shalmaneser III (859-824 B.C.) Relating to His Campaigns to the West* (Koninklijke Brill NV, Leiden, The Netherlands, 2000), p. 274, note 5.
27. R. E. Dupuy and T. N. Dupuy, *The Harper Encyclopedia of Military History: From 3500 BC to the Present*, Fourth Edition (HarperCollins, New York, NY, 1993), p. 9.
28. Ö. Harmansah, op. cit., pp. 188-189.
29. Ibid., pp. 193-195.
30. Ibid., Figure 12, p. 194.
31. L. W. King, *Bronze Reliefs from the Gates of Shalmaneser King of Assyria B.C. 860-825* (The British Museum, London, 1915), p. 31.
32. A. C. Waltham, "The Tigris Tunnel and Birkleyn caves, Turkey," *Bulletin of the British Cave Research Association*, Volume 14, pp. 31-34 (1976).

33. I. Atalay, S. Karadoan, A. Yildirim, "Karstification and ground river system in SE Anatolia: A key study from Birkleyn Cave system," *Proceedings of the 7th Turkey-Romania Geographical Academic Seminar, June 1-9, 2010, Antalya-Turkey*, pp. 81-92; p. 91, Photo 4.
34. Ö. Harmansah, op. cit., pp. 184-185, 196.
35. A. Heidel, *The Gilgamesh Epic*, op. cit., p. 68, Tablet IX, Column v, lines 45-51.
36. Ibid., p. ix.
37. Ibid., p. 64, Tablet VIII, Column v, lines 45-48.
38. L. Reich, "Cornelian cherry: from the shores of ancient Greece" *Arnoldia*, Volume 56, Number 1, pp. 2-7 (Winter 1996); p. 3.
39. R. E. Weaver, Jr., "The cornelian cherries," *Arnoldia*, Volume 36, Number 2, pp. 50-56 (March 1976); p. 52.
40. B. B. Piotrovsky, *The Ancient Civilization of Urartu* (Cowles Book Company, Inc., 488 Madison Avenue, New York, N.Y. 10022), picture on p. 181, caption 113 on p. 216.
41. A. Heidel, *The Gilgamesh Epic*, op. cit., p. 68, Tablet IX, Column v, lines 47-49.
42. R. E. Weaver, Jr., op. cit., pp. 51-52.
43. J. Owen, "Earliest known winery found in Armenian cave," *National Geographic Daily News*, 10 January 2011.
44. *Adam and Eve*, op. cit., Book 1, 66:7-8.
45. Ibid., Book 2, 1:7-8.

11. UTNAPISHTIM'S FLOOD

"Six days and six nights the wind blew, the downpour, the tempest, and the flood overwhelmed the land. When the seventh day arrived, the tempest, the flood, which had fought like an army, subsided in its onslaught. The sea grew quiet, the storm abated, the flood ceased."[1]

11.1 Introduction

The preceding chapter introduced the young king Gilgamesh and his quest for renewed life. In this chapter Gilgamesh travels by boat across a large sea to reach the home of Utnapishtim/Noah and his wife. The story in this chapter may perhaps be considered a sequel to the story of Noah's flood. At one place along the way Gilgamesh has to cross the dangerous "waters of death." Eventually he arrives at the home of Utnapishtim, who narrates to him the story of the great flood. There are many similarities between the flood of Noah and the flood of Utnapishtim. After narrating the flood story to Gilgamesh, Utnapishtim tells him how he might find a supernatural tree that will provide him renewal of life. When a serpent takes the tree from him he abandons his quest and returns to his home city of Uruk.

11. Utnapishtim's Flood

The story here begins in Section 11.2 with the barmaid Siduri telling Gilgamesh how he might reach Utnapishtim, who had previously been granted eternal life. To do this Gilgamesh must cross a large, and in one place toxic, sea. The boatman Urshanabi agrees to take Gilgamesh across the sea to meet Utnapishtim. After his arrival, Utnapishtim tells Gilgamesh the story of a great flood seemingly associated with the same sea, and this story is quoted in Section 11.3. Some of the similarities and differences between Utnapishtim's flood story and the biblical story of Noah's flood are considered in Section 11.4. The flood in the *Gilgamesh Epic* also has parallels in other flood stories from the Middle East, and a few of those are mentioned in Section 11.5. The tree of life in the biblical garden of Eden seems also to have a parallel in the epic, and this possibility is considered in Sections 11.6 and 11.7.

11.2 Crossing the sea
After leaving the garden, Gilgamesh encountered "Sid[uri, the barmaid], who dwells by the edge of the sea,"[2] as in the summary quoted in Section 10.3. He asked her for advice on how he might reach Utnapishtim and find out from him how to obtain eternal life. She replied that only the gods had eternal life. In an appealing admonition she urged Gilgamesh to be content and to enjoy the time that he had left:

> Gilgamesh, whither runnest thou? The life which thou seekest thou wilt not find; for when the gods created mankind, they allotted death to mankind, but life they retained in their keeping. Thou, O Gilgamesh, let thy belly be full; day and night be thou merry; make every

day a day of rejoicing. Day and night do thou dance and play. Let thy raiment be clean, thy head be washed, and thyself be bathed in water. Cherish the little one holding thy hand, and let the wife rejoice in thy bosom. This is the lot of mankind.[3]

This advice is not so different from some biblical teachings, and a few verses are included here for consideration:

> Go, eat your bread with enjoyment, and drink your wine with a merry heart; for God has already approved what you do. Let your garments be always white; let not oil be lacking on your head. Enjoy life with the wife whom you love, all the days of your vain life which he has given you under the sun, because that is your portion in life and in your toil at which you toil under the sun (Ecclesiastes 9:7-9).

When Gilgamesh persisted, Siduri told him that with help he might be able to travel across the sea including the "waters of death" and reach Utnapishtim. She also gave him advice on how to proceed. The essential parts of this story are quoted here.[4]

Tablet X
Column ii

15. [Gilgamesh furthermore] said to her, to the barmaid:
16. "[Now], barmaid, which is the way to Utnapi[shtim]?
17. [What are] the directions? Give me, oh, give me the directions!
18. If it is possible, (even) the sea will I cross!
19. (But) if it is not possible, I will roam over the steppe."
20. The barmaid said to him, to Gilgamesh:
21. "Gilgamesh, there never has been a crossing;
22. And whoever from the days of old has come thus far has not been able to cross the sea.

11. Utnapishtim's Flood

⋮

24. Difficult is the place of crossing (and) very difficult its passage;
25. And deep are the waters of death, which bar its approaches.
26. Where, Gilgamesh, wilt thou cross the sea?
27. (And) when thou arrivest at the waters of death, what wilt thou do?
28. Gilgamesh, there is Urshanabi, the boatman of Utnapishtim.

⋮

30. [Hi]m let thy face behold.
31. [If it is possi]ble, cross over with him; if it is not possible, turn back (home)."

⋮

Column iii

32. Gilgamesh (furthermore) said to him, to [Urshanabi]:
33. "Now, Urshanabi, which [is the road to Utnapishtim]?
34. What are the directions? Give me, oh, give [me the directions]!
35. If it is possible, (even) the sea will I cross; (but) if it is not possible, [I will roam over the steppe]."

⋮

47. Gilgamesh and Urshanabi (then) boarded [the ship].
48. They launched the ship on the billows and [glided along].

Gilgamesh and Urshanabi arrived by boat at a place called the waters of death, survived the crossing of those waters, and eventually reached the home of Utnapishtim and his wife beside the sea. As recalled in Section 10.5, Tiglath Pileser I had traveled near the area of the Tigris Tunnel, and the possibility may be considered that Gilgamesh had in part traveled much the same

Causes And Effects Of The Flood

route many centuries earlier. It also may be inferred from the quotations above that if the very determined Gilgamesh had been unable to cross the sea, he would have attempted to reach Utnapishtim by land ("I will roam over the steppe").

Several locations have been proposed for the dwelling of Utnapishtim. In the quotation earlier in Section 10.3, Heidel clearly considered that there may be a historical basis or at least a geographical basis for the story of Gilgamesh's quest. However, he suggested that Utnapishtim's home may have been at some distance from Lebanon across the Mediterranean Sea. As discussed here previously, the Bible can be interpreted to suggest a focus on the area between Mesopotamia and the land of Ararat for both the Garden of Eden and the flood story. Therefore it seems reasonable to continue here to consider the vicinity of Lake Van as a possible location for the flood of the Gilgamesh story and the later home of Utnapishtim. Roaming over the steppe could then be interpreted as walking around the lake to reach Utnapishtim rather than crossing to him by boat.

One of the characteristics of the waters of the sea to be crossed by Gilgamesh is that somewhere in the crossing they would encounter "the waters of death," a place where the water was apparently highly toxic. In this context it may be appropriate to mention the Dead Sea, which is located on the eastern side of Palestine and also has no outlet (being the lowest point on earth). The maximum length of the Dead Sea is thirty-one miles and its area is about two hundred thirty square miles. That sea is called "dead" because no fish can survive in it. It seems possible that at some specific locations (perhaps especially near Mt. Nemrut) and at some times in the past the waters of Lake Van could have been

11. Utnapishtim's Flood

known as the waters of death, harmful even to the touch. Urshanabi, the boatman of Utnapishtim, agreed to help bring Gilgamesh across the lake. Urshanabi also advised Gilgamesh on the importance of keeping his hands out of the water:

Tablet X
Column iv
1. Urshanabi [said] to him, [to Gilgamesh]:
2. "Press on, Gilgamesh! [Take a pole (for thrusting)...]
3. Let not thy hand touch the waters of death [...].⁵

Such toxicity could have resulted from chemical effects associated with its location in a region of volcanic activity, and some eruptive events may have occurred beneath the surface of the lake.

In a discussion of the general toxicity of the waters of Lake Van for plants along the shoreline in more modern times, Lynch and Oswald discussed their observations of plant death on those occasions when the water level chances to rise. They were especially interested in such effects in the vicinity of lagoons along the margin of the lake:

> On the margin or even in the bed of such lagoons one might often see a group of willows. Some had been immersed a foot or two by the rise in the waters; and, while their neighbours on dry land were green and thriving, these were quite dead. The most notable example was observed by Oswald within the little broken-down crater on the southern shore opposite Akhlat. It receives the lake within its enfolding arms. We have called it Sheikh Ora after a little village of that name which was discovered in its south-east corner. Oswald sailed across to examine this interesting spot while I was busily

engaged at Akhlat. Between the village and the water he came across a small grove of willows upon which the lake had gained. Those above the water line were evidently flourishing; but those which stood in the lake had been killed and their bark withered, so that many of the stems were quite gaunt and bare. The average diameter of the trunks of the dead and the living was not appreciably different. It was therefore not a question of an advance of the lake dating back very many years. On the other hand there had been time for the chemical properties of the water to exercise their destructive effect.[6]

In the Sheikh Ora crater a giant mulberry, which may have been some 500 years old, was standing with half its roots in the water and was already doomed.[7]

It is somewhat striking that what was considered "[t]he most notable example" of these plant poisoning effects occurred in the Sheikh Ora crater "on the southern shore opposite Akhlat." This location, close to Mt. Nemrut, would be near where Gilgamesh, approaching from the southwest, would first have encountered the lake. Further, the largest plant victim mentioned was a giant 500-year-old mulberry, also in the Sheik Ora crater.

Assuming some historicity to the Gilgamesh story of the sea crossing, it is tempting to imagine that the toxic waters at that much earlier time might have been highly acidic because of their association with an active underwater volcanic vent:

Can lakes near volcanoes become acidic enough to be dangerous to people and animals?

Yes. Crater lakes atop volcanoes are typically the most acid, with pH values as low as 0.1 (very strong acid). Normal lake waters, in contrast, have relatively neutral pH values near 7.0. The crater lake at El Chichon volcano in Mexico had a pH of 0.5 in 1983 and Mount Pinatubo's crater lake had a pH of 1.9 in 1992. The acid waters of these lakes are capable of causing burns to human skin but are unlikely to dissolve metal quickly. Gases from magma that dissolve in lake water to form such acidic brews include carbon dioxide, sulfur dioxide, hydrogen sulfide, hydrogen chloride, and hydrogen fluoride.[8]

The presence of such waters supports the idea that the sea beside which Utnapishtim/Noah dwelt and on which the ark had floated was in fact Lake Van with its adjacent volcanoes and its still existing toxic waters. In any case the crossing was said to have been successful.

11.3 The Gilgamesh flood story

After Gilgamesh arrived at the home of Utnapishtim he was impatient to learn the secret of eternal life, but Utnapishtim began his response to Gilgamesh's questioning by relating the story of the flood. To convey a better sense of the form of the flood story in the *Gilgamesh Epic*, a few lines will be quoted here from Heidel's translation. The nature of possible similarities and differences in comparison to the biblical flood will be evident, and those will be discussed in the following section. This segment of the story relates to the period from the start of the actual rainfall associated with the flood to the sacrifices made after the ark had landed. The construction of the ark was described in preceding lines that are not reproduced here. The story is narrated to Gilgamesh by Utnapishtim, the Babylonian representation of Noah. Footnotes of Heidel (and others) are abbreviated and displayed in the text within braces:

Tablet XI

96. As soon as the first shimmer of morning beamed forth,
97. A black cloud came up from out the horizon.
98. Adad {god of storm and rain} thunders within it,
99. While Shullat and Hanish {bodyguards of Adad} go before,
100. Coming as heralds over hill and plain;
101. Irragal {or Nergal, god of the underworld} pulls out the masts;
102. Ninurta {god of war and lord of wells and irrigation works} comes along (and) causes the dikes to give way;
103. The Anunnaki {judges in the underworld} raised (their) torches,
104. Lighting up the land with their brightness {lightning};
105. The raging of Adad reached unto heaven {thunder}
106. (And) turned into darkness all that was light.
107. [....] the land he broke (?) like a po[t(?)].
108. (For) one day the tem[pest blew].
109. Fast it blew and [....].
110. Like a battle [it ca]me over the p[eople].
111. No man could see his fellow.
112. The people could not be recognized from heaven.
113. (Even) the gods were terror-stricken at the deluge.
114. They fled (and) ascended to the heaven of Anu {god of the sky};
115. The gods cowered like dogs (and) crouched in distress(?).
116. Ishtar cried out like a woman in travail;
117. The lovely-voiced Lady of the g[ods] lamented:
118. "In truth, the olden time has turned to clay {is past},
119. Because I commanded evil in the assembly of the gods!

11. Utnapishtim's Flood

120. How could I command (such) evil in the assembly of the gods!
121. (How) could I command war to destroy my people,
122. (For) it is I who bring forth {give birth to} these my people!
123. Like the spawn of fish they {people's bodies} (now) fill the sea!"
124. The Anunnaki-gods wept with her;
125. The gods sat bowed (and) weeping.
126. Covered were their lips . . . . .
127. Six days and [six] nights
128. The wind blew, the downpour, the tempest, (and) the flo[od] overwhelmed the land.
129. When the seventh day arrived, the tempest, the flood,
130. Which had fought like an army, subsided in (its) onslaught.
131. The sea grew quiet, the storm abated, the flood ceased.
135. I opened a window, and light fell upon my face. {line transposed}
132. I looked upon the sea, (all) was silence,
133. And all mankind had turned to clay {all were dead};
134. The . . . . {sea} was as level as a (flat) roof.
136. I bowed, sat down, and wept,
137. My tears running down over my face.
138. I looked in (all) directions for the boundaries of the sea.
139. At (a distance of) twelve (double-hours) there emerged a stretch of land.
140. On Mount Nisir {or Nimush} the ship landed.
141. Mount Nisir held the ship fast and did not let (it) move.
142. One day, a second day Mount Nisir held the ship fast and did not let (it) move.

Causes And Effects Of The Flood

143. A third day, a fourth day Mount Nisir held the ship fast and did not let (it) move.
144. A fifth day, a sixth day Mount Nisir held the ship fast and did not let (it) move.
145. When the seventh day arrived,
146. I sent forth a dove and let (her) go.
147. The dove went away and came back to me;
148. There was no resting-place, and so she returned.
149. (Then) I sent forth a swallow and let (her) go.
150. The swallow went away and came back to me;
151. There was no resting-place, and so she returned.
152. (Then) I sent forth a raven and let (her) go.
153. The raven went away, and when she saw that the waters had abated,
154. She ate, she flew about, she cawed, (and) did not return.
155. (Then) I sent forth (everything) to the four winds and offered a sacrifice.
156. I poured out a libation on the peak of the mountain.
157. Seven and (yet) seven kettles I set up.
158. Under them I heaped up (sweet) cane, cedar, and myrtle.
159. The gods smelled the savor,
160. The gods smelled the sweet savor.
161. The gods gathered like flies over the sacrificer.[9]

While this quotation should convey a sense of the style and scope of the Gilgamesh flood story, many preceding and succeeding parallels are omitted in the above brief excerpt.

11.4 Comparison with the biblical flood story

For some purposes it may be sufficient here to present a summary table of a few of the points of comparison between the biblical and Gilgamesh flood versions. The last column of Table 11.1

11. Utnapishtim's Flood

may be considered to be mostly a condensed version of Heidel's conclusions.[10] Some of the subjects in the table are of particular interest here, and more details relating to those topics are included in the following paragraphs.

Table 11.1. Comparison of the biblical and Gilgamesh flood stories.

Subject	Bible	Gilgamesh Epic, etc.
Author of flood	God	Non-unanimous gods
Reason for flood	Depravity of humans	Noisiness of humans
Position of hero	Noah tenth leader from creation	Utnapishtim tenth king of Babylon
Home of hero	Not identified	Shuruppak, Babylonia
Saving of hero	Righteousness	Sacrifices to gods
Announcing of flood	Warning from God	Prophetic dream
The ark	Boxlike, many rooms	Boxlike, many rooms
Occupants of ark	Noah, family, animals	Utnapishtim, wealth, family, friends, animals
Causes of flood	Rain, earthquakes, underground waters	Rain, earthquakes, overflow of rivers, failure of dikes
Area flooded	Entire earth	Large sea
Duration in ark	About one year	Many days
Landing place of ark	Mountains of Ararat	Mount Nisir
Bird episodes	Raven, three doves	Dove, swallow, raven

Causes And Effects Of The Flood

When the Gilgamesh flood story is compared to the Genesis version, several similarities and differences between the accounts are evident. A few of these are considered in more detail in the subsections below:

A. Reason for the flood

The reason for the flood shown in the preceding table is made very clear in the biblical version:

> The Lord saw that the wickedness of man was great in the earth, and that every imagination of the thoughts of his heart was only evil continually. And the Lord was sorry that he had made man on the earth, and it grieved him to his heart. So the Lord said, "I will blot out man whom I have created from the face of the ground, man and beast and creeping things and birds of the air, for I am sorry that I have made them" (Genesis 6:5-7).

The corresponding cause for the flood in the Gilgamesh story can be inferred from the Atrahasis form of this epic:

> [T]he epic states that, when the people had multiplied and, apparently, had become prosperous, they became so noisy as to deprive Enlil {storm god} of his sleep. In an attempt to quiet them, Enlil sent plague after plague. But, in the final analysis, it was all of no avail; mankind became more numerous (and evidently more noisy) than before. In utter exasperation, Enlil at last sent the flood to destroy them all and to put an end to their unbearable noise.[11]

A different but more plausible reason for the flood is provided by *Jasher*. After the heavy rainfall began, the people

11. Utnapishtim's Flood

who were locked outside of the ark sought to get in so that they wouldn't drown:

> And they called to Noah, saying, open for us that we may come to thee in the ark – and wherefore shall we die?
> And Noah, with a loud voice, answered them from the ark, saying, have you not all rebelled against the Lord, and said that he does not exist ? and therefore the Lord brought upon you this evil, to destroy and cut you off from the face of the earth.[12]

Thus the people who were locked out of the ark had been guilty of what might have been considered the greatest sin of all – declaring that the Lord doesn't exist. In modern terminology it might be said that atheists weren't permitted in the ark.

B. Ancestry of the hero

One vague similarity between the versions concerns the ancestry of the flood hero. In the biblical flood story, Noah was a member of the tenth generation from creation (Genesis 5:1-29). On the other hand, "According to Berossus, the deluge hero was the tenth prediluvian king in Babylonia."[13]

C. Physical causes of the flood

The causes of the flooding in the Bible and in the *Gilgamesh Epic* are of particular interest. Section 4.3 provides some idea of how the normal annual snow melt can provide for a more or less periodic annual inundation in the vicinity of the Tigris and Euphrates Rivers. However, this inundation would usually be incapable of explaining the flood conditions described in Genesis and the *Gilgamesh Epic*. It is possible, though, that the exceptional frost rings noted previously in Section 9.2 could have been accompanied by much higher than normal snow accumulations. It should also be noted that nowhere in either the biblical flood story or the *Gilgamesh Epic* is the flood associated in any way with a river. In fact the epic refers repeatedly to the ark as floating upon a sea, and this may be seen, for example, in lines 123, 131, 132, and 138 of the quotation in Section 11.3. When the storm abated, the nearest land was barely visible in the distance. Then the ark ran aground, and from the description the sea would seem to have been many miles across. As suggested previously, Lake Van is a possible identification for this sea.

D. Duration of the rainfall

It may be considered here also what might have been the cause of the extraordinarily intense rainfall associated with the biblical flood of Noah's time in a region which normally received very little rain. It is helpful first to compare the actual statements concerning the duration of the rainfall in the biblical version of the flood and in the Gilgamesh version as quoted in Section 11.3:

> Bible
> For in seven days I will send rain upon the earth forty days and forty nights; and every living thing that I have made I will blot out from the face of the ground (Genesis 7:4).

> And after seven days the waters of the flood came upon the earth (Genesis 7:10).

> And rain fell upon the earth forty days and forty nights (Genesis 7:12).

> The flood continued forty days upon the earth; and the waters increased, and bore up the ark, and it rose high above the earth (Genesis 7:17).

Gilgamesh Epic
127. Six days and [six] nights
128. The wind blew, the downpour, the tempest, (and) the flo[od] overwhelmed the land.
129. When the seventh day arrived, the tempest, the flood,
130. Which had fought like an army, subsided in (its) onslaught.
131. The sea grew quiet, the storm abated, the flood ceased.

Besides the possibility of flooding resulting directly from the rainfall, there is also the significant possibility of melting snow and ice in mountains subjected to relatively warm rains. The concept of rain-caused flooding was considered previously in Section 7.3.

According to the biblical version, the storm arrived seven days after its final prediction, and when it had arrived it continued for forty days. But forty is a popular number in biblical stories, and in the flood account it is also said that forty days after the mountain tops were visible Noah sent out a raven. On the other hand, according to the *Gilgamesh Epic* the storm lasted for seven days, and seven days after the ship landed Utnapishtim sent forth a dove (as quoted in lines 145-146 in Section 11.3 above). The multiple uses of the period of seven days in both versions and forty days in the biblical version suggest that these time periods have been given familiar schematic values, and it is difficult to say whether any of these numbers may have been interchanged and which of them, if any, might be literally correct. However, with the events in common such as the ark/ship, the flood, the visibility of the mountaintops, the sending out of birds,

Causes And Effects Of The Flood

and the subsequent sacrifice, it would be difficult to argue that these two stories are entirely independent of each other.

In Section 7.3 it was noted that the most rain ever reported for a seven day storm was about 5.4 meters. It was also noted that the Lake Van drainage basin has an area that is about 4.57 times as large as the area of the lake itself. That means that a rainfall of about 5.4 meters would cause a rise in the lake level of about $5.4 \times 4.57 = 24.7$ meters. Though this value is only a rough estimate, it conveys the idea that a week of record heavy rain at Lake Van could easily be sufficient to float Utnapishtim's ship. It should be acknowledged that rainfall at the rate indicated above would not be likely to occur at Lake Van under normal storm conditions. On the other hand, the flood of interest is being associated with severe flooding over other widely separated sites in what has also been considered to be a once in four thousand years weather event. That event is believed to have caused much colder than normal temperatures, so unusual snow accumulations are also a possibility.

E. Fountains of the deep

As another parallel, the biblical account ascribes much of the flooding to waters from the fountains of the deep:

> [O]n that day all the fountains of the great deep burst forth, and the windows of the heavens were opened. And rain fell upon the earth .... (Genesis 7:11-12).

> [T]he fountains of the deep and the windows of the heavens were closed, the rain from the heavens was restrained" (Genesis 8:2).

11. Utnapishtim's Flood

In the biblical account the two main sources of the flood waters are the fountains of the deep and the windows of heaven. Though it might not be stated, the same could have been true of the Gilgamesh version. The concept of underground waters is certainly known in the *Gilgamesh Epic*. In the instructions for the building of the ship, it was securely covered in the way that the earth covers the underground waters:[14] "Cover it [li]ke the subterranean waters."

The explicit meaning of the words "fountains of the deep" may seem a bit mysterious in the Genesis story of the flood, and translations of this phrase vary. Similar language may be found in the *Atrahasis Epic* that dates from the reign of King Ammizaduga (c. 1582-1562 B.C., according to the low or short chronology) of the First Babylonian Dynasty.[15] The name Atrahasis is another designation for Gilgamesh. These additional instances may provide a different perspective on the same underlying usage.

The *Atrahasis Epic* seems to have been written a few centuries after the *Gilgamesh Epic*, and it adds to the story the occurrence of a severe multi-year famine to punish humanity for its noisiness before the ultimate punishment by flood was imposed. One aspect of the famine (real or otherwise) was the discontinuation of all sources of water. The description of this circumstance is quite similar to the stoppage of water sources at the termination of the flood of Noah/Utnapishtim. Reproduced here are some lines relating to this event from the *Atrahasis Epic*:[16]

Fragment No. I

Column i

11. Ab[ove] let Adad {god of storm and rain} make scarce his [rain],

12. [Below] let the springs not flow.
13. [Let the floods not rise] from the source.
14. Let the wind blow
15. . . . .
16. Let the clouds hold back(?)
17. [That rain from heaven] pour not forth.

Fragment No. IV
Column ii

29. Above [Adad made scarce his rain];
30. Bel[ow] (the fountains) were stopped, [so that the flood did not rise at the source].

Column iii

44. Above let Adad make scarce his rain;
45. Below [let (the fountains) be] stopped, so that the flood cannot rise at the source.
   ⋮
54. Above Adad made scarce his rain;
55. Below (the fountains) were stopped, so that the flood could not rise at the source.

Thus, the biblical "fountains of the deep" may in the Atrahasis Epic be represented by "springs" and "fountains," which seem to be a "source" from which "floods" of water rise. These ancient examples help to support the interpretation of the biblical "fountains of the deep" as springs. The floods in these Atrahasis texts may correspond to rivers that were imagined to have started out as springs. At the time of the flood, the flow in these rivers might have been augmented substantially by snow in the mountains that was melted by the rain. Together with direct rainfall, the increased flow in the rivers may have contributed significantly to the flood of Noah's time.

The bursting of the fountains of the deep in the biblical version can perhaps be associated with widespread earthquake-induced collapsing of the ground and freeing of underground rivers and reservoirs or the redirecting of above-ground waters. On this subject it may be appropriate to consider the following line from the *Gilgamesh Epic*, quoted above in Section 11.3:

107. [. . . .] the land he broke (?) like a po[t(?)].

The breaking of the land like a pot might also imply the freeing by earthquake of the waters contained within the ground. The analogy of something being broken like a pot was a familiar concept in the ancient world. A human caused example from Urartu is the following:[17] "The stone walls of the city were destroyed 'like an earthen pot' and razed to the ground 'with iron axes and swords'." A more modern example from Urartu (Ararat) of underground waters being released by the occurrence of an earthquake was discussed previously in Section 7.2.

F. Sacrifices after the flood

The main interest here is still on the circumstances associated with the flood. After the flood had subsided and the occupants of the ark had left it, Noah in the biblical version sacrificed to the Lord, and Utnapishtim in the Gilgamesh version sacrificed to the gods:

Bible:
   Then Noah built an altar to the Lord, and took of every clean animal and of every clean bird, and offered

burnt offerings on the altar. And when the Lord smelled the pleasing odor, the Lord said in his heart, "I will never again curse the ground because of man, for the imagination of man's heart is evil from his youth; neither will I ever again destroy every living creature as I have done. While the earth remains, seedtime and harvest, cold and heat, summer and winter, day and night, shall not cease" (Genesis 8:20-22).

Gilgamesh Epic:
155. (Then) I {Utnapishtim} sent forth (everything) to the four winds and offered a sacrifice.
156. I poured out a libation on the peak of the mountain.
157. Seven and (yet) seven kettles I set up.
158. Under them I heaped up (sweet) cane, cedar, and myrtle.
159. The gods smelled the savor,
160. The gods smelled the sweet savor
161. The gods gathered like flies over the sacrificer.[18]

While there are obvious differences between these two accounts in both content and style, the report of a major sacrifice in both accounts after the departure from the ark is significant. The similarity of the reports of the gods being impressed by the smell of the sacrifice in both accounts is particularly striking. In the biblical version "the Lord smelled the pleasing odor," while in the epic "The gods smelled the sweet savor." Thus, in these descriptions the mention of the smells of the sacrifices supports the idea of the two stories being related. The possibility of an indifferent reaction to sacrifices is also included at a later place in the Bible: "And I will lay your cities waste, and will make your sanctuaries desolate, and I will not smell your pleasing odors" (Leviticus 26:31).

11. Utnapishtim's Flood

11.5 Comparison with other flood stories

It may also be noted that there seem to be several similarities between the events of the Gilgamesh story and occurrences unrelated to Noah's flood.

A. Violence of the storm

The violence of the storm before and during the flood in the Gilgamesh story may be compared to what seem to be eruption-related tempests in Sumer and Egypt. The storm descriptions in the *Gilgamesh Epic* and in the Sumerian Lament could both relate to Noah's flood, while the Tempest Stela relates to an event many centuries later.

> *Gilgamesh Epic*:
> The raging of Adad reached unto heaven {thunder} (and) turned into darkness all that was light. [. . . .] the land he broke(?) like a po[t(?)]. (For) one day the tem[pest blew]. Fast it blew and [. . . .]. Like a battle [it ca]me over the p[eople]. No man could see his fellow. The people could not be recognized from heaven. (Even) the gods were terror-stricken at the deluge. . . . The wind blew, the downpour, the tempest, (and) the flo[od] overwhelmed the land.[19]

> Lament over Sumer:
> The storm was a harrow coming from above, the city was struck (as) by a hoe. On that day, heaven rumbled, the earth trembled, the storm worked without respite.[20]

> Tempest Stela:
> The gods [caused] the sky to come in a tempest of r[ain], with darkness in the western region and the sky being unleashed without [cessation, louder than] the cries of the masses, more powerful than [. . .], [while the rain

raged (?)] on the mountains louder than the noise of the cataract which is at Elephantine {island in the Nile River}.[21]

The reports in the *Gilgamesh Epic* and in the Sumerian Lament from before the flood both have a close similarity to the corresponding report in *Jasher* quoted previously in Section 9.2:

> And on that day, the Lord caused the whole earth to shake, and the sun darkened, and the foundations of the world raged, and whole earth was moved violently, and the lightning flashed, and the thunder roared, and all the fountains in the earth were broken up, such as was not known to the inhabitants before; and God did this mighty act, in order to terrify the sons of men, that there might be no more evil upon earth.
> And still the sons of men would not return from their evil ways, and they increased the anger of the Lord at that time, and did not even direct their hearts to all this. And at the end of seven days, in the six hundredth year of the life of Noah, the waters of the flood were upon the earth.

B. Human casualties

Gilgamesh Epic:
> "How could I {Ishtar} command (such) evil in the assembly of the gods! (How) could I command war to destroy my people, (for) it is I who bring forth {give birth to} these my people! Like the spawn of fish they {people's bodies} (now) fill the sea!"
> I {Utnapishtim} looked upon the sea, (all) was silence, and all mankind had turned to clay {all were dead}.[22]

11. Utnapishtim's Flood

Lament over Sumer:
[...] they piled up in heaps [...] they spread out like sheaves. There were corpses (floating) in the Euphrates. ...[23]

Tempest Stela:
Then every house, every quarter that they (scil. the storm and rain) reached [... their corpses(?)] floating on the water like skiffs of papyrus outside the palace audience chamber for a period of [...] days ....[24]

C. Darkness

The relationship isn't obvious, but perhaps there is another connection between statements already quoted above. This connection involves the darkness that accompanied the storm:

Gilgamesh Epic:
The raging of Adad reached unto heaven {thunder} (and) turned into darkness all that was light. ... No man could see his fellow. The people could not be recognized from heaven.[25]

Lament over Sumer:
The heavens were darkened, they were covered by a shadow, the mountains roared, the sun lay down at the horizon, dust passed over the mountains, the moon lay at the zenith, the people were afraid.[26]

Tempest Stela:
[N]o torch could be lit in the Two Lands {Upper and Lower Egypt}.[27]

All of these statements could be associated with the occurrence of unusual darkness.

D. The storm of Elijah's era

There is also a curious parallel between the Gilgamesh version of the beginning of the storm in the flood story and a biblical account of the beginning of a much later storm during the time of Elijah. The biblical story was quoted previously in Section 9.6, and the Gilgamesh story is quoted above in Section 11.3. They are reproduced together here for easier comparison. A somewhat similar storm representation for the tempest stela in Egypt is also included:

Bible:

And Eli'jah said to Ahab, "Go up, eat and drink; for there is a sound of the rushing of rain." So Ahab went up to eat and to drink. And Eli'jah went up to the top of Carmel; and he bowed himself down upon the earth, and put his face between his knees. And he said to his servant, "Go up now, look toward the sea." And he went up and looked, and said, "There is nothing." And he said, "Go again seven times." And at the seventh time he said, "Behold, a little cloud like a man's hand is rising out of the sea." And he said, "Go up, say to Ahab, 'Prepare your chariot and go down, lest the rain stop you.' " And in a little while the heavens grew black with clouds and wind, and there was a great rain. And Ahab rode and went to Jezreel. And the hand of the Lord was on Eli'jah; and he girded up his loins and ran before Ahab to the entrance of Jezreel (1 Kings 18:41-46).

Gilgamesh Epic:
97. A black cloud came up from out the horizon.
98. Adad {god of storm and rain} thunders within it,
99. While Shullat and Hanish {bodyguards of Adad} go before,
100. Coming as heralds over hill and plain;

101. Irragal {or Nergal, god of the underworld} pulls out the masts;
102. Ninurta {god of war and lord of wells and irrigation works} comes along (and) causes the dikes to give way;
103. The Anunnaki {judges in the underworld} raised (their) torches,
104. Lighting up the land with their brightness {lightning};
105. The raging of Adad reached unto heaven {thunder}.
. . .[28]

Tempest Stela:
  The gods [caused] the sky to come in a tempest of r[ain], with darkness in the western region and the sky being unleashed without [cessation, louder than] the cries of the masses, more powerful than [. . .], [while the rain raged (?)] on the mountains louder than the noise of the cataract which is at Elephantine {island in the Nile River}.[29]

In both the biblical and Gilgamesh versions the storm began as a cloud or darkness rising from the sea or some other horizon. In both the biblical and tempest stories the storm and darkness are said to have come from the west.

11.6 The tree of life

An important aspect of the biblical story of the garden of Eden is the presence in the garden of the tree of life. According to the biblical story, a person who eats from that tree will live forever:

And the Lord God planted a garden in Eden, in the east; and there he put the man whom he had formed. And out

Causes And Effects Of The Flood

of the ground the Lord God made to grow every tree that is pleasant to the sight and good for food, the tree of life also in the midst of the garden, and the tree of the knowledge of good and evil (Genesis 2:8-9).

Then the Lord God said, "Behold, the man has become like one of us, knowing good and evil; and now, lest he put forth his hand and take also of the tree of life, and eat, and live for ever"– therefore the Lord God sent him forth from the garden of Eden, to till the ground from which he was taken (Genesis 3:22-23).

The story of the tree of life is an important part of the biblical account of the garden of Eden, so it is natural to enquire whether that story might have any parallel in the Gilgamesh story. Gilgamesh had sought out Utnapishtim in the first place in an effort to learn the secret of eternal life, and on his return voyage he tried a method told him by Utapishtim. In this approach Gilgamesh was to dive down to the bottom of the sea (apparently not so toxic in that region of the sea) and bring back a prickly plant or small tree. Eating of the fruit of that tree would restore him to his youth:

Tablet XI
263. Utnapishtim [said] to him, [to] Gilgamesh:
264. "Gilgamesh, thou hast come hither, thou hast become weary, thou hast exerted thyself;
265. What shall I give thee (wherewith) thou mayest return to thy land?
266. Gilgamesh, I will reveal (unto thee) a hidden thing,
267. Namely, a [secret of the gods will I] tell thee:
268. There is a plant like a thorn [. . . .].
269. Like a rose(?) its thorn(s) will pr[ick thy hands].

270. If thy hands will obtain that plant, [thou wilt find new life]."
271. When Gilgamesh heard that he opened [. . . .].
272. He tied heavy stones [to his feet];
273. They pulled him down into the deep, [and he saw the plant].
274. He took the plant, (though) it pr[icked his hands].
275. He cut the heavy stones [from his feet],
276. (And) the . . . . threw him to its shore.[30]

The tree in this story clearly has a similar role to that of the tree of life in the garden of Eden (Genesis 2:9, 3:22-23). Eating of the tree of life would assure a person of renewed life. From lines 272 to 276 one understands that Gilgamesh dived with extra weight to be released later, and buoyancy helping to speed him back to the surface. He was successful in obtaining the plant, and he intended to bring it back with him to his royal city of Uruk in Sumer so that he could eat of it in his old age. Unfortunately, at a rest stop on the way back, a serpent came out of the water to take and eat the plant. It then shed its skin (an imagined restoration of youth), and swam away, leaving a very sad Gilgamesh:

Tablet XI
284. After thirty (additional) double-hours they stopped for the night.
285. Gilgamesh saw a pool with cold water;
286. He descended into it and bathed in the water.
287. A serpent perceived the fragrance of the plant;
288. It came up [from the water] and snatched the plant,
289. Sloughing (its) skin on its return. {By eating this magic plant, the serpent gained the power to shed its old skin and thereby to renew its life.}
290. Then Gilgamesh sat down (and) wept,
291. His tears flowing over his cheeks.

292. [. . . .] of Urshanabi, the boatman:
293. "[For] whom, Urshanabi, have my hands become weary?
294. For whom is the blood of my heart being spent?
295. For myself I have not obtained any boon.
296. For the 'earth-lion' {serpent} have I obtained the boon.³¹

One might inquire whether the serpent in the epic could be associated in any way with the serpent in the garden of Eden. There the first man and woman had access to the tree of life, and by eating of its fruit they could live forever. However, they were warned by God that if they ate the fruit from the tree of the knowledge of good and evil they would be condemned to die (Genesis 2:9,16-17). In the Genesis story the "serpent" seems initially to have been a creature that could walk and talk, and that furthermore was determined to get the first man and woman in trouble with God (Section 2.4). The serpent did this by persuading them to eat of the forbidden fruit (Genesis 3:1-13), and as a consequence God drove them from the garden. The serpent was also punished. His ability to walk was taken from him and he was condemned to slither on his belly (Genesis 3:14). In other words he seems to have become a more ordinary venomous snake (Section 5.2).

Many features of the Eden and various flood stories may seem too fantastic to be literal factual history. Examples of such features include the seemingly impossible lifetimes, a magical tree (prickly in the epic), and an evil serpent (walking and talking in the garden of Eden). Nevertheless, it may be of interest to consider the possibility of a single unifying story, factual or otherwise, underlying the accounts considered here. Like the

11. Utnapishtim's Flood

original texts, the possibilities suggested below should not all be understood as serious proposals.

While a particular shedding of a serpent's skin can be interpreted to imply a degree of rejuvenation, it does not, of course, mean that the serpent would live forever. The Gilgamesh story could, however, be creatively interpreted to suggest that perhaps a descendant of the serpent in the garden of Eden had retained some of its ancestor's wily ways. By taking away Gilgamesh's magical tree (or tree of life), the later serpent also took away Gilgamesh's last chance for eternal life. If the original serpent in the garden of Eden were imagined to be an ancestor of the serpent of the epic, the ancestor would probably have been proud of his descendant. In both stories the actions of the serpent brought an end to the hopes of the respective heroes (Adam and Gilgamesh) for eternal or at least extended life. This hypothetical scenario would be of greatest interest if the garden of Eden and flood stories were related geographically, and such a possibility will be considered here.

One might wonder why in the Gilgamesh story the tree of life was beneath the surface of the sea, while in the biblical story the tree of life was in the above-sea-level garden of Eden. First it should be emphasized that these two stories do not necessarily refer to the very same tree, since the garden of Eden story is imagined to have taken place a few centuries before the flood of Noah/Utnapishtim.[32] However, if the tree of life was in any sense a real tree, then it could have had similar descendant trees (and descendant tree stories) in the same vicinity.

If the sea in the Gilgamesh story corresponds to modern Lake Van in Turkey (as suggested here), there is an easy explanation

Causes And Effects Of The Flood

for the underwater location of the tree. At the time of the flood the water level in Lake Van increased dramatically enough to float the ark and save the lives of its occupants. This rising water could also have submerged much of the original garden, especially if it was not far from the original shore of the lake. A likely location would have been in the vicinity of the Arjish/Ercis branch of the lake, as discussed in Section 4.4.

Gilgamesh may be assumed to have arrived soon enough after the flood that Utnapishtim was still alive, assuming that he was a mortal of normal lifespan and not immortal as claimed in the epic. Perhaps at that time the water level had not yet returned to its pre-flood value, and a "tree of life" was still edible but submerged by the higher post-flood water level. Most such changes at Lake Van seem to decay back to "normal" in a period of a few years. Thus, the tree might well have been underwater when Gilgamesh arrived, but it had not been there too long for it to still be edible by the serpent in the story. The chronology of this serpent story is in reasonable agreement with the considerations in Section 10.2, where it was suggested that Gilgamesh's visit may have occurred at some time between the flood in about 2035 B.C. and the approximate composition date of the epic in about 2000 B.C., after Gilgamesh's return to his home. Thus, Gilgamesh would have visited Noah/Utnapishtim shortly after 2035 B.C. or before (for example) about 2030 B.C.

If there was a tree of life in the garden of Eden and perhaps another one at the time and place of Noah/Utnapishtim, it would be of interest to know more about them. There is actually only a little that can be said with confidence; but, as the reader will have noticed, speculation is always possible. One conclusion is of the negative sort. There seems to be no evidence in the Bible or elsewhere that the tree of life ever provided anyone with eternal life. Adam and Eve were apparently expelled from the garden before any data on this subject could be obtained. God sent him (them) out the garden "lest he put forth his hand and take also of the tree of life, and eat, and live for ever" (Genesis 3:22). The conclusion of Adam's very long life is reported in the typical perfunctory manner: "Thus all the days that Adam lived were nine hundred and thirty years; and he died" (Genesis 5:5).

In the *Gilgamesh Epic*, Utnapishtim and his wife are said to have been granted eternal life, but not because of anything they ate. The story of what happened after the flood is related by Utnapishtim as follows:

Tablet XI
189. "Then Enlil {warrior god} went up into the ship.
190. He took my hand and caused me to go aboard.
191. He caused my wife to go aboard (and) to kneel down at my side.
192. Standing between us, he touched our foreheads and blessed us:
193. 'Hitherto Utnapishtim has been but a man;
194. But now Utnapishtim and his wife shall be like unto us gods.
195. In the distance, at the mouth of the rivers, Utnapishtim shall dwell!'
196. So they took me and caused me to dwell in the distance, at the mouth of the rivers.[33]

It is of interest that when Utnapishtim and his wife were granted eternal life they were made to live "at the mouth of the rivers." In fact it seems that they were physically placed at the

Causes And Effects Of The Flood

mouth of the rivers: "So they took me and caused me to dwell in the distance, at the mouth of the rivers." Adam also had originally been given no choice of where to live, and he was placed in the garden of Eden:

> And the Lord God planted a garden in Eden, in the east; and there he put the man whom he had formed (Genesis 2:8).

> The Lord God took the man and put him in the garden of Eden to till it and keep it (Genesis 2:15).

Later, however, Adam and Eve were expelled from the garden of Eden because of their disobedience, as reviewed in Section 2.4. Since both the biblical garden of Eden and Utnapishtim's home "at the mouth of the rivers" are associated with a tree of life and an evil serpent, it seems natural to wonder whether the garden of Eden could have been located at some place that could be identified as "the mouth of the rivers."

As discussed in Section 4.4, the garden of Eden may have been near the northern end of what is now called the Ercis gulf of Lake Van. Several rivers have their mouths in this region of the lake. Also, as noted in Section 5.2 the so-called mountain of snakes is near the northern end of the Ercis gulf. Thus, this general region remains a viable candidate for the location of the garden of Eden and for the post-flood home of Noah/Utnapishtim. It may be noted also that a location near "the mouth of the rivers" would have had the least toxic water associated with the lake, and it is only in those rivers and their estuaries that any fish may be said today to live in Lake Van, as noted in Section 5.3. Certainly, Gilgamesh would not have chosen to dive in the area of waters that he had been instructed not to touch. As noted in Section 11.2, the boatman Urshanabi had warned Gilgamesh about the poisonous water at one place in the lake that they had to cross on their journey to Utnapishtim: "Let not thy hand touch the waters of death."

In contrast to Utnapishtim, his biblical counterpart Noah is not said to have lived forever. Furthermore, there seems to be no indication (beyond the statements in the epic) that Utnapishtim and his wife actually did experience extended lives. Thus, the only hint in the Noah/Utnapishtim stories that could have been construed as successful immortality-by-eating is the report mentioned above of the serpent shedding its skin.

11.7 Identifying the tree

Supernatural or otherwise, it might still be of interest to know something about the physical characteristics of what in the Bible was called the tree of life. As a logical starting point for this seemingly doubtful subject, it may be noted that even today there are trees that go by the name arborvitae, which is Latin for tree of life. They are said to have received this name originally because tea made from their bark and leaves was found to cure scurvy. The name remains relevant because of the still important medicinal properties of oil made from arborvitae.

> One of the first documented uses of indigenous medicine in North America was the cure in the winter of 1536 of Jacques Cartier's crew from a disease he called "Scorbut" (scurvy). Cartier's second voyage (1535-1536) was undertaken at the command of King Francois 1[er] to complete the discovery of the western lands under the

same climate and parallels as in France. At Stadaconna, now Quebec City, Cartier's crew was cured from scurvy [scorbut] by ascorbic acid (vitamin C) obtained as a decoction from the Iroquois. It was prepared by boiling winter leaves and the bark from an evergreen tree. The tree, identified as "Annedda", became known as the "tree of life" or "arbre de vie" because of its remarkable curative effects. In the winter, scurvy was the most prevalent disease among the Iroquois. This was due to the lack of food and vitamin C.[34]

The actual identity of the evergreen tree of life "Annedda" was not well documented and became controversial. Candidate trees include certain species of cedar, spruce, pine, fir, hemlock, and juniper. The common juniper (*Juniperus communis*) could be of particular interest:

> *Juniperus communis*, the common juniper, is a species of conifer in the genus Juniperus, in the family Cupressaceae. It has the largest geographical range of any woody plant, with a circumpolar distribution throughout the cool temperate Northern Hemisphere from the Arctic south in mountains to around 30° N latitude in North America, Europe and Asia.[35]

Scurvy was a dreaded disease that tended to affect sailors who were at sea for long periods of time, as well as people who experienced long cold winters:

> Scurvy is an acute chronic illness caused by a dietary deficiencey of ascorbic acid (vitamin C). . . . Exposure to long periods of cold temperatures can lead to ascorbic-acid insufficiency. . . .
> Scurvy is characterized principally by anemia, hemorrhagic manifestations in the skin (ecchymoses and

11. Utnapishtim's Flood

perifollicular haemorrhage), and in the musculoskeletal system (haemorrhage into periosteum and muscles). The gums start to bleed. Teeth are loosened. With no vitamin C intake, the symptoms of scurvy would occur after one to three months. Unless treated, scurvy is fatal.

At Stadaconna (46° 49' N, 71° 13' E) and in November 1535, Canada's cold struck with its entire rigor, and ice thickened to two fathoms. In December, over 50 of the Iroquois died from an unknown sickness (scurvy). The sickness began to spread to Cartier's crews in all three of his ships. By mid-February 1536, of the 110 member crews, 8 were already dead and more than 50 past all hope of recovery.

Land-based scurvy is well documented as occurring during times of food shortages. Late springs and low levels of vitamin C in stored grain contributed to endemics of sub-clinical scurvy. As for the candidate trees of life, food would include seeds, buds, inner bark, cambium and sap of trees.[36]

Pierre Belon (1517-1564), one of the early investigators of the treatment of scurvy, visited Turkey and found there a tree which may have been a tree of life:[37] "When Belon visited Turkey, he found a tree similar to the one at Fontainebleau in France, which was brought from Canada and called 'Arbre de Vie'." *Juniperus Communis* is native to both Canada and Turkey, and it is possible that this is the species that Belon observed. As noted above, scurvy tends to occur during long periods of cold temperatures. Lake Van and the associated garden of Eden as considered here would have been somewhat farther south than the area of what is now known as Quebec where Cartier landed with his ships. On the other hand, the elevation of the lake is more than a mile above sea level, and it is situated among the much higher snow-capped peaks of the mountains of Ararat. Modern explorers have been surprised by how cold it is when the wind blows down on the north shore of the lake (Section 6.3): "To our amazement, the Ardjîch [Ercis] plain that extends to our feet is covered with thick snow. Its valley is undoubtedly more open to the winds from the North and is therefore colder." Thus, it is possible that the winters at the location of Lake Van in what is now Turkey could have been as cold as those near sea level in Quebec. If that were the case, then the danger from scurvy could have been as great for the early inhabitants of the garden of Eden as it was for the much later permanent or temporary residents of Quebec. That the winter temperatures near Lake Van were probably as cold as those encountered by the Iroquois and their visitors can be confirmed in two similar quotations: "The winters in the drainage basin of Lake Van are severe. At least 3 months in the year, the average monthly temperature drops below 0° C."[38] "The world's largest soda lake, Lake Van, is situated on the eastern Anatolian high plateau of Turkey at about 43° E and 38.5° N. At this latitude, winter is very severe with temperatures frequently under 0° C during at least three months of the year."[39]

The juvenile junipers have sharp needles, and some of them (including *Juniperus communis*) retain such prickly needles at maturity. With others the needles are replaced by scale-like leaves. This is of interest because in the *Gilgamesh Epic*, as quoted in Section 11.6, the then-underwater tree that was said to rejuvenate its consumer was prickly to the touch.

Tablet XI
266. Gilgamesh, I will reveal (unto thee) a hidden thing,
267. Namely, a [secret of the gods will I] tell thee:

11. Utnapishtim's Flood

268. There is a plant like a thorn [. . . .].
269. Like a rose(?) its thorn(s) will pr[ick thy hands].
270. If thy hands will obtain that plant, [thou wilt find new life]."

The tree of life identification would not necessarily be inappropriate for juniper trees. Like the modern arborvitae, the juniper has long been used in medicine, and a single illustration should suggest the possibilities:[40] "Florida's Seminole Indians had many medical uses for the Eastern Red Cedar [*Juniperus virginiana*]. They used it to treat cold symptoms, swollen joints, stiff neck or back, swollen legs, eye diseases, fever, headache, dizziness and diarrhea." Other species of juniper were also used by native Americans.

Both the arborvitae and juniper trees have berry-like cones. Juniper berries may be brown or orange, but in most species (including *Juniperus communis*) they are blue. The berries are usually edible and aromatic, and sometimes they are used as a spice. The strong aroma of the berries is significant here, because the serpent was said to have been attracted to the rejuvinating tree by its aroma:[41]

Tablet XI
287. A serpent perceived the fragrance of the plant;
288. It came up [from the water] and snatched the plant,

The genus name Thuja of the arborvitae comes from a Greek word for perfume. It was suggested in Section 10.7 that carnelian and lapis lazuli plants in the *Gilgamesh Epic* could have been named for the colors of their fruit. Besides the plants mentioned

Causes And Effects Of The Flood

previously, the "berries" of Juniperus Communis could also be considered to be the same blue color as lapis lazuli.

It isn't possible to know with certainty which if any juniper species might have been present near Lake Van at the time of the garden of Eden or the flood. The potential usefulness of these trees is also shown by the Utah Juniper or *Juniperus osteosperma*, about which much is known:

> Juniper berries were used as food by the Indian inhabitants of the Southwest and the Great Basin. The shredded bark was smoked in cane cigarettes as a tobacco substitute, and was woven into sandals and other clothing. Rope was made from it. Juniper bark torches have been found at Tonto National Monument, Arizona, and in the prehistoric salt mine near Montezuma Castle National Monument, Arizona, which was in use from A.D. 100–1450. Juniper firewood was a favored cooking fuel of Great Basin and Southwestern Indians. Its wood was also much used for carving utensils and other objects, because it is fine-textured, fairly soft, and easily worked.[42]

As mentioned above, the Utah juniper was widely used by Southwestern Indians for carving utensils, and that use could be consistent with the idea, suggested in Section 10.7, of Gilgamesh on the other side of the world and several millennia earlier seemingly carving a bowl from the wood of a "lapis lazuli" plant. Considering the past usefulness to humans of junipers and other trees of the cypress family, it would seem appropriate to find that one of these trees might also have been identified in ancient times as a "tree of life."

It is curious that the French explorers of the early fifteen hundreds together with their Iroquois colleagues in what is now known as Quebec should have chosen to call their scurvy-curing plants "tree of life." The choice of this name would seem to suggest that these people were well aware of the biblical story of the garden of Eden (and or the later Gilgamesh parallel) in which a plant that if eaten would lead to eternal life. The identification in Quebec of a particular family of plants that had "remarkable curative effects" in extended periods of cold weather and food shortages suggests that these plants might have been the ones that were said in the ancient past to have made possible eternal life. However, while the plants could have provided dramatic improvements in health and life expectancy, they would not in this interpretation actually have ensured eternal life.

11.8 Conclusion

With the beginning of this chapter, Gilgamesh arrived at a large sea (Lake Van), which he had to cross to reach Utnapishtim and perhaps learn from him the secret of eternal life. The crossing was successful, and Utnapishtim began by relating to Gilgamesh the story of a great flood of perhaps the same sea that Gilgamesh had just crossed. This flood story has many features in common with the biblical flood of Noah's time. Eventually, Utnapishtim described to Gilgamesh a plant that seems similar to the tree of life that had been in the garden of Eden but that after the flood was beneath the waters of the sea. Gilgamesh recovered the plant, but unfortunately a serpent snatched it away and ate of it himself. This occurrence was so discouraging to Gilgamesh that he abandoned his quest and returned to his home.

Causes And Effects Of The Flood

---

1. A. Heidel, *The Gilgamesh Epic and Old Testament Parallels* (The University of Chicago Press, Chicago, 1963), pp. 85-86, Tablet XI, lines 127-131 (translator indications deleted, full text in Section 11.3).
2. Ibid., p. 68, Tablet IX, Column vi, line 37.
3. Ibid., p. 70, Tablet X, A. The Old Babylonian Version, Column iii, lines 1-14 (translator indications omitted).
4. Ibid., B. The Assyrian Version, pp. 73-74, Tablet X, Column ii, lines 15-22, 24-28, 30-31; pp. 76-77, Tablet X, Column iii, lines 32-35, 47-48.
5. Ibid., p. 77, Tablet X, Column iv, lines 1-3.
6. H. F. B. Lynch, *Armenia Travels and Studies*, Volume 2 (Longmans, Green, and Co., London, 1901), p. 48.
7. Ibid., p. 52
8. "Volcano Hazards FAQs," https://www2.usgs.gov/faq/categories/9820/3195. Web. 28 May 2017.
9. A. Heidel, op. cit., pp. 84-87, Tablet XI, lines 96 to 161.
10. Ibid., Chapter IV, "The story of the flood," pp. 224-269.
11. Ibid., pp. 225-226.
12. *The Book of Yashar*, Translated from the Hebrew and Published by Mordecai Manuel Noah (Hermon Press, New York, 1972), (cited here in the form *Jasher*); *Jasher* 6:18-19.
13. A. Heidel, op. cit., p. 227.
14. Ibid., p. 81, Tablet XI, line 31; and comments on p. 234.
15. Ibid., pp. 106-107.
16. Ibid., pp. 107-108, 112, 114, 115.
17. B. P. Piotrovsky, *The Ancient Civilization of Urartu*, Translated from the Russian by J. Hogarth (Cowles Book Company, Inc., New York, 1969), p. 110.
18. A. Heidel, op. cit., p. 87, Tablet XI, lines 155-161.
19. Ibid., p. 85, Tablet XI, lines 105-113, 128.

11. Utnapishtim's Flood

20. P. Michalowski, *The Lamentation over the Destruction of Sumer and Ur* (Eisenbrauns, Winona Lake, 1989), p. 41, lines 80ß – 81.
21. R. K. Ritner, "A new translation of the Tempest Stele of Ahmose," Appendix A in K. P. Foster, R. K. Ritner, and B. R. Foster, "Texts, storms, and the Thera eruption," *Journal of Near Eastern Studies*, Volume 55, Number 1, pp. 1-14 (January 1996); p. 11, lines (8)-(10).
22. A. Heidel, op. cit., pp. 85-86, Tablet XI, lines 120-123, 132-133.
23. P. Michalowski, op. cit., p. 43, lines 93-94.
24. R. K. Ritner and N. Moeller, "The Ahmose 'Tempest Stela', Thera and comparative chronology," *Journal of Near Eastern Studies*, Volume 73, Number 1, pp. 1-19 (April 2014); pp. 6-7.
25. A. Heidel, op. cit., p. 85, Tablet XI, lines 105-106, 111-112.
26. P. Michalowski, op. cit., p. 41, lines 82-84.
27. R. K. Ritner and N. Moeller, op. cit., p. 7.
28. A. Heidel, op. cit., pp. 84-85, Tablet XI, lines 97 to 105.
29. R. K. Ritner, op. cit.
30. A. Heidel, op. cit., p. 91, Tablet XI, lines 263-276.
31. Ibid., p. 92, Tablet XI, lines 284-296.
32. This chronology will be considered further in Chapter 12, "From Adam to Abram."
33. A. Heidel, op. cit., p. 88, Tablet XI, lines 189-196.
34. D. J. Durzan, "Arginine, scurvy and Cartier's 'tree of life'," *Journal of Ethnobiology and Ethnomedicine* (2 February 2009), http://ethnobiomed.biomedcentral.com/articles/10.1186/1746-4269-5-5.
35. *Juniperus communis*, Wikipedia. Web. 29 April 2017.
36. D. J. Durzan, op. cit.
37. Ibid.

38. E. T. Degens, H. K. Wong, S. Kempe, and F. Kurtman, "A geological study of Lake Van, Eastern Turkey," *A Geologische Rundschau,* Volume 73, Number 2, pp. 701-734 (June 1984); p. 716.
39. M. Kadioglu, Z. Sen, and E. Batur, "The greatest soda-water lake in the world and how it is influenced by climatic change." *Annales Geophysicae*, Volume 15, pp. 1489-1497 (1997); p. 1489.
40. G. Deane, "Junipers," http://www.eattheweeds.com/junipers. Web. 21 March 2017.
41. A. Heidel, op. cit., p. 92, Tablet XI, lines 287-288.
42. R. M. Lanner, *Trees of the Great Basin: A Natural History* (University of Nevada Press, Reno, 1984), p. 114.

12. FROM ADAM TO ABRAM

"This is the book of the generations of Adam. When God created man, he made him in the likeness of God. Male and female he created them, and he blessed them and named them Man when they were created. When Adam had lived a hundred and thirty years, he became the father of a son . . ." (Genesis 5:1-3).

12.1 Introduction

One purpose of this book has been to develop a chronology of the events described in the Bible from the time of the biblical creation story to the time of Abram. Thus, this work may be considered to complement a previous study that focused on chronological patterns from the New Testament period back in time to Abram.[1] The present chapter attempts to combine and summarize the various aspects of the chronologies for this earlier time period that have been investigated in the preceding chapters. Another objective here is to associate calendar years with some of the more important biblically-reported events prior to the time of Abram.

It would seem appropriate to approach this chronological summary starting from the latest and perhaps most reliably documented time period prior to the time of Abram. An especially significant and potentially datable biblical event for the time before Abram is the flood that has been associated with Noah and also with the corresponding Babylonian hero Utnapishtim. Many of the previous pages have been devoted to understanding this flood, including its location, extent, and especially the date of its occurrence. Thus, the first step here will be to attempt to establish an approximate historical chronology for the time between the flood and Abram.

For times before Noah, several relative biblical and Bible-related chronologies have been summarized in Chapter 3. It was found that there is close agreement among the chronologies concerning the numbers and names of the people represented, but there are many substantial differences between the chronologies concerning the years that are associated with particular individuals. The chronologies also include many implausibly high ages of the patriarchs when their sons are said to have been born. Possible modifications of the early chronologies before and after Noah are considered in this chapter. These modifications tend to make the ages of the ancient patriarchs similar to the ages of typical modern historical people. The modifications are also compatible with the more plausible historical dates for the flood of Noah's era that have been inferred in previous chapters.

The chronology of the time period between Noah and Abram is considered in Section 12.2. The most reasonable of several different biblical chronologies for this period (the Masoretic version) is taken as a starting point. As discussed in Section 12.3, there is some ambiguity about the immediate descendants of Noah. In Section 12.4 an approximate averaging procedure is employed for replacing a few of the least-likely-appearing patriarchal ages with a more reasonable approximate average value found for a sequence of eight consecutive postdiluvian patriarchs. The results agree well with previously obtained dates for the time periods associated with Noah and Abram. The period from Adam to Noah is summarized in Section 12.5. In the absence of a more appealing procedure, the ages of the antediluvian patriarchs at the births of their sons are replaced by the average age that had already been suggested for the post-flood period considered in Section 12.2. As discussed in Section 12.6, a similar averaging procedure has been applied previously to the chronology of China for about the same period of time as that of interest in the study of the flood in Noah's time.

12.2 From Noah to Abram

The primary purpose of this section is to develop a plausible chronology for times between Noah and Abram. There are important Bible stories relating to the time before Abram, and it is perhaps appropriate to focus initially on Abram's more immediate predecessors. For times after Abram and his family, the various fundamental versions of the Bible have fewer differences in chronology and fewer instances where the numbers given seem conspicuously doubtful. Generally, however, the Bible provides a wealth of chronological information; and the implications of this information for times after Abram have been emphasized previously. However, for the period from Noah to Abram, there are some basic concerns that must be addressed. The biblical data for this period are summarized in Table 12.1.

Causes And Effects Of The Flood

Table 12.1. Biblical chronology of the patriarchs from Noah to Abram. The numbers shown are the ages of these patriarchs at the births of some of their biblically most important sons.

Patriarch	Verses	Masoretic	Samaritan	Septuagint
Noah	Genesis 5:32	502	502	502
Shem	Genesis 11:10	100	100	100
Arpachshad	Genesis 11:12	35	135	135
Kainan	Genesis 11:13	(30)		130
Shelah	Genesis 11:14	30	130	130
Eber	Genesis 11:16	34	134	134
Peleg	Genesis 11:18	30	130	130
Reu	Genesis 11:20	32	132	132
Serug	Genesis 11:22	30	130	130
Nahor	Genesis 11:24	29	79	79
Terah	Genesis 11:26	70	70	70
Abram	Genesis 16:16	86	86	86

One obvious problem with the information in the table is that the various versions of the Bible sometimes disagree significantly with each other. Thus, for the time from Arpachshad (the grandson of Noah) to Nahor (the grandfather of Abram), the ages of the patriarchs at the births of their principal successors were greater by exactly one hundred years (fifty years for Nahor) in the Samaritan and Septuagint versions of the Bible than in the more plausible Masoretic version.[2] It is clear that to develop a useful and unique reference chronology, some assumptions are needed to address this and other discrepancies.

Since the Masoretic numbers for this period mostly seem more reasonable than those of the Samaritan and Septuagint texts (at least to the author), the Samaritan and Septuagint numbers will be discarded for this section of the chronology. Thus, in the following discussion only the Masoretic numbers are included, supplemented by the number thirty (in parentheses) for the age of Kainan. It may be noticed that the Masoretic and Samaritan versions of this genealogy in Table 12.1 do not actually include Kainan, though he is present in the Septuagint (Genesis 11:13), the New Testament (Luke 3:36), and the Book of Jubilees (Jubilees 8:1-5). The indicated age of thirty years is one hundred less than the Septuagint value of one hundred and thirty years, by analogy with the version differences of other names in this era that already have been mentioned.

12.3 Noah and his sons

It may be observed in Table 12.1 that the ages of Noah and Shem (at the top of the table) and Terah and Abram (at the bottom) when their principal sons were born are higher than would be expected for typical human beings. Curiously, these particular ages also agree among the versions. Perhaps there has been an inclination to assign to the more notable characters in the chronology lifetimes that are more impressive. The most striking example in this list is Noah himself, who is said to have been six hundred years old at the time of the flood (Genesis 6:6, 11) and to have lived a few centuries after that. His total lifespan is said to have been nine hundred fifty years, one of the longest lifetimes in the Bible. The tendency to assign long lifetimes to heroes would certainly not be unique to the Bible. Thus, in an extreme case Utnapishtim (the Babylonian counterpart of Noah) is said in the *Gilgamesh Epic* to have been granted immortality (Section 11.6).

Causes And Effects Of The Flood

The age of Noah at the birth of his son Shem is indicated in Table 12.1 to be 502 years. This value is sometimes given as 500 years instead of 502,[3] and this minor discrepancy may warrant a few comments. As a starting point one may consider the verse indicated in the table: "After Noah was five hundred years old, Noah became the father of Shem, Ham, and Japheth" (Genesis 5:32). If this verse is considered in isolation, it might seem clear that Noah was five hundred years old when Shem was born. However, this reading would also seem to suggest that Noah became the father of three sons in one year. To clarify this situation, special attention must be given to the word "after" at the beginning of the verse quoted above. It would seem that each of Noah's three sons could have been born at any time after Noah had become 500 years old. Unfortunately, the word "after" is also translation-dependent. It occurs in many, but not all, translations of this verse; and, for example, it is replaced by the more restrictive word "when" in some translations.

The question of the specific age of Noah when Shem was born may, however, be resolved by a consideration of two additional verses:

> "Noah was six hundred years old when the flood of waters came upon the earth" (Genesis 7:6).

> "When Shem was a hundred years old he became the father of Arpach'shad two years after the flood" (Genesis 11:10).

According to these verses, Shem was a hundred years old two years after the flood. In other words Shem was a hundred years old when Noah was six hundred and two. This implies that Shem

12. From Adam to Abram

was born when Noah was five hundred and two years old, and this is the figure included in Table 12.1. The same arithmetic was also carried out long ago in *Jasher*: "And Noah was five hundred and two years old when [his wife] Naamah bare Shem."[4]

There may also be uncertainty about the order in which the sons of Noah were born. The most popular idea seems to be that they were born in the order Shem, Ham, and Japheth; and there is mixed textual support for this arrangement in the Bible. The first of the following verses was quoted already above:

> After Noah was five hundred years old, Noah became the father of Shem, Ham, and Japheth (Genesis 5:32).

> And Noah had three sons, Shem, Ham, and Japheth (Genesis 6:10).

> On the very same day Noah and his sons, Shem and Ham and Japheth, and Noah's wife and the three wives of his sons with them entered the ark . . . (Genesis 7:13).

> The sons of Noah who went forth from the ark were Shem, Ham, and Japheth (Genesis 9:18).

> He also said, "Blessed by the Lord my God be Shem; and let Canaan [son of Ham] be his slave. God enlarge Japheth, and let him dwell in the tents of Shem; and let Canaan be his slave" (Genesis 9:26-27).

Chapter 10 of Genesis provides a more extended listing of the descendants of Noah. However, there is also a curious ambiguity in that chapter concerning the order in which the sons of Noah were born. The opening verse Genesis 10:1 (as the earlier verses in Genesis quoted above seem to indicate somewhat relentlessly) states that the birth order was Shem, Ham, and Japheth.

Surprisingly however, the immediately following and more extended discussion of the descendants of Noah mostly presents his sons in exactly the opposite order of Japheth, Ham, and Shem:

> These are the generations of the sons of Noah, Shem, Ham, and Japheth; sons were born to them after the flood (Genesis 10:1).
>
> The sons of Japheth: Gomer, Magog, Madai, Javan, Tubal, Meshech, and Tiras. . . . These are the sons of Japheth in their lands, each with his own language, by their families, in their nations (Genesis 10:2-5).
>
> The sons of Ham: Cush, Egypt, Put, and Canaan. . . . These are the sons of Ham, by their families, their languages, their lands, and their nations (Genesis 10:6-20).
>
> To Shem also, the father of all the children of Eber, the elder brother of Japheth, children were born. The sons of Shem: Elam, Asshur, Arpachshad, Lud, and Aram. . . . These are the sons of Shem, by their families, their languages, their lands, and their nations (Genesis 10:21-31).
>
> These are the families of the sons of Noah, according to their genealogies, in their nations; and from these the nations spread abroad on the earth after the flood (Genesis 10:32).

The first sentence in Genesis 10:21 has the appearance of a late and somewhat awkward addition to the text, and no corresponding sentence appears in the preceding paragraphs concerning the descendants of Japheth and Ham. The wording of this sentence isn't entirely clear, but it seems to have two purposes. One of these is to remind the reader that in Genesis 10:1 (and the verses quoted previously above) and in spite of the

12. From Adam to Abram

organization of Genesis 10:2-32 Shem was older than his brothers. Also, Eber is sometimes considered to be the eponymous ancestor of the Hebrews, and another purpose may be to emphasize that this same Shem who was being claimed as the eldest son of Noah was also a direct ancestor of Eber.

This difference in order could be imagined to have important implications relating to the sons and their birthrights:

> While all a man's sons had his protection and some benefits, the first-born son had a special inheritance. His was the principal inheritance of property and name. Through him the family line was continued.[5]

One might imagine that the more comprehensive genealogical information given in Genesis 10:2-32 would correspond to the original historical reality, while the various short listings in Genesis 10:1 and in the other preceding verses quoted above could represent very intentional later additions and editing. Putting Shem first in many places could improve the apparent birthright status of the Hebrew descendants of Shem, some of whom may themselves have participated in the editing of the ancient documents.

Some listings concerning the sons of Noah are also given in *Jasher*. These listings consistently identify Japheth as Noah's first son and Shem as his last. Thus, the ambiguity seen in the biblical texts appears to be absent in *Jasher*. In one seemingly damaged version (statement on Ham missing), one finds again that Japheth was born before Shem:

> And Noah was four hundred and ninety eight years old, when he took Naamah for a wife.

> And Naamah conceived and bare a son, and he called his name Japheth, saying, God has enlarged me in the earth; [and she conceived again . . . Ham . . . (postulating missing text)], and she conceived again and bare a son, and he called his name Shem, saying, God has made me a remnant, to raise up seed in the midst of the earth (*Jasher* 5:16-17).

In the more complete form of the list of Noah's descendants found in Chapter 7 of *Jasher*, one finds clearly that Noah's sons were born in the order Japheth, Ham, and Shem:

> And these are the names of the sons of Noah: Japheth, Ham and Shem; and children were born to them after the flood, for they had taken wives before the flood (*Jasher* 7:1).
> These are the sons of Japheth; Gomer, Magog, Madai, Javan, Tubal, Meshech and Tiras, seven sons. . . . [T]hese are the sons of Japheth according to their families, and their numbers in those days were about four hundred and sixty men (*Jasher* 7:2-9).
> And these are the sons of Ham; Cush, Mitzraim, Phut and Canaan, four sons. . . . These are the sons of Ham, according to their families, and their numbers in those days were about seven hundred and thiry men (*Jasher* 7:10-14).
> And these are the sons of Shem; Elam, Ashur, Arpachshad, Lud and Aram, five sons. . . .These are the sons of Ham [error, should be Shem], according to their families; and their numbers in those days were about three hundred men (*Jasher* 7:15-18).

Similar information is given by Josephus[6] and in Jubilees (Jubilees 7:8-10), but in these sources the order of birth of Noah's sons is first given as Shem, Japheth, and Ham. Later, however,

12. From Adam to Abram

the order in Josephus seems to be either Japheth, Ham, and Shem (as in *Jasher*) or perhaps Japheth, Shem, and Ham[7]

There is a somewhat similar ambiguity concerning the sons of Shem. In one location the text would seem to indicate that Arpach'shad was the firstborn son of Shem:

> When Shem was a hundred years old, he became the father of Arpach'shad two years after the flood; and Shem lived after the birth of Arpach'shad five hundred years, and had other sons and daughters (Genesis 11:10-11).

On the other hand, in a more formal listing of the descendants of Noah, Arpach'shad is said to have been the third of Shem's five sons.

> The sons of Shem: Elam, Asshur, Arpach'shad, Lud, and Aram. The sons of Aram: Uz, Hul, Gether, and Mash (Genesis 10:22-23).

In a much later text, the verses above have perhaps been improperly merged:

> The sons of Shem: Elam, Asshur, Arpach'shad, Lud, Aram, Uz, Hul, Gether, and Meshech (1 Chronicles 1:17).

It would seem that Arpach'shad was a son of Shem, even as Shem was a son of Noah. Putting Arpach'shad first could be imagined to improve the birthright status of his Hebrew descendants. A comprehensive statement of the position of Arpach'shad among the descendants of Shem is also provided in Jubilees and *Jasher:*

> And these are the sons of Shem: Elam, and Asshur, and Arpach'shad – this (son) was born two years after the flood – and Lud, and Aram (Jubilees 7:18).
>
> And these are the sons of Shem; Elam, Ashur, Arpachshad, Lud and Aram, five sons. . . (*Jasher* 7:15).

Abram was a descendant of Arpach'shad, and the preceding paragraphs might provide some perspective concerning his place within the family of Noah.

12.4 Toward a more realistic chronology

It should be recalled that the ages given in Table 12.1 do not necessarily represent the ages of the patriarchs at the births of their first sons. As seen above, the people represented in the table could perhaps be better described as biblically important sons rather than eldest sons. The Masoretic ages associated with the patriarchs from Arpach'shad to Nahor (including thirty years for Kainan) at the birth of a typical child in the table (not necessarily the youngest or the eldest) would probably not be particularly unreasonable, even in more modern times.

If the indicated numbers in the table are reasonable, then it should be possible to use the data in the table to approximate any unknown or seemingly unreasonable data in estimating multigenerational time intervals. The average of the ages for the patriarchs between Arpach'shad and Nahor is about thirty-one years. Therefore, as a rough rationalization of the table, it is suggested that Shem may have been born when Noah was about thirty-one years old (rather than five hundred and two), and Arpach'shad may have been born when Shem was about thirty-one years old (rather than one hundred). Similarly, it is suggested that Terah may have been about thirty-one years old when Abram was born, and Abram may have been about thirty-one years old when his Egyptian wife Hagar gave birth to Ishmael.

With the modifications just summarized, an alternative approximate biblical chronology for the period from Noah to Abraham is given in Table 12.2. In this version of the table, the qualitatively-estimated thirty-one year intervals are included in parentheses.

Table 12.2. Approximate biblical chronology for the period from Noah to Abram. The Biblical Age column corresponds to the Masoretic ages shown previously in Table 12.1. The column Adjusted Age indicates the approximate ages of the patriarchs at the births of their sons, and the column Date indicates (except for the flood line) the dates of birth of those sons.

Patriarch	Biblical Age	Adjusted Age	Date
Noah	502	(31)	2064 B.C.
flood			2035
Shem	100	(31)	2033
Arpachshad	35	35	1998
Kainan	30	30	1968
Shelah	30	30	1938
Eber	34	34	1904
Peleg	30	30	1874
Reu	32	32	1842
Serug	30	30	1812
Nahor	29	29	1783
Terah	70	(31)	1752
Abram	86	(31)	1721

The dates given in the Date column are chosen to be compatible with the chronology in the Adjusted Age column, and those dates also agree with the few absolute dates that have been tentatively inferred in these investigations. Thus, the first line in the table suggests that Noah may have been about 31 years of age in 2064 B.C. when his son Shem was born. The first line also

12. From Adam to Abram

implies that Noah himself was born in about 2095 B.C. The tentatively adopted date for Noah's flood is 2035 B.C.,[8] and thus Noah was about sixty years old at the onset of the flood (rather than six hundred years old as indicated in Genesis 7:6,11). Similarly, Arpachshad is indicated to have been born in the year 2033 B.C. This number is consistent with the following biblical statement: "[Shem] became the father of Arpach'shad two years after the flood" (Genesis 11:10). It is also indicated in Table 12.2 that Abram became the father of Ishmael in about the year 1721 B.C. This date is nearly the same as that suggested previously in a chronology for the period between the journeys to Egypt by Abram and Jacob.[9]

The chronology in Table 12.2 also provides a possible approach to finding approximate historical dates for other events associated with specific biblical characters. As a starting point one may consider what the Bible says about the person Peleg, whose name appears in Table 12.2: "To Eber were born two sons: the name of the one was Peleg [That is Division], for in his days the earth was divided, and his brother's name was Joktan" (Genesis 10:25, see also 1 Chronicles 1:19). This "division" of the earth is often associated with the scattering of people in the tower-of-Babel story after their language had been "confused:"

> "Come let us go down, and there confuse their language, that they may not understand one another's speech." So the Lord scattered them abroad from there over the face of all the earth, and they left off building the city. Therefore its name was called Ba'bel, because there the Lord confused the language of all the earth; and from there the Lord scattered them abroad over the face of all the earth (Genesis 11:7-9).

Causes And Effects Of The Flood

"The name [Peleg] is explained as 'division,' because in his days the earth was divided – probably a reference to the Tower of Babel story."[10]

Contrary to the above, however, an alternative meaning of this "division" of the earth is provided in Jubilees:

> And she [Mu'ak] bare him [Shelah] a son in the fifth year thereof, and he called his name Eber: and he took unto himself a wife, and her name was 'Azurad, the daughter of Nebrod, in the thirty-second jubilee, in the seventh week, in the third year thereof. And in the sixth year thereof, she bare him a son, and he called his name Peleg; for in the days when he was born the children of Noah began to divide the earth amongst themselves: for this reason he called his name Peleg. And they divided (it) secretly amongst themselves, and told it to Noah. And it came to pass in the beginning of the thirty-third jubilee that they divided the earth into three parts, for Shem and Ham and Japheth, according to the inheritance of each, in the first year in the first week, when one of us ["i.e., one of the angels"], who had been sent, was with them. And he called his sons, and they drew nigh to him, they and their children, and he divided the earth into the lots, which his three sons were to take in possession, and they reached forth their hands and took the writing out of the bosom of Noah, their father . . . (Jubilees 8:7-11).

Thus, according to Jubilees the earth was divided up among the sons of Noah, and it was this division that began in about 1904 B.C., near the time of the birth of Peleg as indicated in Table 12.2. The detailed discussion of this division goes on in Jubilees for the many remaining verses of Chapter 8. The subdivision of the earth among the grandchildren of Noah occupies Chapter 9. Noah's death and burial are also recorded: "And Noah slept with his fathers, and was buried on Mount Lubar in the land of Ararat" (Jubilees 10:15).

Earlier in Jubilees, Lubar also is said to have been the mountain on which the ark had landed: "And the ark went and rested on the top of Lubar, one of the mountains of Ararat" (Jubilees 5:28). After the writing of Jubilees, the mountain on which the ark landed was referred to by Nicolaus of Damascus (c. 64 - 4 B.C.) as Mount Baris, as quoted in Section 9.5. Perhaps there is some relationship between the names Lubar and Baris.

In Jubilees the tower of Babel story refers to a somewhat later event than the dividing of the earth. The relating of that event occupies verses 18 to 27 in Chapter 10 of Jubilees. The first part of the story is the following:

> And in the three and thirtieth jubilee, in the first year in the second week, Peleg took to himself a wife, whose name was Lomna the daughter of Sina'ar, and she bare him a son in the fourth year of this week, and he called his name Reu; for he said: "Behold the children of men have become evil through the wicked purpose of building for themselves a city and a tower in the land of Shinar." For they departed from the land of Ararat eastward to Shinar; for in his days they built the city and the tower, saying, "Go to, let us ascend thereby into heaven." And they began to build, and in the fourth week they made brick with fire, and the bricks served them for stone, and the clay with which they cemented them together was asphalt which cometh out of the sea, and out of the fountains of water in the land of Shinar. And they built it: forty and three years were they building it. . . . (Jubilees 10:18-21).

The above verses explain how it is that the ancestors of Abram came to be in Babylonia. It seems that it was Reu and his family (and perhaps others) that moved from Ararat to Shinar. The meaning of Shinar has been given in part as follows: "Name for Babylonia occurring eight times in the Old Testament."[11] Thus, the move by Abram's ancestors is said to have occurred after they had received their allotment of land in the division of the earth that had begun near the time of the birth of Peleg. The building of the tower of Babel may have begun later in about 1874 B.C., near the time of the birth of Reu as indicated in Table 12.2. Thus, the dividing of the earth among the descendants of Noah is said in Jubilees to have happened earlier than the confusion of languages and the scattering of those people who were participating in the building of the tower of Babel (Jubilees 10:26-27).

12.5 From Adam to Noah

There could be intrinsic interest in collecting and comparing the various genealogies for any place and any period of time. Given that information, one can attempt to identify, for example, the earliest or the most reliable of the available chronologies. Unfortunately, for the earliest biblical periods many of the reported lifetimes associated with the persons of interest are vastly longer than the typical lifetimes known for more recent times. This situation was already encountered for the period from Noah to Abram discussed in the previous section. In Table 12.2 eight consecutive ages in the Masoretic list (including an inferred value for Kainan) seemed not to be unreasonable and were retained as found. On the other hand, the two first and the two last ages in the list seem very high and were replaced by 31 years, the average of the eight reasonable-seeming values.

12. From Adam to Abram

The suggestion implicit in this interpretation is that actual human lifetimes have always been about the same, as have the ages of fathers at the births of their sons. However, the earliest biblical records have been subjected to the most transcriptions and thus have the greatest probability of discrepancies. Also, it is possible that the intentional lengthening of ages upon transcription could be intended to increase in the reader a sense of awe at the supernatural content that appears to be present in a story. Unfortunately, it seems to be at least as likely today that skepticism concerning impossible-sounding ages in an important story might cause a reader to dismiss the story as fiction.

The practice of assigning extremely long time periods seems to be common for the era of interest in the vicinity of Mesopotamia. Perhaps the best-known example is provided by what is known as the Sumerian King List, especially for the period before the occurrence of a great flood. One of the eight antediluvian kings on the list is Enmenlu-Anna, and he is said to have reigned for 43,200 years. This is far longer than the greatest biblical age, Methuselah's nine hundred and sixty-nine years (Genesis 5:27).

It is important to not overstate any seeming analogy between the Sumerian and biblical antediluvian leaders. First of all, these two groups of leaders would seem to be associated with two different geographical regions. The Sumerians lived in Mesopotamia ("between the rivers" Tigris and Euphrates). Thus, any flooding in Mesopotamia would probably have been directly associated with those rivers. On the other hand, the biblical personages from Adam to Noah probably lived near Lake Van, and as discussed in previous sections the biblical flood should probably be associated with this lake and its surrounding lands in Urartu (Ararat). The reported Sumerian kings are not all from the same city, so their reign lengths should not necessarily be summed to obtain overall time intervals. Also, the reported names of the members of the Sumerian and biblical groups seem to have nothing in common. Furthermore, the Sumerian kings are each assigned a reign length in years, while the antediluvian patriarchs are usually assigned ages at the births and deaths of their sons. The only particularly likely link between the biblical and Sumerian stories could occur if the flooding of Lake Van and of the Tigris and Euphrates Rivers were caused by the same underlying weather event.

In spite of the absence of a direct connection between the very long antediluvian Sumerian and biblical ages or reigns, it has been of interest to consider how in both cases such long time periods might be interpreted. An appealing discussion of the Sumerian reign lengths has been suggested by Harrison.[12] The relevant data from Harrison's interpretation is summarized in Table 12.3.

12. From Adam to Abram

Table 12.3. Reign lengths of the antediluvian rulers of the Sumerian King List. The columns labeled actual reign are calculated from the stated reigns as described in the text below.

City	Ruler	Stated reign years	Actual reign years	Actual reign months
Eridu	Alulim	28,800	8	0
	Alalgar	36,000	10	0
Badtibira	Enmenlu-Anna	43,200	12	0
	Enmengal-Anna	28,800	8	0
	divine Dumuzi	36,000	10	0
Larak	Ensipazi-Anna	28,800	8	0
Sippar	Enmendur-Anna	21,000	5	10
Shuruppak	Ubar-Tutu	18,600	5	2

The stated reigns in Table 12.3 are vastly longer than any ordinary human lifetimes. It has been suggested that these seemingly very long reigns might be an artifact of the still poorly understood Sumerian sexagesimal numbering system. In that system the number 60 and its multiples are very important. Specifically, the actual reign lengths for the rulers listed in the table can perhaps be obtained by scaling down the stated lengths by the factor 60 raised to the second power (or 3600). This operation leads to the "Actual reign" shown in Table 12.3. Such a procedure is made especially attractive by the fact that all of the results can be expressed in integer numbers of years and months. As an illustration of this procedure, we may consider the case of the king Enmendur-Anna who is said to have reigned for 21,000 years. To find the actual reign of this king, we may divide this number by 3600: 21,000/3,600 years is equal to 70/12 years, which can be read as 70 months or 5 years and 10 months as shown in Table 12.3.

The purpose of this system for expressing the length of a king's reign is not yet fully understood:[13]

> So little insight has been gained into the theoretical dynamics of Sumerian mathematics that it is impossible to say with certainty what the reason was for employing base-60 squared as a constant, assuming that this was its actual function in the King List, as seems eminently probable. It was certainly integral to the structure of the various recorded reigns, unlike some constants in modern mathematics that grace an equation but are not indispensable entities.[14] Why base-60 should have been squared in order to perform its function satisfactorily is also problematical. Perhaps, after all, base-60 was intended to serve as a symbol of relative power and importance, which the compilers of the ancient Sumerian King List associated with those men whose reigns they recorded.
>
> Regardless of the immediate answers to these queries, it seems clear that base-60 squared should be recognized as an "ideal" constant, which, however, must be factored out once it has been isolated so that it is not reckoned as part of the overall calculation. . . .

12. From Adam to Abram

It is now appropriate to consider, at least briefly, the possible ages of the antediluvian biblical patriarchs at the births of their sons. As noted in Section 3.3, the Samaritan and closely similar Jubilees chronologies for these patriarchs provide the most plausible (or least implausible) of the bible-related chronologies for this era. Table 12.4 includes a summary of the Samaritan chronology from Table 3.2. Like the Sumerian chronology just discussed, the ages here seem unreasonably large, but at least they are not nearly so large as the uncorrected Sumerian values.

It may be recalled from Section 12.2 that for the time between Noah and Abram the first two ages and the last two ages in Table 12.2 were replaced by the average of the more reasonable intervening eight ages. That average was about thirty-one years. A similar circumstance arises in the Samaritan chronology between Adam and Shem as shown in Table 3.1. Thus, in Table 12.4 the first two ages (Adam and Seth) and the last two ages (Noah and Shem) seem significantly larger than the ages of the intervening patriarchs. In this case the average ages of the intervening patriarchs at the births of their sons would be about 67.4 years. While this age (or each of the component ages) seems high for the births of the sons, it could be about right for the ages of the patriarchs at the times of their deaths. However that may be, the final column assumes that the ages of the patriarchs at the births of their sons are all about thirty-one years, as assumed previously for the period between Noah and Abram.

Table 12.4. Approximate biblical chronology of the antediluvian patriarchs from Adam to Shem. The column Adjusted Age indicates the estimated approximate ages of the patriarchs at the births of their listed sons.

Patriarch	Biblical Age	Adjusted Age	Date
Adam	130	(31)	2343 B.C.
Seth	105	(31)	2312
Enosh	90	(31)	2281
Kenan	70	(31)	2250
Mahalalel	65	(31)	2219
Jared	62	(31)	2188
Enoch	65	(31)	2157
Methuselah	67	(31)	2126
Lamech	53	(31)	2095
Noah	502	(31)	2064
flood			2035
Shem	100	(31)	2033

Adam is, of course, said in the Bible to have been created rather than born, and he is said to have been moved east from his place of creation to the garden of Eden (Section 2.4). On the other hand, some readers would be inclined to believe that Adam was actually born to ordinary human parents. With this alternative possibility his birth might have been in about 2374 B.C. (2343 + 31), with an uncertainty of at least several years.

12. From Adam to Abram

The location of Adam's origin is also of interest. In this context one may recall the creation stories from Ebla that were introduced previously in Section 1.4. Thus, the first part of an Eblaite creation hymn is the following:

> "Lord of heaven and earth:
> the earth was not, you created it,
> the light of the day was not, you created it,
> the morning light you had not [yet] made exist."[15]

This short hymn is similar in important respects to the creation story in Genesis. Thus, the creation events in Genesis may have been recorded at about the same time that written descriptions of a similar creation story appeared in the archives of Ebla. In the non-biblical interpretation mentioned in Section 2.4, one could speculate that Adam and Eve could have been taken from their place of origin before being relocated east to the garden of Eden.

The biblical creation story can be used to obtain an approximate estimate of the place from which Adam and Eve may have come in their journey to the garden of Eden. From the biblical account one finds the following: "And the Lord God planted a garden in Eden, in the east; and there he put the man whom he had formed" (Genesis 2:8). Thus, it would seem that the garden of Eden was east of Adam's place of origin. On the other hand, Ebla was one of the most advanced civilizations of its time, and it was located mostly west from the location that has been suggested here for the garden of Eden. Thus, it could be possible in principle that Adam and Eve were moved to the garden from somewhere in the domain of Ebla. A few words on the chronology of Ebla would help in determining whether the chronology of the life of Adam could in any way be compatible with the archive period of Ebla:

> The city (of Ebla) was excavated starting in 1964, and became famous for the Ebla tablets, an archive of about 20,000 cuneiform tablets found there, dated to around 2350 BC. . . . The archive period, which is designated "Mardikh IIB1", lasted from c. 2400 BC until c. 2300 BC. The end of the period is known as the "first destruction", mainly referring to the destruction of the royal palace and much of the acropolis.
>
> The written archives do not date from before Igrish-Halam's reign, which saw Ebla paying tribute to Mari, and an extensive invasion of Eblaite cities in the middle Euphrates region led by the Mariote king Iblul-II. Ebla recovered under King Irkab-Damu in about 2340 BC; becoming prosperous and launching a successful counter-offensive against Mari.[16]

The first man mentioned in the Bible was named Adam, and there are also indications that the name Adam may be associated more directly with the word "man": "This is the book of the generations of Adam. When God created man, he made him in the likeness of God" (Genesis 5:1).

> Till now the only attestation of *'ādām*, "man, Adam," outside the Bible appeared in Old Akkadian texts from the period of Sargon the Great (circa 2350 B.C.) in the form of the personal names *A-da-mu*, *'A-da-mu*, and *A-dam-u*. Now from Ebla comes the personal name *A-da-mu*, one of the 14 governors of the provinces under King Igriš-Halam.[17]

An approximate chronology for the kings of Ebla during the archive period has been given as shown in Table 12.5.[18]

12. From Adam to Abram

Table 12.5. Approximate chronology of the rulers of Ebla during the archive period.

Ruler	Approximate Reign	Comment
Kun-Damu	2400 B.C.	
Adub-Damu	2380	Short reign
Igrish-Halam	2360	Ruled 12 years
Irkab-Damu	2340	Ruled 11 years
Isar-Damu	2320	Ruled about 35 years
Ir'ak-Damu		A prince, might have ascended the throne for a short period

As implied above, the Eblaite word for man seems to have been some form of the personal name A-da-mu, and this name is said to have been known from the reign of Sargon the Great (c. 2350 B.C.). The same name is known in Ebla in the reign of Igriš-Halam, and from Table 12.5 that ruler reigned in c. 2360 B.C. These dates are approximately compatible with the date of roughly 2374 B.C. inferred earlier in this section for the origin of

Adam in a location west of Eden. It appears that this person may have anticipated the soon-to-occur destruction of Ebla near the end of the archive period and thus migrated (as Adam) to the east with Eve to what became known as the garden of Eden. He previously could have learned from the archives of Ebla, the world's first library, the Eblaite story of the creation of the world. That creation story is quoted briefly earlier in this section, and it has been discussed in more detail in Section 1.5.

12.6 Other lifetime estimates

In this chapter several assumptions have been made in estimating approximate dates for biblical events between Adam and Abram. One assumption is that normal life spans of human beings haven't changed much in the past few thousand years. The age of a person at the birth of a child is particularly important in developing relative and absolute chronologies for a family. The age of a person at death or at some other event after the birth of a child of interest would, of course, be dependent on the incidence of war, disease, and the quality of medical care. However, such ages are usually of less importance in the context of chronology.

Sometimes the age of a parent at the birth of a child is not available, or for various reasons does not appear to be reliable. In such a circumstance it might be necessary to consider some approximate method for estimating ages. In this section we have adopted the expedient of using the average of a series of reasonable-appearing parental ages at the births of their children as a substitute for unreasonable appearing ages elsewhere in the genealogy. Specifically, a series of such ages appearing between Noah and Abram yielded about thirty-one years as the average

12. From Adam to Abram

age of a father at the birth of a son. This period of thirty-one years was employed as a replacement for a few seemingly excessive ages in the period from Noah to Abram, and then it was used further as the sole basis for estimating ages in the period from Adam to Noah.

For comparison it is found that chronological records for almost the same era have been found in China. These records give the appearance of relatively good internal consistency, and even good agreement with astronomically-based data for the same period. The first known instruction "to begin systematic astronomical observations and calendar making was as follows: 'Thereupon (King Yao) commissioned Hsi and Ho reverently to follow the august Heaven, and calculate and delineate (the movements of) the sun, moon and (other) celestial bodies, and respectfully give the seasons to the people.' "[19] The records referred to here include in particular the reports of the extraordinary floods mentioned previously in Section 9.3. Those floods in China may have been at about the same time as the flood associated with Noah and his Bablyonian counterpart Utnapishtim. A tentative date of about 2035 B.C. for this flood has been suggested previously (Section 12.4).

An important application of the ancient astronomical data is the combining of those data with ancient genealogies to establish the typical length in China of one generation: "The traditional Chinese view is that there are 30 years between two generations."[20] The thirty year generation period found for times near the flooding in China is of course close to the approximate thirty-one year generation period inferred here for times near the corresponding flooding in the vicinity of Mesopotamia.

12.7 Conclusion

The main subject of this chapter has been the development of an approximate absolute chronology for the people and events between the Genesis creation story and the life of Abram. For this purpose it has been necessary to use an averaging procedure based on what seem to be the more plausible recorded ages to develop substitutes for ages that are less plausible. This is, of course, a subjective process, but it seems to be satisfactory for times after Noah's flood. In particular, it provides good agreement with dates inferred previously for the flood event itself. Some insights have been suggested for the chronology before Noah, but uncertainties may be significant.

---

1. L. W. Casperson, *Patterns of Biblical Chronology* (Westbow Press, Bloomington, Indiana, 2012).
2. S. J. De Vries, "Chronology of the OT," *The Interpreter's Dictionary of the Bible, an Illustrated Encyclopedia*, Edited by G. A. Buttrick (Abingdon Press, Nashville, 1962), Volume 1, A-D, pp. 580-599; p. 581.
3. See for example, S. J. De Vries, op. cit.
4. *The Book of Yashar*, Translated from the Hebrew and Published by Mordecai Manuel Noah (Hermon Press, New York, 1972), (cited here in the form *Jasher*); *Jasher* 5:18.
5. O. J. Baab, "Birthright," *The Interpreter's Dictionary of the Bible, an Illustrated Encyclopedia*, op. cit., p. 440.
6. F. Josephus, *The Life and Works of Flavius Josephus*, Translated by William Whiston (Holt, Rinehart and Winston, New York), *The Antiquities of the Jews*, Book I, 4:1, p. 39.
7. Ibid., Book I, 6:1-4, pp. 40-41.

8. L. W. Casperson, op. cit., Chapter 21, "Cause of the famine," p. 542, reference 24; see also Chapter 9 of this study.
9. L. W. Casperson, op. cit., Section 24.5, "Revised chronology," pp. 609-610, Table 24.3.
10. S. Cohen, "Peleg," *The Interpreter's Dictionary of the Bible, An Illustrated Encyclopedia*, Edited by G. A. Buttrick (Abingdon Press, Nashville, 1962), Volume 3, K-Q, p. 709.
11. E. G. Hirsch and G. A. Barton, "Shinar," *The Jewish Encyclopedia* (Funk and Wagnalls Company, New York, 1905), Volume XI, Samson-Talmid Hakam, pp. 290-291.
12. R. K. Harrison, "Reinvestigating the antediluvian Sumerian king list," *Journal of the Evangelical Theological Society*, Volume 36, Number 1, pp. 3-8 (March 1993).
13. R. K. Harrison, op. cit., p. 7.
14. Reference from quote: "Thus, in Einstein's famous equation $E = mc^2$, where E represents energy, m stands for mass, and c^2 is the velocity of light squared, the constant (c^2) is often dispensed with by modern astrophysicists and others."
15. G. Pettinato, *The Archives of Ebla, An Empire Inscribed in Clay* (Doubleday & Company, Inc., Garden City, New York, 1981), p. 244.
16. "Ebla," Wikipedia, Retrieved from the internet on 1 September 2017.
17. G. Pettinato, op. cit., Afterword by M. Dahood, "Ebla, Ugarit, and the Bible," pp. 271-321; p. 274.
18. "Ebla," Wikipedia, Retrieved on 3 September 2017.
19. K. D. Pang, "Extraordinary floods in early Chinese history and their absolute dates," *Journal of Hydrology*, Volume 96, 139-155 (1987); p. 140.
20. Ibid., p. 145.

PART 3
ADVENTURES IN EGYPT

13. ARRIVAL IN EGYPT

"Now there was a famine in the land. So Abram went down to Egypt to sojourn there, for the famine was severe in the land" (Genesis 12:10).

13.1 Introduction

The focus of this chapter is on the period of time when the early Hebrews first interacted with the Egyptians. In the biblical records these encounters began when Abram, together with his extended family and servants, travelled to Egypt in response to a famine in Canaan. Further interactions then continued until the exodus. Some aspects of the Hebrew and Egyptian chronologies for this period have been discussed in a previous study.[1] Because of space limitations other subjects were not included, and some of those are introduced here as Chapters 13-17. During this period of relatively intense involvement it should not be surprising to find that there are literary or archeological indications within each of these communities of involvement with the other group. The time period of particular interest in this chapter begins with the first arrival of Abram and perhaps his Hyksos colleagues in Egypt. Activities during times of Hyksos

13. Arrival in Egypt

domination will be considered in Chapter 14, and times closer to the exodus will be treated in Chapters 15 to 17.

Actual dates have been proposed previously for the events associated with the initial arrival in Egypt of both Abram and the Hyksos. This information is recalled, summarized, and extended in Section 13.2. Several of the most important dates for the entire Hyksos era are suggested in Table 13.1, and at least a partial rationale for most of those dates accompanies the table. Interactions between Abram, his wife Sarai, and the king of Egypt are discussed in Section 13.3. Particular topics of this section include the king's identity, his desire to marry Sarai, and his hope of establishing a political alliance with Abram. Section 13.4 indicates the apparent ease with which Abram acquired great wealth in his various homes of Ur, Haran, Egypt, and finally Canaan. In connection with his activities in Egypt, Abram is sometimes said to have been very knowledgeable and skillful in teaching the math and science that he had learned from the Chaldeans. Evidence relating to such claims is also considered in Section 13.5

13.2 Chronology of the Hyksos era

As noted above, direct interactions between the Hebrews and the Egyptians began when Abram traveled to Egypt: "Now there was a famine in the land [of Canaan]. So Abram went down to Egypt to sojourn there, for the famine was severe in the land" (Genesis 12:10). Assuming that Abram's journey to Egypt coincided with the journey of the Hyksos from Canaan to Egypt, the general date of that occasion had been estimated to be in the range of 1730 B.C. to 1720 B.C.[2] Like the later famine of Joseph's time, the

Adventures In Egypt

famine that brought Abram, his family and associates, and probably the Hyksos into Egypt may have been the consequence of a volcanic eruption. In fact, there was a major eruption in this date range, and from eruption-related information it was inferred that Abram's journey to Egypt probably occurred in about 1727 B.C.[3] Further consideration of the possible identity of this volcano has been included in Section 8.2.

An approximate chronology of the Hebrews from the time of Abram's entry into Egypt until Jacob's entry into Egypt was given previously.[4] An extended version of this chronology is given here as Table 13.1.[5] This table attempts to represent the dates of several of the important events during the period from Abram's entry into Egypt until the accession of the first post-Hyksos Egyptian kings. The entry date of 1727 B.C. is included in the approximate chronology shown in Table 13.1. Dates in the table should be understood as best estimates given the biblical and other data available, and supporting arguments for them are given. Dates marked with circa (c.) are invented to be physically reasonable in context, but errors in these values may be as large as several years. The unmarked dates have a more specific basis and should usually be more accurate.

13. Arrival in Egypt

Table 13.1. Approximate chronology for the period from the entry of Abram into Canaan to the accession of Amenophis I (Eighteenth Dynasty).

Event	Date
Abram moves from Haran to Canaan	1729 B.C.
Abram journeys to Egypt	1727
Tanis established, Abram returns to Canaan	1722
Marriage of Abram to Hagar	1719
Birth of Ishmael	1718
Birth of Isaac	1704
Marriage of Isaac to Rebekah	1684
Birth of Esau and Jacob	1683
Conquest by Hyksos, Accession of Salatis/Sheshi	1674
Birth of Joseph	1664
Accession of Khyan (Hyksos)	1661
Accession of Auserre Apophis I (Hyksos)	1651
Joseph taken to Egypt	1647
Joseph interprets dreams of butler and baker	1636
Joseph promoted, first year of plenty, birth of Apophis II	1634
Eruption of Thera volcano	1628
First year of famine	1627
Jacob's journey to Egypt	1625
End of famine	1620
Death of Jacob in Egypt	1608
Accession of Aqenenre Apophis II (Hyksos)	c. 1593
Accession of Khamudy (Hyksos)	c. 1578
Accession of Ahmose I (Eighteenth Dynasty)	1570
Expulsion of Hyksos	1567
Eruption that caused tempest	1554
Tempest damage reconstruction	1549
Accession of Amenophis I (Eighteenth Dynasty)	1546

Most of the newly introduced dates in the table will be discussed here in the next few paragraphs. The table indicates that Abram moved to Hebron (in Canaan) in 1729 B.C. and Tanis

Adventures In Egypt

(in Egypt) in about 1722 B.C. The basis for these additions is the following story in Jubilees:

> And he [Abram] removed from thence and went towards the south [cf. Genesis 12:9], and he came to Hebron, and Hebron was built at that time, and he dwelt there two years, and he went (thence) into the land of the south, to Bealoth [a town in southern Judah (Joshua 15:24)], and there was a famine in the land. And Abram went into Egypt [cf. Genesis 12:10]. . . , and he dwelt in Egypt five years before his wife was torn away from him. Now Tanais [i.e., Zoan] in Egypt was at that time built – seven years after Hebron [cf. Numbers 13:22]. And it came to pass when Pharaoh seized Sarai, the wife of Abram, that the Lord plagued Pharaoh and his house with great plagues because of Sarai, Abram's wife. And Abram was very glorious by reason of possessions in sheep, and cattle, and asses, and horses, and camels, and menservants, and maidservants, and in silver and gold exeedingly. And Lot also, his brother's son, was wealthy. And Pharaoh gave back Sarai, the wife of Abram, and he sent him out of the land of Egypt, and he journeyed to the place where he had pitched his tent at the beginning, to the place of the altar, with Ai on the east, and Bethel on the west, and he blessed the Lord his God who had brought him back in peace (Jubilees 13:10-15).

The more familiar biblical version of much of the above story is included in Genesis 12:9-13:4. However, Jubilees adds specific information on the dates of the establishment of the cities of Hebron and Tanis (Zoan, San el-Hagar). These dates can't be directly inferred from the related biblical statement in Numbers 13:22. In Jubilees Hebron is said to have been built two years before the famine and Abram's entry into Egypt (in 1727 B.C.),

13. Arrival in Egypt

whereas Tanis was built after Abram had been in Egypt for five years and seemingly just before he left to return to Canaan. The dates of these two events are included in the table.

It should be noted that in contrast to the time intervals suggested above, Artapanus reports a longer period of time for Abram's stay in Egypt:

> Artabanus in his Jewish History says that the Jews were called Ermiuth, which when interpreted after the Greek language means Judaeans, and that they were called Hebrews from Abraham. And he, they say, came with all his household into Egypt, to Pharethothes the king of the Egyptians, and taught him astrology; and after remaining there twenty years, removed back again into the regions of Syria: but that many of those who had come with him remained in Egypt because of the prosperity of the country.[6]

Thus, Artapanus suggests that the Jews were called Hebrews from Abraham rather than from Eber as mentioned in Section 12.3 and that they remained in Egypt for twenty years rather than five years as reported in Jubilees. However, the schematic twenty-year period seems long and less plausible for the urgent story line in Jubilees of Pharaoh having "sent him out of the land of Egypt." The Genesis version also has the king being very blunt in expelling Abram from Egypt:

> So Pharaoh called Abram, and said, "What is this you have done to me? Why did you not tell me that she was your wife? Why did you say, 'She is my sister,' so that I took her for my wife? Now then, here is your wife, take her, and be gone" And Pharaoh gave men orders concerning him;

and they set him on the way, with his wife and all that he had (Genesis 12:18-20).

In any case, the five-year interval is incorporated in Table 13.1 for the departure from Egypt. Artapanus states "that many of those who had come with him remained in Egypt because of the prosperity of the country," and this may be a representation of the initial immigration by the Hyksos.

Other dates have also now been incorporated into Table 13.1. Possible values for the reign lengths of Salatis and Khyan will be discussed in Section 14.5. The first king of the Eighteenth Dynasty was Ahmose I, and in the previously adopted Egyptian chronology Ahmose I came to the throne in 1570 B.C.[7] As inferred by Hayes, the expulsion of the Hyksos occurred in 1567 B.C., "when the last Hyksos was driven from Egypt by the founder of the New Kingdom [Ahmose I].[8]

The last Hyksos king is usually considered to have been Khamudy, but there hasn't been agreement on the length of his reign before the expulsion of the Hyksos. The period of eleven years is adopted here:

> Another reign length can be inferred from the note on the verso of the Rhind Mathematical Papyrus whereby in the 11th regnal year of the ruling king, Heliopolis has been conquered, and "he of the South" has attacked and taken Sile. Since "he of the South" must denote the Theban ruler Ahmose, the regnal year 11 can only be assigned to the successor of the Hyksos king Apapi: Khamudi. The Hyksos capital Avaris will have fallen to Ahmose not much later.[9]

13. Arrival in Egypt

If Khamudy came to the throne eleven years before the expulsion of the Hyksos, then his accession would have been in about 1578 B.C., as shown in the table.

For completeness a few additional dates estimated previously are also included at the end of Table 13.1. The volcanic eruption that led to the tempest documented on the Tempest Stela is suggested in Section 8.4 to have occurred in 1554 B.C., with the reconstruction beginning in about 1549 B.C. After serving a few years as co-regent with Ahmose I, Amenophis I became sole regent in about 1546 B.C.[10] The above dates account for most of the additions included in Table 13.1 above. It probably isn't necessary to emphasize, but any chronological conclusions relating to the still poorly understood Hyksos period can at best be regarded as reasonable inferences. Furthermore, all of the dates in Table 13.1 depend at least in part on choices made for the underlying Egyptian chronology. Some estimates for the chronology of this time period differ significantly from the relatively popular model employed here.

13.3 Alliance or dalliance
If Abram entered Egypt in about 1727 B.C., then from the assumed coincidence of the Egyptian and biblical four-hundred-year intervals the god/king Seth may have begun his reign in Egypt at about the same time. It seems possible that the Seth of the Four Hundred Year Stela was actually the same person as the known mortal king Meribre Seth associated with the Thirteenth Dynasty, and this possibility was considered previously.[11] This king is thought to have usurped the throne of his predecessor Sehotepkare Intef IV, and thus he may also have arrived in Egypt

451

at about the time of the Hyksos.[12] He seems to have had various problems during and following his reign. If he were king of Egypt during Abram's visit, then he may have sought to marry Abram's wife Sarai, who had pretended she was Abram's sister (Genesis 12:10-20). In punishment "God put a stop to his unjust inclinations, by sending upon him a distemper, and a sedition against his government."[13]

After his reign King Seth was anathematized by his successors on the throne.[14] Whether this treatment was due to his ancestry, religious practices, political shortcomings, or some other factor is difficult to know; and this subject will be mentioned again in Section 13.4. In any case, his reputation may have been rehabilitated by the time of the construction of the Four Hundred Year Stela. In any case, King Seti I (also named for the god Seth) celebrated the four-hundred-year anniversary of the entry of the god/king Seth into Egypt, and the stela was emplaced a few years after the anniversary by Seti's son Ramesses II to commemorate that celebration. Other migrations of the Hyksos to Egypt also occurred, and a discussion of a later group of immigrants will be included in Section 14.2.

For completeness a few further aspects of Abram's visit to Egypt are recalled here. One important point, mentioned above, is that the king of Egypt during Abram's visit sought to marry Abram's wife Sarai, who said that she was his sister (Genesis 12:10-15). The king claimed that he had intended thereby to establish a marriage alliance with Abram:

> And when he [the king] had found out the truth, he excused himself to Abram, that supposing the woman to be his sister, and not his wife, he set his affections on her,

13. Arrival in Egypt

as desiring an affinity with him by marrying her, but not as incited by lust to abuse her.[15]

The possibility of a marriage alliance between the king of Egypt and Abram isn't so far-fetched as it might sound. In most accounts Hagar seems to have been a maidservant whom Sarai had acquired in Egypt: "Now Sar'ai, Abram's wife, bore him no children. She had an Egyptian maid whose name was Hagar . . ." (Genesis 16:1). According to *Jasher*, on the other hand, Hagar was actually a daughter of the king of Egypt: "And the king took a maiden whom he begat by his concubines, and he gave her to Sarai for a handmaid."[16] In view of Sarai's previous years of childlessness, it may not have been long before Hagar became Abram's wife (as suggested in Table 13.1), thus establishing a marriage alliance. However, instead of the king marrying Abram's supposed sister, Abram may have married the king's daughter.

It is of interest to recall from the previous section that the building (or rebuilding) of Tanis in the Nile delta may be associated with the presence of Abram in Egypt. Abram had entered Egypt with his Hyksos allies in about 1727 B.C. as indicated in Table 13.1. It should be mentioned, however, that some building is said to have taken place at Tanis as early as the beginning of the Twelfth Dynasty in the twentieth century B.C.[17] It may also be mentioned that the site of Tanis, associated with Abram in Jubilees, is now often referred to as San el-Hagar.

The Bible has examples of other special marriage relationships besides the one suggested above with respect to Abram. Infertility seems typically to have been considered the fault of the wife:

> When Rachel saw that she bore Jacob no children, she envied her sister; and she said to Jacob, "Give me children, or I shall die!" Jacob's anger was kindled against Rachel, and he said, "Am I in the place of God, who has withheld from you the fruit of the womb?" Then she said, "Here is my maid Bilhah; go in to her, that she may bear upon my knees, and even I may have children through her." So she gave him her maid Bilhah as a wife; and Jacob went in to her. And Bilhah conceived and bore Jacob a son (Genesis 30:1-3).
>
> When Leah saw that she had ceased bearing children, she took her maid Zilpah and gave her to Jacob as a wife. Then Leah's maid Zilpah bore Jacob a son (Genesis 30:9-10).

Actual marriage alliances are also reported in the Bible. Thus, many centuries after Abram another king of Egypt established a marriage alliance with a Hebrew king:

> So the kingdom was established in the hand of Solomon. Solomon made a marriage alliance with Pharaoh king of Egypt; he took Pharaoh's daughter, and brought her into the city of David, until he had finished building his own house and the house of the Lord and the wall around Jerusalem (1 Kings 2:46-3:1).
>
> And he made the Hall of the Throne where he was to pronounce judgment, even the Hall of Judgment; it was finished with cedar from floor to rafters. His own house where he was to dwell, in the other court back of the hall, was of like workmanship. Solomon also made a house like this hall for Pharaoh's daughter whom he had taken in marriage (1 Kings 7:7-8).

13. Arrival in Egypt

Pharaoh king of Egypt had gone up and captured Gezer and burnt it with fire, and had slain the Canaanites who dwelt in the city, and had given it as dowry to his daughter, Solomon's wife; so Solomon rebuilt Gezer (1 Kings 9:16-17).

But Pharaoh's daughter went up from the city of David to her own house which Solomon had built for her (1 Kings 9:24).

Solomon brought Pharaoh's daughter up from the city of David to the house which he had built for her, for he said, "My wife shall not live in the house of David king of Israel, for the places to which the ark of the Lord has come are holy" (2 Chronicles 8:11).

The significance of this alliance may have diminished over time as Solomon came to have a great many other wives:

> Now King Solomon loved many foreign women: the daughter of Pharaoh, and Moabite, Ammonite, E'domite, Sido'nian, and Hittite women, from the nations concerning which the Lord had said to the people of Israel, "You shall not enter into marriage with them, neither shall they with you, for surely they will turn away your heart after their gods"; Solomon clung to these in love. He had seven hundred wives, princesses, and three hundred concubines; and his wives turned away his heart. For when Solomon was old his wives turned away his heart after other gods; and his heart was not wholly true to the Lord his God, as was the heart of David his father (1 Kings 11:1-4).

One could imagine that Pharaoh might not have been pleased with Solomon, and eventually Pharaoh even formed a marriage alliance with one of Solomon's enemies:

> And the Lord raised up an adversary against Solomon, Hadad the E'domite; he was of the royal house in Edom. . . . And Hadad found great favor in the sight of Pharaoh, so that he gave him in marriage the sister of his own wife, the sister of Tah'penes the queen (1 Kings 11:14, 19).

13.4 Acquiring wealth

Having considered briefly the subjects of surrogate wives and marriage alliances with Egypt, other topics relating to Abram's visit to Egypt may also be addressed. The royal treatment that he sometimes received in Ur, Haran, and Egypt is reflected in part by his increasing level of prosperity. Abram was said to have been given great wealth before he left the land of Shinar on his way to Haran:

> "And all the kings, princes and servants gave Abram many gifts of silver and gold and pearl, and the king and his princes sent him away, and he went in peace. And Abram went forth from the king in peace, and many of the king's servants followed him, and about three hundred men joined him. And Abram returned on that day and went to his father's house, he and the men that followed him, and Abram served the Lord his God all the days of his life, and he walked in his ways and followed his law. And from that day forward Abram inclined the hearts of the sons of men to serve the Lord."[18]

> "And Terah took his son Abram and his grandson Lot, the son of Haran, and Sarai his daughter-in-law, the wife of his son Abram, and all the souls of his household and went with them from Ur Casdim to go to the land of Canaan. And when they came as far as the land of Haran they remained there, for it was exceeding good land for pasture, and of sufficient extent for those who

13. Arrival in Egypt

accompanied them. And the people of the land of Haran saw that Abram was good and upright with God and men, and that the Lord his God was with him, and some of the people of the land of Haran came and joined Abram, and he taught them the instruction of the Lord and his ways; and these men remained with Abram in his house and they adhered to him."[19]

It is clear from quotations such as these that (to the extent that *Jasher* is historically reliable) Abram had already become at least moderately wealthy during his stay in Ur and before he went to Haran.

The biblical account agrees that Abram's resources increased in Haran before he traveled farther to the land of Canaan:

> Now the Lord said to Abram, "Go from your country and your kindred and your father's house to the land that I will show you. And I will make of you a great nation, and I will bless you, and make your name great, so that you will be a blessing" (Genesis 12:1-2).
> So Abram went, as the Lord had told him; and Lot went with him. . . . . And Abram took Sar'ai his wife, and Lot his brother's son, and all their possessions which they had gathered, and the persons that they had gotten in Haran; and they set forth to go to the land of Canaan (Genesis 12:4-5).

While Abram and Lot were clearly already wealthy at that time, there is no implication that they were too prosperous for the geographical and economic situation in any place at which they had stopped. Thus they had apparently fit in comfortably (politically and agriculturally) in Haran. As noted explicitly above, Haran is said in *Jasher* to have had "exceeding good land for pasture, and of sufficient extent for those who accompanied them"[20] The people he encountered admired him, and many of them joined him in his journeys. While Abram was in Haran, he was apparently contacted by the Chaldeans, who urged him to return to Ur. Perhaps they wished him to take up a leadership position like that of his father Terah (Section 13.5 below). To help Abram know what to do, he turned to God in prayer:

> And he said, "Shall I return unto Ur of the Chaldees who seek my face that I may return to them, or am I to remain here in this place? The right path before Thee prosper it in the hands of Thy servant that he may fulfil (it) and that I may not walk in the deceitfulness of my heart, O my God." And he made an end of speaking and praying, and behold the word of the Lord was sent to him through me, saying: "Get thee up from thy country, and from thy kindred and from the house of thy father unto a land which I shall show thee, and I shall make thee a great and numerous nation" (Jubilees 12:21).

The land to which Abram was sent was Canaan. However, after a few years a severe famine struck the land of Canaan (Jubilees 13:1-11). As a result Abram and his family and associates then moved farther south into Egypt. Their trip seems to have been motivated only by the urgency of the famine rather than by any shortage of land: "Now there was a famine in the land. So Abram went down to Egypt to sojourn there, for the famine was severe in the land" (Genesis 12:10).

When Abram and his entourage returned from Egypt to Canaan a few years later, it seems that their wealth may again have increased dramatically:

13. Arrival in Egypt

> And Abram was very glorious by reason of possessions in sheep, and cattle, and asses, and horses, and camels, and menservants, and maidservants, and in silver and gold exceedingly. And Lot also, his brother's son, was wealthy (Jubilees 13:14).

> So Abram went up from Egypt, he and his wife, and all that he had, and Lot with him, into the Negeb. Now Abram was very rich in cattle, in silver, and in gold. . . . And Lot, who went with Abram, also had flocks and herds and tents, so that the land could not support both of them dwelling together; for their possessions were so great that they could not dwell together. . . . (Genesis 13:1-2, 5-6).

Much of this wealth was acquired while Abram and his extended family were in Egypt. According to *Jasher*, Pharaoh seems to have been very generous:

> And the king approached Sarai and said to her, what is that man to thee who brought thee hither? and she said, he is my brother. And the king said, it is incumbent upon us to make him great, to elevate him and to do unto him all the good which thou shalt command us; and at that time the king sent to Abram silver and gold and precious stones in abundance, together with cattle, men servants and maid servants; and the king ordered Abram to be brought, and he sat in the court of the king's house, and the king greatly exalted Abram on that night.[21]

Thus, Sarai may have asked for and received much wealth for Abram. However, as discussed above, the king's plans for Sarai didn't work out as he had hoped, and perhaps surprisingly his response was to further reward both Abram and Sarai:

> Now therefore [Abram] here is thy wife, take her and go from our land lest we all die on her account. And Pharaoh took more cattle, men servants and maid servants, and silver and gold, to give to Abram, and he returned unto him Sarai his wife. And the king took a maiden whom he begat by his concubines, and he gave her to Sarai for a handmaid.[22]

Josephus also reports that Pharaoh, the king of Egypt, gave Abram much money, but many other privileges were granted to him as well:

> He [Pharaoh] also made him [Abram] a large present in money, and gave him leave to enter into conversation with the most learned among the Egyptians; from which conversation his virtue and his reputation became more conspicuous than they had been before.
> For whereas the Egyptians were formerly addicted to different customs, and despised one another's sacred and accustomed rites, and were very angry one with another on that account, Abram conferred with each of them, and, confuting the reasonings they made use of, every one for their own practices, demonstrated that such reasonings were vain and void of truth: whereupon he was admired by them in those conferences as a very wise man, and one of great sagacity, when he discoursed on any subject he undertook; and this not only in understanding it, but in persuading other men also to assent to him. He communicated to them arithmetic, and delivered to them the science of astronomy; for before Abram came into Egypt they were unacquainted with those parts of learning; for that science came from the Chaldeans into Egypt, and from thence to the Greeks also.[23]

Besides its brief information on one of the source's of Abram's great wealth, another interesting aspect of the quotation above is

13. Arrival in Egypt

its statement that Abram sought persistently to modify and clarify the religious teachings of the Egyptians, their "sacred and accustomed rites." This concept is compatible with the archeological evidence as summarized by Hayes: "Over the Hyksos bridge there flowed new blood strains, new religious and philosophical concepts, and new artistic styles and media."[24] It seems likely, however, that any religious teachings that may have been advocated by Abram were soon forgotten.[25] It is also possible that king Seth was anathematized by his successors because of his tolerance of untraditional religious beliefs, or his excessive attraction to beautiful women, or perhaps his giving away of a substantial portion of Egypt's royal wealth.

Given the recorded eagerness, even of people whom Abram had wronged, to provide him with land, cattle, servants, silver, and gold, it would seem that by the time of his return from Egypt to Canaan he and Lot had become exceedingly wealthy. Then for perhaps the first time it seems to have become impossible for Abram and Lot and their possessions to be accommodated within the same land:

> So Abram went up from Egypt, he and his wife, and all that he had, and Lot with him, into the Negeb. Now Abram was very rich in cattle, in silver, and in gold. And he journeyed on from the Negeb as far as Bethel, to the place where his tent had been at the beginning, between Bethel and Ai, to the place where he had made an altar at the first; and there Abram called on the name of the Lord. And Lot, who went with Abram, also had flocks and herds and tents, so that the land could not support both of them dwelling together; for their possessions were so great that they could not dwell together, and there was strife between the herdsmen of Abram's cattle and the

herdsmen of Lot's cattle. At that time the Canaanites and the Per'izzites dwelt in the land.

Then Abram said to Lot, "Let there be no strife between you and me, and between your herdsmen and my herdsmen; for we are kinsmen. Is not the whole land before you? Separate yourself from me. If you take the left hand, then I will go to the right; or if you take the right hand, then I will go to the left.". . . So Lot chose for himself all the Jordan valley, and Lot journeyed east; thus they separated from each other. Abram dwelt in the land of Canaan, while Lot dwelt among the cities of the valley and moved his tent as far as Sodom (Genesis 13:1-13).

Additional wealth came to Abram after his return from Egypt to Canaan and after separating from Lot. In a near repeat of the previous story of Sarai, Abram, and the king of Egypt, there then occurred the subsequent story of Sarah, Abraham, and the king of Gerar:

From there Abraham journeyed toward the territory of the Negeb, and dwelt between Kadesh and Shur; and he sojourned in Gerar. And Abraham said of Sarah his wife, "She is my sister." And Abim'elech king of Gerar sent and took Sarah. . . . Then God said to him . . . , restore the man's wife. . . . Then Abim'elech took sheep and oxen, and male and female slaves, and gave them to Abraham, and restored Sarah his wife to him. And Abim'elech said, "Behold, my land is before you; dwell where it pleases you." To Sarah he said; "Behold, I have given your brother a thousand pieces of silver; it is your vindication in the eyes of all who are with you; and before every one you are righted" (Genesis 20:1-18).

An extended version of the same story is included in *Jasher*.[26]

13.5 Arithmetic and astronomy

The possible role of Abram in conveying to the Egyptians Chaldean methods in arithmetic and astronomy is also of interest. A quotation from Josephus in the preceding section claims that before Abram's arrival in Egypt the Egyptians were unacquainted with the sciences of arithmetic and astronomy. This claim might not be entirely false, but it seems at least to be an overstatement of the actual case. That Abram could have been highly educated in the math and sciences of the Chaldeans is, however, likely to be true. His family is sometimes said to have been closely associated with the royal family of Ur in Mesopotamia:

> "[A]nd Nahor begat Terah, and Terah was thirty-eight years old, and he begat Haran and Nahor. . . "[27]

> And Terah the son of Nahor, prince of Nimrod's host, was in those days very great in the sight of the king and his subjects, and the king and princes loved him, and they elevated him very high. And Terah took a [another] wife, and her name was Amthelo the daughter of Cornebo; and the wife of Terah conceived and bare him a son in those days. Terah was seventy years old when he begat him, and Terah called the name of his son that was born to him Abram, because the king had raised him in those days, and dignified him above all his princes that were with him.[28]

On the other hand, several now famous Egyptian mathematical papyri have been discovered that date from before the arrival of Abram in Egypt. Thus the Kahun Mathematical Fragments, the Berlin Papyrus, and the Reisner Papyrus all seem to have been written during the Twelfth Dynasty (c. 1990-1780 B.C.),[29] whereas Abram is understood here to have arrived in

Adventures In Egypt

Egypt in about 1727 B.C., during the Thirteenth Dynasty (c. 1780-1620 B.C.). The largest and perhaps most important ancient mathematical treatise is the Rhind Mathematical Papyrus. It was copied in year thirty-three of Apophis I from a now-lost text of the Twelfth Dynasty. Similarly, the Moscow Mathematical Papyrus, also of the Thirteenth Dynasty is believed to be adapted from some earlier work of the Twelfth Dynasty. The Mathematical Leather Roll in the British Museum is, however, thought to date from the Seventeenth Century B.C.[30]

From the above information it would be difficult to argue convincingly that Abram contributed significantly to the development or propagation of the understanding of mathematics in Egypt. Most early evidence of Egyptian mathematical skills dates from many years before the arrival of Abram, with no conspicuously novel methods having been discerned in texts from the years soon after his visit. On the other hand, the copying of mathematical documents confirms at least that there was a continuing interest in mathematics during the Hyksos era. More specifically, the copying of the Rhind Mathematical Papyrus (in the thirty-third year of Apophis) would from Table 13.1 have occurred in about 1619 B.C., just after the end of the seven-year famine in Egypt.[31] At that time Joseph, if identified as Chancellor Hur,[32] would perhaps still have been second in command of Egypt.

It may be appropriate to turn now to the question of whether Abram might have contributed to the Egyptian knowledge of astronomy. As noted previously in Section 13.2, Artapanus had stated that Abram taught Pharaoh astrology: "And he [Abram], they say, came with all his household into Egypt, to Pharethothes

13. Arrival in Egypt

the king of the Egyptians, and taught him astrology".[33] It would seem that this statement might refer to astronomy rather than astrology, since astrology would be inconsistent with Abram's religious beliefs. It should be noted especially that according to Josephus as quoted above: "He communicated to them arithmetic, and delivered to them the science of astronomy; for before Abram came into Egypt they were unacquainted with those parts of learning; for that science came from the Chaldeans into Egypt, and from thence to the Greeks also".[34] Abram's possible teaching of arithmetic has been considered above, but a few comments about his possible teaching of astronomy might be appropriate here.

As a starting point, it should be acknowledged that many of the inhabitants of Mesopotamia did indeed seem to have a fascination with astrology. This is evident to Bible readers from the story of the birth of Jesus: "Now when Jesus was born in Bethlehem of Judea in the days of Herod the king, behold, wise men from the East came to Jerusalem, saying, 'Where is he who has been born king of the Jews? For we have seen his star in the East, and have come to worship him. . .' " (Matthew 2:1-2). A similar story is reported by *Jasher* concerning the birth of Abram and the astrologers of the king:

> And it was in the night that Abram was born, that all the servants of Terah, and all the wise men of Nimrod, and his conjurors came and ate and drank in the house of Terah, and they rejoiced with him on that night. And when all the wise men and conjurors went out from the house of Terah, they lifted up their eyes toward heaven that night to look at the stars, and they saw, and behold one very large star came from the east and ran in the

heavens, and he swallowed up the four stars from the four sides of the heavens. And all the wise men of the king and his conjurors were astonished at the sight, and the sages understood this matter, and they knew its import.[35]

According to these sources astrology might seem to have been an important part of Mesopotamian culture for many centuries.

On the other hand, Abram himself is said to have inquired into the power of the sun, moon, and stars; and he convinced himself that they were in fact powerless. One account of his conclusions is included in Jubilees, and in this version his insights occurred after his arrival with his father in Haran:

> And in the sixth week, in the fifth year thereof, Abram sat up throughout the night on the new moon of the seventh month to observe the stars from the evening to the morning, in order to see what would be the character of the year with regard to the rains, and he was alone as he sat and observed. And a word came into his heart and he said: "All the signs of the stars, and the signs of the moon and of the sun are all in the hand of the Lord. Why do I search (them) out?
> > If He desireth, he causeth it to rain, morning and evening;
>
> And if He desireth, He withholdeth it,
> And all things are in His hand."[36]

In the somewhat similar *Jasher* version, the revelations came while Abram was still in Ur:

> And the Lord gave Abram an understanding heart, and he knew all the works of that generation were vain, and that all their gods were vain and were of no avail.
> And Abram saw the sun shining upon the earth, and Abram said unto himself surely now this sun that shines

13. Arrival in Egypt

upon the earth is God, and him will I serve. And Abram served the sun in that day and he prayed to him, and when evening came the sun set as usual, and Abram said within himself, surely this cannot be God?

And Abram still continued to speak within himself, who is he who made the heavens and the earth? who created upon earth? where is he? And night darkened over him, and he lifted up his eyes toward the west, north, south and east, and he saw that the sun had vanished from the earth, and the day became dark. And Abram saw the stars and moon before him, and he said, surely this is the God who created the whole earth as well as man, and behold these his servants are gods around him; and Abram served the moon and prayed to it all that night. And in the morning when it was light and the sun shone upon the earth as usual, Abram saw all the things that the Lord God had made upon earth.

And Abram said unto himself, surely these are not gods that made the earth and all mankind, but these are the servants of God.[37]

According to both of these accounts, Abram was very interested in the sun, moon, and stars; but he concluded that they were creations rather than creators or gods. If these stories reflect a historical reality, it seems clear that Abram would not have been an advocate for or a teacher of astrology.

On the other hand, it is possible in principle that Abram might have attempted to teach the Egyptians the "science of astronomy" as quoted above from Josephus in Section 13.4. It would be of interest to see if this possibility is supported in any way by archeology. Some hints on the subject may be found in the second tomb of Senmut (sometimes written Sen-ne-mut or Senenmut), Hatshepsut's chief steward:[38]

Except for its foundation deposits, which we have already discussed, Sen-ne-mut's unfinished second tomb, under the forecourt of the Deir el Bahri temple, produced no portable objects which can be associated directly with its owner. The glory of this tomb is its inscribed antechamber which boasts, among other interesting features, the earliest and one of the finest astronomical ceilings now known – a great chart of the heavens complete with a decan table and a list of monthly festivals, laid out and "drawn by the most skilful penmen of the mid-Eighteenth Dynasty."[39]

Hatshepsut reigned during about 1503-1483 B.C.[40] Thus the earliest astronomical ceiling known was created during the period of the Hebrews' stay in Egypt, and Senmut is sometimes considered to have been an astronomer.[41] It has been suggested that the positions of the planets drawn on the ceiling correspond to the actual night sky in about May of 1534 B.C.[42] With the chronology employed here, this date could have been during the early lifetime of Senmut

With the above background and Josephus's explicit claim, it would seem that Abram himself, coming from Mesopotamia, could have been a knowledgeable teacher of astronomy during his visit to Egypt. In possible further support, it may be noted that Seti I created the Four Hundred Year Stela to celebrate the beginning of the reign in Egypt of the god/king Seth (and possibly the arrival of the Hyksos as well).[43] As noted previously in this chapter, that arrival probably included Abram.[44] Seti's unusual personal interest in astronomy, could in part be a recognition of Abram as an astronomy teacher arriving at the same time as the Hyksos. That interest is suggested in a description of Seti's tomb and its favorable comparison with the "vast hypostyle hall" of the Temple of Amun in Karnak:

> Equally amazing in its own way and infinitely superior in the quality of its sculptured decoration is Sethy's huge rock-cut tomb (No. 17) in the Valley of the Tombs of the Kings. Here a series of passages and chambers, adorned with painted reliefs of great beauty and delicacy, leads gradually downward for a distance of more than three hundred feet to a pillared sepulchral hall and vaulted crypt, where, under a painted astronomical ceiling, the king's body once rested in a magnificent alabaster sarcophagus.[45]

Thus, Seti may have intended in death to lie under a star-filled sky for eternity.

13.6 Conclusion

This chapter commenced with a summary of the chronology of the approximately one hundred sixty year time period between the entry of Abram and the Hyksos into Egypt (1727 B.C.) until the Hyksos were expelled from Egypt (1567 B.C.). The emphasis of the following discussions was on the activities of Abram himself in his travels to and from Egypt, and his few-year sojourn there. He seems to have interacted with the highest levels of the government and probably the religious and scientific establishments as well. There are hints that he may have attempted to teach members of these groups, but his degree of success in such endeavors is not easy to quantify. The king was attracted to Abram's wife Sarai, and he also seems to have done everything he could to form an alliance with Abram. Beyond the accomplishment of just surviving in the complex Hyksos/Egyptian environment, Abram also managed to acquire great wealth during his visit.

---

1. L. W. Casperson, *Patterns of Biblical Chronology* (Westbow Press, 2012). Brief summaries of chronological conclusions may be found in Table 14.1 (p. 333), Table 16.1 (p. 408), Table 23.1 (p. 579), Table 24.3 (p. 610), and associated texts.
2. Ibid., Section 24.3, "The 400 year stela," and references, pp. 598-605.
3. Ibid., Section 24.4, "Abram and the famine," Section 24.5, "Revised chronology," pp. 605-610.
4. Ibid., Table 24.3, p. 610.
5. Reproduced in part from L. W. Casperson, op. cit., Table 23.1, p. 579 and Table 24.3, p. 610; with several added events and their possible dates.
6. L. W. Casperson, op. cit., "Appendix A: Artapanus," Paragraph 1, pp. 622-623.
7. Ibid., Section 7.3, "Absolute chronology of the Eighteenth Dynasty," pp. 139-145.
8. W. C. Hayes, "Egypt: From the death of Amenemes III to Seqenenre II," *The Cambridge Ancient History, Third Edition, Volume II, Part I, History of the Middle East and the Aegean Region c. 1800-1380 B.C.*, Edited by I. E. S. Edwards, C. J. Gadd, N. G. L. Hammond, and E. Sollberger (Cambridge University Press, Cambridge, 1973), Chapter 2, p. 58.
9. T. Schneider, "The Relative Chronology of the Middle Kingdom and the Hyksos Period (Dyns. 12-17)" in Erik Hornung, Rolf Krauss, and David A. Warburton (editors), *Ancient Egyptian Chronology*, Handbook of Oriental Studies, Section 1, The Near and Middle East; Volume

83, (Koninklijke Brill NV, Leiden, The Netherlands, 2006), pp. 194-195.
10. L. W. Casperson, op. cit., Section 7.3, "Absolute chronology of the Eighteenth Dynasty," pp. 139-145; Table 7.1, p. 145.
11. Ibid., Section 24.6, "Abram in Canaan and Egypt," pp. 610-618.
12. Darrell D. Baker, *The Encyclopedia of the Egyptian Pharaohs, Volume 1, Predynastic to the Twentieth Dynasty 3300-1069 B.C.* (Bannerstone Press, Oakville, CT, 2008), p. 406.
13. F. Josephus, *The Life and Works of Flavius Josephus*, Translated by William Whiston (Holt, Rinehart, and Winston, New York), *Antiquities of the Jews,* Book 1, 8:1, p. 43.
14. L. W. Casperson, op. cit., Section 24.6, "Abram in Canaan and Egypt," pp. 610-618; and K. S. B. Ryholt and A. Bülow-Jacobsen, *The Political Situation in Egypt during the Second Intermediate Period c. 1800-1550 B.C.* (Museum Tusculanum Press, University of Copenhagen, 1997), pp. 284-285.
15. F. Josephus, op. cit., *Antiquities of the Jews*, Book 1, 8:1, p. 43.
16. *The Book of Yashar*, Translated from the Hebrew and Published by Mordecai Manuel Noah (Hermon Press, New York, 1972), (cited here and below in the form *Jasher*); *Jasher* 15:31.
17. N. Grimal, *A History of Ancient Egypt* (Barnes & Noble Books, New York, 1997), p. 160.
18. *Jasher* 12:40-43.
19. *Jasher* 13:1-2.
20. *Jasher* 13:1.
21. *Jasher* 15:21-22.
22. *Jasher* 15:30-31.

23. F. Josephus, op. cit., *Antiquities of the Jews*, Book 1, 8:1-2, p. 43.
24. W. C. Hayes, *The Scepter of Egypt, A Background for the Study of the Egyptian Antiquities in The Metropolitan Museum of Art, Part II: The Hyksos Period and the New Kingdom (1675-1080 B.C.)* (Harvard University Press, Cambridge, Massachusetts, 1959), p. 4.
25. L. W. Casperson, op. cit., Section 24.6, "Abram in Canaan and Egypt," pp. 610-618; p. 614.
26. *Jasher* 20.
27. *Jasher* 7:22.
28. *Jasher* 7:49-51.
29. M. Clagett, *Ancient Egyptian Science, A Source Book*, Volume Three, Ancient Egyptian Mathematics (American Philosophical Society, Philadelphia, 1999), Part 2: Documents, pp. 112-279.
30. Ibid., pp. 255-260.
31. L. W. Casperson, op. cit., Section 23.4, "Joseph and Apophis," pp. 575-581, Table 23.1; p. 579.
32. Ibid., Section 23.5, "Finding Joseph," pp. 581-589.
33. Ibid., "Appendix A: Artapanus," Paragraph 1, pp. 622-623.
34. F. Josephus, op. cit., *Antiquities of the Jews*, Book 1, 8:1-2, p. 43.
35. *Jasher* 8:1-3.
36. Jubilees 12:16-18.
37. *Jasher* 9:12-19.
38. W. C. Hayes, *The Scepter of Egypt*, op. cit., pp. 111-112.
39. H. E. Winlock, "The Egyptian expedition 1925-1927: The Museum's excavations at Thebes" *The Metropolitan Museum of Art Bulletin*, Volume 23, Number 2, Part 2 (February 1928), pp. 3-58; p. 37.
40. L. W. Casperson, op. cit., Section 14.2, "Manetho's list of kings," pp. 329-336, Table 14.1.

41. B. Novakovic, "Senenmut: An ancient Egyptian astronomer," *Publications of the Astronomical Observatory of Belgrade*, No. 85 (2008), pp. 19-23.
42. O. V. Spaeth, "Dating the Oldest Egyptian Star Map," *Centaurus*, Volume 42, Issue 3, pp. 159-179 (July 2000).
43. L. W. Casperson, op. cit., Section 24.3, "The 400 year stela," pp. 598-605.
44. Ibid., Section 24.5, "Revised chronology," pp. 609-610, Table 24.3.
45. W. C. Hayes, *The Scepter of Egypt*, op. cit., pp. 327-328.

14. CONQUEST OF EGYPT

"There was a king of ours whose name was Timaus. Under him it came to pass, I know not how, that God was averse to us, and there came, after a surprising manner, men of ignoble birth out of the eastern parts, and had boldness enough to make an expedition into our country, and with ease subdued it by force, yet without our hazarding a battle with them."[1]

14.1 Introduction

As discussed previously, the Hyksos were a group of foreigners who entered Egypt peacefully in about 1730-1720 B.C., perhaps as Abram did in response to famine in his country of origin.[2] This first arrival was the beginning of a multi-year process by which the Hyksos eventually gained control of most of Egypt. The initially peaceful occupation seems to have developed into a destructive military engagement. The Hyksos succeeded in their conquest, and they then maintained a major presence in Egypt for about a century. The purpose of this chapter is to seek a better understanding of the conquest process and perhaps identify the initial Hyksos leader in this conflict.

The actual invasion that led to the Hyksos takeover of Egypt has been said to have begun in about 1674 B.C. One of the principal sources of information on this subject is provided by the historian Manetho, who wrote in Egypt during the third century B.C. His claims relating to the Hyksos are sometimes questioned, but they are recalled here in Section 14.2. The identity of the leader of the Hyksos during and after their invasion is also of interest, and the possibility that this leader might have been an Arab is considered in Section 14.3. There are also several indications that some of the descendants of Abram might have come to be known as Arabs. A particular Arab of interest for the kingship position is Ishmael, Abram's eldest son, and this possibility is mentioned in Section 14.4. The case for Ishmael as the "Sultan" of the Arabs (seemingly implied by Manetho and other sources) may be regarded as interesting, but the reader should reach his or her own conclusions. The interval of time between the conquest and the reign of Apophis I (of Joseph's time) is also of interest, and a possible chronology for that period is inferred in Section 14.5.

14.2 Invasion to domination

The Hyksos appear to have been mostly Semitic, and originally they came to Egypt from northern Sinai and Palestine:

> In Egypt we can recognize two principal stages in the Hyksos rise to power, the first of which had its origin in the northeastern Delta around 1720 B.C. This was the time of the Asiatic occupation of the town of . . . Avaris, and the founding there of a temple to the Hyksos counterpart of the god Seth, an event commemorated on a New Kingdom stela [the Four Hundred Year Stela] as having taken place four hundred years before the reign of Hor-em-heb, the last king of the Eighteenth Dynasty. For

> approximately forty-five years (about 1720-1675 B.C.) the first waves of Asiatic princes appear to have consolidated their position and extended their holdings in northern Egypt; but they have left us no monuments, inscribed or otherwise, which can with assurance be assigned to this initial phase of the Hyksos domination.[3]

A similar summary by Grimal refers to this same period of time:

> The Hyksos seizure of Avaris took place around 1730-1720 B.C. . . . Hyksos control over northern Egypt evolved in a number of stages. Starting from Avaris they gradually moved towards Memphis, following the eastern edge of the Delta. They established centres at Farasha, Tell el-Sahaba (at the mouth of the Wadi Tumilat), Bubastis, Inshas and Tell el-Yahudiya (about 20 kilometres north of Heliopolis). This progression took place over a period of almost half a century, until about 1675 B.C. The Thirteenth Dynasty had by then reached its thirty-third or thirty-fourth king, Dedumesiu I. If this king is to be identified with Manetho's Tutimaius, then it would have been during his reign that the Hyksos became rulers of Egypt. This identification would appear to be confirmed by the fact that Dedumesiu is the last known king of the Thirteenth Dynasty in the inscriptions on Theban monuments at Thebes, Deir el-Bahri and Gebelein.[4]

Abram's visit would have been part of the first phase of this migration. It was after about 1675 B.C. that the occupation became a more aggressive and perhaps violent activity.

The actual conquest of Egypt by the Hyksos is thought to have occurred in about 1674 B.C.,[5] and this event could not so easily be described as peaceful. At about that time there seems to have been a major and more sudden invasion, following which

14. Conquest of Egypt

the Hyksos installed their own king of Egypt. Several aspects of the conquest will be considered in this chapter. One of the most authoritative early writers on the subject of the Hyksos conquest of Egypt was the historian and priest Manetho of about the third century B.C. Among his resources he may have had access to the then developing Library of Alexandria. Extensive quotations from Manetho's *History of Egypt* have been left for us by the Jewish historian Flavius Josephus in his work *Against Apion*, and several of those quotations were considered previously.[6] A few words relating to the origin of the Hyksos are quoted again here:

> I shall begin with the writings of the Egyptians; not indeed of those that have written in the Egyptian language, which it is impossible for me to do. But Manetho was a man who was by birth an Egyptian, yet had he made himself master of the Greek learning, as is very evident; for he wrote the history of his own country in the Greek tongue, by translating it, as he saith himself, out of their sacred records; he also finds great fault with Herodotus for his ignorance and false relations of Egyptian affairs. Now this Manetho, in the second book of his Egyptian History, writes concerning us in the following manner. I will set down his very words, as if I were to bring the very man himself into a court for a witness: "There was a king of ours whose name was Timaus. Under him it came to pass, I know not how, that God was averse to us, and there came, after a surprising manner, men of ignoble birth out of the eastern parts, and had boldness enough to make an expedition into our country, and with ease subdued it by force, yet without our hazarding a battle with them. So when they had gotten those that governed us under their power, they afterwards burnt down our cities, and demolished the temples of the gods, and used all the inhabitants after a most barbarous

Adventures In Egypt

manner; nay, some they slew, and led their children and their wives into slavery. At length they made one of themselves king, whose name was Salatis; he also lived at Memphis, and made both the upper and lower regions pay tribute, and left garrisons in places that were the most proper for them. He chiefly aimed to secure the eastern parts, as foreseeing that the Assyrians, who had then the greatest power, would be desirous of that kingdom, and invade them; and as he found in the Saite Nomos, [Seth-roite,] a city very proper for this purpose, and which lay upon the Bubastic channel, but with regard to a certain theologic notion was called Avaris, this he rebuilt, and made very strong by the walls he built about it, and by a most numerous garrison of two hundred and forty thousand armed men whom he put into it to keep it. Thither Salatis came in summer time, partly to gather his corn, and pay his soldiers their wages, and partly to exercise his armed men, and thereby to terrify foreigners. When this man had reigned thirteen years, after him reigned another. . . . This whole nation was styled Hycsos, that is, Shepherd-kings; for the first syllable Hyc, according to the sacred dialect, denotes a king, as is Sos a shepherd; but this according to the ordinary dialect; and of these is compounded Hycsos: but some say that these people were Arabians."[7]

14.3 King of the Arabs

The Jewish historian Artapanus lived and wrote in Alexandria in about the 2nd century B.C., and thus he also is likely to have had access to the Library of Alexandria. Several excerpts from his work *Concerning the Jews* were preserved by Eusebius,[8] and those excerpts were numbered by paragraph and reproduced previously.[9] The first three paragraphs relating to the movement

14. Conquest of Egypt

of Abraham and Joseph to Egypt and a few of their activities are reproduced here:

Chapter 18

1. Artabanus in his *Jewish History* says that the Jews were called Ermiuth, which when interpreted after the Greek language means Judaeans, and that they were called Hebrews from Abraham. And he, they say, came with all his household into Egypt, to Pharethothes the king of the Egyptians, and taught him astrology; and after remaining there twenty years, removed back again into the regions of Syria: but that many of those who had come with him remained in Egypt because of the prosperity of the country.

Chapter 23

2. Artapanus says, in his book *Concerning the Jews*, that Joseph was a descendant of Abraham and son of Jacob: and because he surpassed his brethren in understanding and wisdom, they plotted against him. But he became aware of their conspiracy, and besought the neighbouring Arabs to convey him across to Egypt: and they did what he requested; for the kings of the Arabians are offshoots of Israel, being sons of Abraham [by Hagar and Ketura], and brethren of Isaac. And when he had come to Egypt and been commended to the king, he was made administrator of the whole country. And whereas the Egyptians previously occupied the land in an irregular way, because the country was not divided, and the weaker were unjustly treated by the stronger, he was the first to divide the land, and mark it out with boundaries, and much that lay waste he rendered fit for tillage, and allotted certain of the arable lands to the priests.

3. He was also the inventor of measures, and for these things he was greatly beloved by the Egyptians. He married Aseneth a daughter of the priest of Heliopolis, by

whom he begat sons. And afterwards his father and his brethren came to him, bringing much substance, and were set to dwell in Heliopolis and Sais, and the Syrians multiplied in Egypt.[10]

A few points suggested in the above text may merit additional comments. The quotation from Manetho in the previous section ended with the following statement about the Hyksos: "[S]ome say that these people were Arabians." As seen from previous discussions, Abram and his extended family that visited Egypt were probably included within the group called Hyksos.[11] Thus, Manetho's quotation may be understood to imply an association between Abram's family and the Arabs. Artapanus as quoted above provides us with a direct link between the family of Abram/Abraham and the Arabs: "The kings of the Arabians are offshoots of Israel, being sons of Abraham, and brethren of Isaac."[12] Brothers (or half-brothers) of Isaac would seem only to be interpretable to include Ishmael (son of Abram and Hagar, Genesis 16:15) and perhaps also the sons of Abraham and his third wife Keturah: "Abraham took another wife, whose name was Keturah. She bore him Zimran, Jokshan, Medan, Midian, Ishbak, and Shuah" (Genesis 25:1-2).

It is of interest here to contemplate possible implications of Ishmael being a king of the Arabs. One conclusion is that during his lifetime Ishmael's sons would have been considered to be princes. With this thought in mind, one might review what is said in Genesis about Ishmael's sons:

> As for Ish'mael, I have heard you; behold, I will bless him and make him fruitful and multiply him exceedingly;

14. Conquest of Egypt

he shall be the father of twelve princes, and I will make him a great nation (Genesis 18:20).

> These are the descendants of Ish'mael, Abraham's son, whom Hagar the Egyptian, Sarah's maid, bore to Abraham. These are the names of the sons of Ish'mael, named in the order of their birth: Neba'ioth, the firstborn of Ish'mael; and Kedar, Adbeel, Mibsam, Mishma, Dumah, Massa, Hadad, Tema, Jetur, Naphish, and Ked'emah. These are the sons of Ish'mael and these are their names, by their villages and by their encampments, twelve princes according to their tribes (Genesis 25:12-16).

The only other princes mentioned in Genesis seem to be "the princes of Pharaoh" (Genesis 12:15), Abraham himself (Genesis 23:6), and Hamor the Hivite (Genesis 34:2). The fact that Ishmael's sons are called princes would be entirely compatible with the idea of him being a king. An important question to ask could be the following: "Of which people might Ishmael have been considered to be the king?" This question will be considered further a few paragraphs below.

The story of Joseph being sold into slavery in Egypt is somewhat different in the version given by Artapanus from the biblical version. In the Bible we are told that Joseph was cast into a pit before being sold to traders heading toward Egypt. The text includes ambiguity about whether the traders were Ishmaelites, Midianites, or both:

> Then they [Joseph's brothers] sat down to eat; and looking up they saw a caravan of Ish'maelites coming from Gilead, with their camels bearing gum, balm, and myrrh, on their way to carry it down to Egypt. Then Judah

said to his brothers, "What profit is it if we slay our brother and conceal his blood? Come, let us sell him to the Ish'maelites, and let not our hand be upon him, for he is our brother, our own flesh." And his brothers heeded him. Then Mid'ianite traders passed by; and they drew Joseph up and lifted him out of the pit, and sold him to the Ish'maelites for twenty shekels of silver; and they took Joseph to Egypt (Genesis 37:25-28).

Meanwhile the Mid'ianites had sold him in Egypt to Pot'i-phar, an officer of Pharaoh, the captain of the guard (Genesis 37:36).

Now Joseph was taken down to Egypt, and Pot'i-phar, an officer of Pharaoh, the captain of the guard, an Egyptian, bought him from the Ish'maelites who had brought him down there (Genesis 39:1).

Thus, it isn't clear from these texts alone the details of how Joseph was sold to or kidnapped by Ishmaelites/Midianites. However, the biblical story would seem to be in partial agreement with the version of Artapanus, since in that version the Ishmaelites and Midianites are both said to be Arabs. As mentioned above, Midian was a son of Abraham by his third wife Keturah. A similar definition of Arabs is provided in Jubilees:

And Ishmael and his sons, and the sons of Keturah and their sons, went together and dwelt from Paran to the entering in of Babylon in all the land which is towards the East facing the desert. And these mingled with each other, and their name was called Arabs, and Ishmaelites (Jubilees 20:12-13).

The previous paragraphs mentioned the partial agreement between the biblical texts and Artapanus concerning the identity

14. Conquest of Egypt

of the traders who took Joseph to Egypt. This terminology of "partial agreement" is used because there remains a discrepancy between these texts. In the version of the Joseph story provided by Artapanus, Joseph became aware of the plotting by his brothers. He then escaped to Egypt with the assistance of Arab traders, and he joined the service of the king without having served as a slave.

There were, of course, other interactions between the Arabs and the descendants of Isaac. One such interaction involved Moses shortly before the exodus of the Hebrews from Egypt. A quotation from Josephus indicates that the land of Midian on the Sinai Peninsula is named for the half-brother of Isaac mentioned above:

> [B]ut when he [Moses] had learned beforehand what plots there were against him, he went away privately; and because the public roads were watched, he took his flight through the deserts, and where his enemies could not suspect he would travel; and, though he was destitute of food, he went on, and despised that difficulty courageously; and when he came to the city of Midian, which lay upon the Red Sea, and was so denominated from one of Abraham's sons by Keturah, he sat upon a certain well, and rested himself there after his laborious journey, and the affliction he had been in.[13]

A quotation from Artapanus provides a further indication that the Midianites were considered either to be Arabs or at least closely related to them:

> So he [Moses] made his escape into Arabia, and lived with Raguel [Reuel, Hobab, Jethro] the ruler of the district [Midian], having married his daughter [Zipporah].

483

> And Raguel wished to make an expedition against the Egyptians in order to restore Moses, and procure the government for his daughter and son-in-law; but Moses prevented it, out of regard for his own nation; and Raguel forbidding him to march against the Arabs, ordered him to plunder Egypt.[14]

Raguel (Jethro) wouldn't want Moses, and presumably his followers, to attack the Arabs, if Raguel as a Midianite were himself an Arab.

14.4 Salatis and Ishmael

As noted above in Section 14.2, the Hyksos are believed to have undertaken the conquest phase of their occupation of Egypt in about 1674 B.C. One of the first things they needed was a king:

> At length they made one of themselves king, whose name was Salatis; he also lived at Memphis, and made both the upper and lower regions pay tribute, and left garrisons in places that were the most proper for them. He chiefly aimed to secure the eastern parts, as foreseeing that the Assyrians, who had then the greatest power, would be desirous of that kingdom, and invade them; and as he found in the Saite Nomos, [Seth-roite,] a city very proper for this purpose, and which lay upon the Bubastic channel, but with regard to a certain theologic notion was called Avaris, this he rebuilt, and made very strong by the walls he built about it, and by a most numerous garrison of two hundred and forty thousand armed men whom he put into it to keep it. Thither Salatis came in summer time, partly to gather his corn, and pay his soldiers their wages, and partly to exercise his armed men, and thereby to terrify foreigners.[15]

14. Conquest of Egypt

At this point it may be useful to review briefly what a few more modern writers have had to say about the origins of the Hyksos:

> Looking further at the scanty monuments left by the Hyksos themselves, we discover a few vague but nevertheless significant hints as to the character of these strange invaders, whom tradition called Arabians and Phoenicians; and contemporary monuments designated as "Asiatics," "barbarians" and "rulers of foreign countries."[16]

Comments by Hayes include the following:

> About 1675 B.C. a Hyksos prince whom Manetho calls Salatis ("the Sultan"?) ousted the contemporary Egyptian ruler from the capital city of Memphis, occupied most of Middle Egypt, and appears to have extended his control southward to include, for the time being at least, the whole of Upper Egypt and Nubia. With Salatis begins the succession of six Hyksos sovereigns – the so-called "Great Hyksos" – who comprised Manetho's Fifteenth Dynasty....
>
> The Manethonian "Salatis" (or "Saites"), unattested elsewhere, was, as already indicated, probably a title or epithet applied to an early Hyksos ruler of sufficient power and importance to have been recognized by posterity as the founder of the Fifteenth Dynasty. Just such a ruler was King Ma-yeb-Re Sheshi, whose seals and seal impressions, of early Hyksos types, are both numerous and widely distributed, examples of the latter having been found as far south as the Middle Kingdom trading post at Kermeh, near the Third Cataract of the Nile. Our own collection includes twenty-seven glazed steatite scarabs of this king, on which his names are preceded by the pharaonic titles the "Good God" or the

"Son of Re" and followed by the phrases "given life" or "May he live forever." Eleven examples bear the Egytianizing throne name "Ma-yeb-Re" ("Just-is-the-heart-of-Re") and sixteen the personal name "Sheshi" – probably, though not certainly, belonging to the same ruler.[17]

The corresponding comments by Grimal are as follows:

> The founder of the first Hyksos dynasty, Manetho's Fifteenth Dynasty, was a man called Salitis, who was probably the same person as the Sheshi mentioned on several seals found at Kerma, suggesting perhaps that the Nubians allied themselves with the Hyksos against the Thebans from the very beginning of the Hyksos period. Salitis is probably also to be identified with a ruler called Sharek whose name appears at this time in Memphis. . . . For twenty years Salitis/Sheshi/Sharek, probably based at Memphis, ruled a kingdom comprising both the Delta and the Nile valley down to Gebelein, as well as the desert trade routes that allowed the Hyksos to make contact with their Nubian allies. This state of affairs lasted until the reign of Apophis I, who delegated part of his authority to another Hyksos family of vassal princes, which Manetho incorrectly describes as the Sixteenth Dynasty.[18]

Perhaps the "Hyksos family of vassal princes" mentioned by Grimal should be imagined to include important Hebrews if their names could have been written as Jacob-El (Jacob?), Bnon (Benoni/Benjamin?), and Hur (Joseph?) as mentioned previously.[19]

An important point in these and other quotations is that the first Hyksos king was identified by Manetho as Salatis or Salitis, but he is also associated with other names given variously as

14. Conquest of Egypt

Shalek, Sharek, or Sheshi. In addition, there has been the inclination to associate the Hyksos with the Arabs, as suggested previously in Section 14.3. Furthermore, it has sometimes been considered that Manetho's version of Salitis is (or became) more of a title than a personal name. In particular Salitis has often been thought to correspond to the familiar Arabic title of Sultan:

> [T]he title *Siltonim* is applied in the Book of Daniel to the lords of the Assyrian monarchy, and is traceable to the same root as Sultan, *Salat*, which signifies to rule, both in Hebrew and Arabic. . . . Manetho, quoted by Josephus, gives *Salatis* as the name of the first of the *Hyksos* or Shepherd Kings of Egypt, a name which has been supposed to be derived from the same root. In one of the Khorsabad inscriptions, translated by Oppert and Menant, the title appears in the following passage: "Hanon roi de Gaza et Lebech *Sultan* d'Egypte se reunirent à Rapih pour y livrer combat et bataille." The title is also said to have been applied to the Governor of Babylon while it was subject to Assyria, and used in the sense of royal on Babylonian weights.[20]

Another similar interpretation is the following:

> Some time appears to have elapsed before the Semitic hordes consolidated themselves under the rule of a single prince, to whom the name of Salatis or Shaladh, "the Sultan," is given, and who established his court at Memphis.[21]

The same interpretation of the title Sultan was given above by Hayes, but he included a question mark, presumably indicating a degree of uncertainty.

The available information seems to suggest that the first king of the Hyksos may have been an Arab who was identified by Manetho with the Arabic title Sultan. As discussed above, however, "the kings of the Arabs are offshoots of Israel, being sons of Abraham, and brethren of Isaac."[22] The father of the Arabs is considered to be Ishmael, who was the eldest son of Abraham. Thus it appears that Ishmael may have been the first king of the Hyksos, and was given the title Sultan or Salitis. If this were true, one could wonder if his personal name Sheshi and throne name Ma-yeb-Re could somehow have evolved from (or to) Ishmael. Also, of course, if Ishmael had been an actual king of Egypt, that would explain entirely why his twelve sons were called princes as indicated previously in Section 14.3.

A reasonable question at this point is why the senior member of a small family group of people who would later be called Arabs could be qualified to become the first Hyksos king of Egypt. It is also of interest to consider whether there is any more specific information that might link Ishmael in particular to what is known about the Hyksos. There could be several possible answers to such questions. First, as noted previously, Ishmael's father Abram had been treated like royalty when he visited Egypt. The king sought to marry Abram's wife, believing her to be his sister. Also, Abram was given a vast fortune when he departed from Egypt, and according to *Jasher* he and his wife Sarai were given Hagar, the daughter of Pharaoh as their maidservant:

> And the king [of Egypt] took a maiden [Hagar] whom he begat by his concubines, and he gave her to Sarai for a handmaid. And the king said to his daughter, it is better

for thee my daughter to be a handmaid in this man's [Abram's] house than to be a mistress in my house, after we have beheld the evil that befel us on account of this woman.[23]

It is clear that the authors of *Jasher* considered Ishmael to be the grandson of the person who had been king of Egypt at the time of Abram's visit, and it is also possible that the king of Egypt at the time of the visit was Meribre Seth. Such an ancestry could, except among traditional anti-Hyksos Egyptian royalty, have strengthened Ishmael's claim to the kingship of Egypt.

It is of interest to consider whether any other known aspects of Ishmael's life would seem to be compatible with his possible role as a Hyksos king. According to *Jasher*, when Ishmael was of age Hagar went with him to Egypt to find him a suitable wife:

> [A]nd Hagar went with her son [Ishmael] to the wilderness, and they dwelt in the wilderness of Paran with the inhabitants of the wilderness, and Ishmael was an archer, and he dwelt in the wilderness a long time. And he and his mother afterward went to the land of Egypt, and they dwelt there, and Hagar took a wife for her son from Egypt, and her name was Meribah.[24]

If this account is historically accurate, it shows that Hagar had retained considerable authority in Egypt and that both she and Ishmael were comfortable visiting there. Ishmael's mother Hagar is said to have been an Egyptian princess, and his wife was also said to be an Egyptian. One could imagine that Ishmael's father Abram may have been remembered favorably, at least by the Hyksos branch of Ishmael's Egyptian constituency. These could have been sufficient credentials for Ishmael to be accepted as king (or sultan) in Egypt. In Manetho's account quoted in Section 14.2, the Hyksos invaders were very destructive in Egypt before they chose Salatis to be their king. According to Manetho they had burned cities, destroyed temples, and murdered or enslaved many of the inhabitants. With the establishment of the king known in part as Salatis, the Hyksos rule seems to have been tolerable to most Egyptians.

The statement in *Jasher* quoted above also indicates that "Ishmael was an archer," and this same claim is repeated in Genesis and Jubilees:

> And God was with the lad, and he grew up; he lived in the wilderness, and became an expert with the bow. He lived in the wilderness of Paran; and his mother took a wife for him from the land of Egypt (Genesis 21:20-21).

> And the child grew and became an archer, and God was with him; and his mother took him a wife from among the daughters of Egypt (Jubilees 17:13).

Thus, from his childhood Ishmael had apparently become very skilled in archery. Given this expertise, it is of interest that the composite bow was one of the most important inventions introduced into Egypt by the Hyksos:

> Through their Hyksos adversaries the Egyptians probably first became acquainted with the composite bow, bronze daggers and swords of improved types, and other advances in the equipment and technique of war, as well as with some of the important western Asiatic innovations in the arts of peace which we encounter in Egypt for the first time under the Eighteenth Dynasty. Represented as an unmitigated disaster by native historians of later times,

the Hyksos domination appears actually to have provided the Egyptians with both the incentive and the means towards 'world' expansion and so laid the foundations and, to a great extent, determined the character of the New Kingdom, or, as it is often called, 'the Empire'.[25]

One can imagine that this new bow design may have come into Egypt with the first Hyksos king, perhaps Ishmael, who was "an expert with the bow."

One could also assume that a reasonably high economic level would have been helpful to the acceptance and performance of an outsider as king of Egypt. Some insight into social status can be inferred from a person's accustomed manner of dress. Little is known on that subject, but *Jasher* provides an example for one important event in the life of Ishmael's younger half-brother Isaac. As is well known to Bible readers, Abraham's faith and obedience were sorely tested when he was asked by God to sacrifice his son Isaac:

> After these things God tested Abraham, and said to him, "Abraham!" And he said, "Here am I." He said, "Take your son, your only son Isaac, whom you love, and go to the land of Mori'ah, and offer him there as a burnt offering upon one of the mountains of which I shall tell you" (Genesis 22:1-19).

Abraham passed the test, and Isaac was saved.

A more detailed and touching version of this sacrifice story is recorded in *Jasher*. In that version Sarah was not told why Isaac and his father Abraham were going on a journey. Nevertheless, like any good mother, she was very concerned about Isaac's well-being and the care that Abraham was likely to provide along the way. Among other things, she insisted that Isaac wear nice clothes:

> And Sarah said to Abraham, O my lord, I pray thee take heed of thy son, and place thine eyes over him, for I have no other son nor daughter but him. O forsake him not. If he be hungry give him bread, and if he be thirsty give him water to drink; do not let him go on foot, neither let him sit in the sun. Neither let him go by himself in the road, neither force him from whatever he may desire, but do unto him as he may say to thee.
> And Sarah wept bitterly the whole night on account of Isaac, and she gave him instructions till morning. And in the morning Sarah selected a very fine and beautiful garment from those garments which she had in the house, that Abimelech [king of the Philistines] had given to her. And she dressed Isaac her son therewith, and she put a turban upon his head, and she enclosed a precious stone in the top of the turban, and she gave them provision for the road, and they went forth, and Isaac went with his father Araham, and some of their servants accompanied them to see them off the road.
> And Sarah went out with them, and she accompanied them upon the road to see them off, and they said to her, return to the tent. And when Sarah heard the words of her son Isaac she wept bitterly, and Abraham her husband wept with her, and their son wept with them a great weeping; also those who went with them wept greatly. And Sarah caught hold of her son Isaac, and she held him in her arms, and she embraced him and continued to weep with him, and Sarah said, who knoweth if after this day I shall ever see thee again?[26]

Sarah's love for Isaac was reciprocated. Shortly before the intended sacrifice, Isaac had his own instructions for Abraham, and those included the following:

14. Conquest of Egypt

> And Isaac still said to his father, O my father, when thou shalt have slain me and burnt me for an offering, take with thee that which shall remain of my ashes to bring to Sarah my mother, and say to her, this is the sweet smelling savor of Isaac; but do not tell her this if she should sit near a well or upon any high place, lest she should cast her soul after me and die.[27]

Although Isaac was saved from being sacrificed, Sarah's joy on learning the good news caused her to expire:

> [W]hen she heard the word her joy was so exceedingly violent on account of her son, that her soul went out through joy; she died and was gathered to her people.[28]

While this might be considered only a beautiful (but very sad) story, one reason for its inclusion here is the information that it suggests on the attire of Abraham's sons in general and Isaac in particular. On this occasion at least, Isaac wore a very fine and beautiful garment, and he had a turban on his head that held a precious stone at the top. This attire would be consistent with the clothing that one might also expect a more modern sultan to wear, and thus it is also consistent with the idea that, at least on special occasions, Abraham's sons may have been treated and dressed like Arab royalty. Abraham himself could have been considered to be the father of the Arabs, and more specifically it would not be surprising that Isaac's brother Ishmael, if he were a leader of the Hyksos, could also have been associated with some form of the title Sultan.

There are also several Bible-related references to the wife and children of Ishmael:

These are the descendants of Ish'mael, Abraham's son, whom Hagar the Egyptian, Sarah's maid, bore to Abraham. These are the names of the sons of Ish'mael, named in the order of their birth: Neba'ioth, the firstborn of Ish'mael; and Kedar, Adbeel, Mibsam, . . . (Genesis 25:12-13).

So when Esau saw that the Canaanite women did not please Isaac his father, Esau went to Ish'mael and took to wife, besides the wives he had, Ma'halath the daughter of Ish'mael Abraham's son, the sister of Neba'ioth (Genesis 28: 8-9).

These are the descendants of Esau (that is, Edom). Esau took his wives from the Canaanites: Adah the daughter of Elon the Hittite, Oholiba'mah the daughter of Anah the son of Zib'eon the Hivite, and Bas'emath, Ish'mael's daughter, the sister of Neba'ioth (Genesis 36:1-3).

[A]nd his mother [Hagar] took him [Ishmael] a wife from among the daughters of Egypt. And she bare him a son, and he called his name Nebaioth; for she said, "The Lord was nigh to me when I called upon him" (Jubilees 17:13-14).

And he [Ishmael] and his mother afterward went to the land of Egypt, and they dwelt there, and Hagar took a wife for her son from Egypt, and her name was Meribah. And the wife of Ishmael conceived and bare four sons and two daughters, and Ishmael and his mother and his wife and children afterward went and returned to the wilderness [of Paran].[29]

And Ishmael took a wife from the land of Egypt, and her name was Ribah, the same is Meribah. And Ribah bare unto Ishmael Nebayoth, Kedar, Adbeel, Mibsam and their sister Bosmath [Bas'emath]. And Ishmael cast away his wife Ribah, and she went from him and returned to Egypt to the house of her father, and she dwelt there, for she had been very bad in the sight of Ishmael, and in the sight of his father Abraham. And Ishmael afterward took a wife from the land of Canaan, and her name was Malchuth . . .[30]

Jubilees and *Jasher* are in agreement that Ishmael's first wife was an Egyptian. According to the more detailed account in *Jasher*, Ishmael remained in Egypt for the birth of his first four sons and perhaps two daughters (cf. Genesis 28:8-9, Genesis 36:1-3, *Jasher* 21:17-18). After that, he left Egypt and returned to the wilderness of Paran. If the *Jasher* version is correct, it would seem likely that Ishmael had remained in Egypt for perhaps ten or more years. The references and speculations of this section do not prove that Ishmael is the same person as Salatis. They do suggest at least that Ishmael and his family may have had close connections with Egyptian royalty at the same time that the Hyksos were beginning their leadership role in that country.

14.5 From Salatis to Apophis

Having considered in some detail the title and a possible identity of the first Hyksos king, it would perhaps be appropriate to consider also how long that king might have reigned. Perhaps the oldest source of Manetho's chronology is provided by Josephus: "When this man [Salatis] had reigned thirteen years, after him reigned another. . . ."[31] Thus, Manetho reports that the first Hyksos king reigned for thirteen years. This result seems to be reinforced by the Turin Canon. As understood by Hayes, the Canon says the following: "In the Turin Canon (column x, 15)

Adventures In Egypt

the first Hyksos ruler of the Fifteenth dynasty is ascribed a reign of [1]3 (or [2]3?) years. . . ."[32] A reign length of thirteen years would have been sufficient for Ishmael, if he were Salatis, to father four sons and two daughters as suggested in the previous section. To be specific, the name Sheshi and the reign length of thirteen years will be employed here for this king.

As in Table 13.1, Sheshi's first year of reign has been assumed here and by others to have been about 1674 B.C. Thus, Sheshi's last year and his successor's first year would have been thirteen years later in about 1661 B.C., as shown in the table. It is now often accepted that the king who reigned immediately before Apophis was Khyan: "Khayan was, however, succeeded by Apophis who was apparently a usurper."[33] But from Table 13.1, Apophis came to the throne in about 1651 B.C. If these dates are correct, it is implied that Khyan reigned for a period of about ten years. This is consistent with a previous inference: "Since the immediate predecessor of Apophis is ascribed a reign of at least ten years in the Turin King-list, it is therefore inevitable that he should be identified with Khayan."[34]

The early Hebrews in Canaan were closely associated with the Amorites as discussed previously. Thus, following his visit to Egypt, Abram is found to have established a military alliance with Amorites then living in Canaan (Genesis 14:13). He is also likely to have entered Egypt previously with Amorites of Syria-Palestine who became known as the Hyksos.[35] The name Khyan of the Hyksos successor to Salatis has been "interpreted as Amorite Hayanu which the Egyptian form represents perfectly, and this is in all likelihood the correct interpretation."[36] Khyan's importance seems to have extended far beyond Egypt, and

14. Conquest of Egypt

objects with his name have been found or originated in at least Canaan, Knossos (Crete), Hattusha (Turkey), and Babylon (Iraq).

To provide a summary, it may be helpful to recall a statement by Hayes concerning the period during which the Hyksos controlled much of Egypt:

> In 1674 B.C. began the succession of six important Hyksos rulers, whom Manetho calls the Fifteenth Dynasty and who, according to the Turin Canon (column x, 15-21), reigned for a total of 108 years. Since this would bring us down to 1567 B.C., when the last Hyksos was driven from Egypt by the founder of the New Kingdom, it is probable that the numerous other Hyksos 'kings' of the same period were merely chiefs of the many different Asiatic tribes banded together under the leadership of the Great Hyksos. In this category would fall the seventy-five 'shepherd kings' assigned by Africanus to the Sixteenth and Seventeenth dynasties, the eight foreign (?) names listed at the end of column x of the Turin Canon (= the Sixteenth Dynasty?), and the quantity of unplaced Hyksos rulers mentioned on scarabs and other small monuments.[37]

This quotation from Hayes suggests another difficulty in trying to make sense of the chronology of the Hyksos occupation of Egypt. Between the date of the first Hyksos migration into Egypt (estimated here as about 1727 B.C.) and their expulsion (about 1567 B.C.) is a period of about one hundred sixty years. Within that relatively brief period must fit much of the Thirteenth Dynasty and all of the Fourteenth through the Seventeenth Dynasties. Thus, if the dynasties were consecutive, each one would last an average of between thirty and forty years. In contrast the Eighteenth Dynasty alone lasted about two hundred seventy-five years,[38] and the Twelfth Dynasty was about two hundred ten years in duration.[39] A partial explanation of this seeming discrepancy is that the different dynasties of the Hyksos era may have overlapped in time while perhaps being somewhat distinct in geographical focus.

Even with such a temporal overlap between dynasties, there still seem often to be too many kings for each dynasty. The answer to this problem is that not all of the people identified with the titles, scarabs, or other honorific records that one might associate with a king actually held that office in any form. In contemplating the scarab evidence for the existence of many of the Hyksos kings, Van Seters has remarked as follows: "These scarabs yield so many names that they cannot all be considered as primary rulers of Egypt. . . . Even if there were only one-half the number of names, there would still be too many kings for the period."[40]

Having considered the possibility that Ishmael might have had a leadership role among the Hyksos, perhaps even as the first Hyksos king of Egypt, it may be worthwhile in conclusion to recall again that other early members of the family of Abraham might also have held such positions. It isn't possible to be certain of any such identifications, but it may be of interest to mention a few possibilities. Some of these can be identified (tentatively) by a similarity of their activities or their names, or perhaps of the meanings of their names.[41] Thus, Jacob-el was said to have been a Hyksos king at about the same time that the Hebrew Jacob was present in Egypt with his family. Similarly, Bnon was said to have been a Hyksos king, at about the time that Jacob's highly-favored son Benoni/Benjamin (Genesis 35:18, 43:26-34, 45:22)

14. Conquest of Egypt

was also in Egypt. Chancellor Hur in Egypt had the same leadership role that the Bible ascribes to Joseph. Both the name Hur and the names that Joseph gave his children seem to reflect the background of Joseph.

The idea of a person naming his child to recall his own past is, of course, not unique to Joseph, and Moses is said to have done the same:

> Now the priest of Midian had seven daughters: and they came and drew water, and filled the troughs to water their father's flock. And the shepherds came and drove them away: but Moses stood up and helped them, and watered their flock. And when they came to Reuel their father, he said, How is it that ye are come so soon to day? And they said, An Egyptian delivered us out of the hand of the shepherds, and also drew water enough for us, and watered the flock. And he said unto his daughters, And where is he? why is it that ye have left the man? call him, that he may eat bread. And Moses was content to dwell with the man: and he gave Moses Zipporah his daughter. And she bare him a son, and he called his name Gershom: for he said, I have been a stranger in a strange land (Exodus 2:16-22).

The concluding phrase in this quotation, "Stranger in a Strange Land," became the title of a science fiction novel.[42]

14.6 Conclusion

One of the most important ways to establish dates for events described in the Bible is to find connections between biblically-recorded occurrences and historical events known from non-biblical sources. A good place in the early biblical accounts to look for confirmatory historical evidence is in the stories relating to the sojourn of the Hebrews in Egypt. Chapter 13 focused especially on the time period from Abram's journey to Egypt in about 1727 B.C. until the country of Egypt was taken over by the Hyksos in about 1674 B.C. The emphasis in this chapter has been on some details of the conquest itself until the time of Apophis I, possibly the Hyksos king with whom Joseph interacted. Of special interest were the indications that the descendants of Abram may, according to Egyptian records, have included the people that became known as Arabs, as well as those who became known as Hebrews. The Hyksos king at the time of the conquest seems to have been identified by the Egyptian historian Manetho with the Arabic title Salatis or Sultan. Since the Arabs are said to have been descendants of Abram, it is conceivable that the Sultan may also have been such a descendant. Abram's son Ishmael had an Egyptian mother, sometimes said to be the daughter of an Egyptian king. Ishmael is also said to have had an Egyptian wife, and he was the father of twelve princes. Sheshi is often considered to be the personal Egyptian name for Salatis, and the chronology of Sheshi and his successor Khyan have been estimated here.

---

1. "Flavius Josephus Against Apion," *The Life and Works of Flavius Josephus*, Translated by William Whiston, Introductory Essay by H. Stebbing (Holt, Rinehart and Winston, New York), Book 1, Chapter 14, pp. 863-864; L. W. Casperson, "Appendix B: Manetho," *Patterns of Biblical Chronology* (Westbow Press, Bloomington IN, 2012), pp. 630-645; p. 631.

2. L. W. Casperson, op. cit., Section 24.3, "The 400 year stela," pp. 598-605.
3. W. C. Hayes, *The Scepter of Egypt, A Background for the Study of the Egyptian Antiquities in The Metropolitan Museum of Art, Part II: The Hyksos Period and the New Kingdom* (1675-1080 B.C.) (Harvard University Press, Cambridge, Massachusetts, 1959), p. 4.
4. N. Grimal, *A History of Ancient Egypt* (Barnes and Noble, New York, 1997), p. 185.
5. L. W. Casperson, op. cit., Section 23.3, "Chronology of the Hyksos," pp. 569-574.
6. Ibid., "Appendix B: Manetho," pp. 630-645.
7. Ibid., pp. 630-632.
8. Eusebius, *Preparation for the Gospel*, Translated from a revised text by E. H. Gifford, Part 1 (Oxford at the Clarendon Press, 1903), Book 9, Chapters 18, 23, 27; quoting Alexander Polyhistor, who was quoting Artapanus.
9. L. W. Casperson, op. cit., "Appendix A: Artapanus," pp. 622-629.
10. Ibid., pp. 622-623.
11. Ibid., Section 24.3, "The 400 year stela," pp. 598-605.
12. Ibid., "Appendix A: Artapanus," Paragraph 2, p. 623.
13. F. Josephus, *The Life and Works of Flavius Josephus*, Translated by William Whiston (Holt, Rinehart and Winston, New York), *The Antiquities of the Jews*, Book II, 11:1, p. 79.
14. L. W. Casperson, op. cit., "Appendix A: Artapanus," Paragraph 23, p. 626.
15. Ibid., "Appendix B: Manetho," Chapter 14, p. 631.
16. J. H. Breasted, *A History of Egypt from the Earliest Times to the Persian Conquest* (Charles Scribner's Sons, New York, 1905), Book Four, "The Hyksos: The Rise of the Empire," pp. 217-219.
17. W. C. Hayes, op. cit., pp. 4-5.

18. N. Grimal, op. cit., pp. 185, 187.
19. L. W. Casperson, op. cit., Section 23.5, "Finding Joseph," pp. 581-589.
20. T. E. Colebrooke, "On imperial and other titles," *Journal of the Royal Asiatic Society of Great Britain and Ireland*, New Series, Volume 9, Number 2, pp. 314-420 (April, 1877); p. 370.
21. A. H. Sayce, *The Ancient Empires of the East*, (Macmillan and Co., London, 1884), p. 34.
22. L. W. Casperson, op. cit., "Appendix A: Artapanus," Paragraph 2, p. 623.
23. *The Book of Yashar*, Translated from the Hebrew and Published by Mordecai Manuel Noah (Hermon Press, New York, 1972), (cited here and below in the form *Jasher*); *Jasher* 15:31-32.
24. *Jasher* 21:16-17.
25. W. C. Hayes, "Egypt: From the death of Amenemes III to Seqenenre II," *The Cambridge Ancient History, Third Edition, Volume II, Part 1, History of the Middle East and the Aegean Region c. 1800-1380 B.C.*, Edited by I. E. S. Edwards, C. J. Gadd, N. G. L. Hammond, and E. Sollberger (Cambridge University Press, London, 1973), Chapter 2, p. 57.
26. *Jasher* 23:10-18.
27. *Jasher* 23:62.
28. *Jasher* 23:86.
29. *Jasher* 21:17-18.
30. *Jasher* 25:15-18.
31. L. W. Casperson, op. cit., "Appendix B: Manetho," p. 631.
32. W. C. Hayes, "Egypt: From the death of Amenemes III to Seqenenre II," op. cit., Chapter 2, p. 59.
33. K. S. B. Ryholt, *The Political Situation in Egypt during the Second Intermediate Period c. 1800-1550 B.C.* (Museum Tusculanum Press, Copenhagen, 1997), p. 256.

34. Ibid., p. 120.
35. L. W. Casperson, op. cit., Section 24.6, "Abram in Canaan and Egypt," pp. 610-618.
36. K. S. B. Ryholt, op. cit., p. 128.
37. W. C. Hayes, "Egypt: From the death of Amenemes III to Seqenenre II," op. cit., Chapter 2, p. 58.
38. L. W. Casperson, op. cit., Table 7.1, p. 145.
39. W. C. Hayes, "Chronological Table," *The Scepter of Egypt, A Background for the Study of the Egyptian Antiquities in The Metropolitan Museum of Art, Part I: From the Earliest Times to the End of the Middle Kingdom* (Harvard University Press, Cambridge, Massachusetts, 1953), p. 2.
40. J. V. Van Seters, *The Hyksos: A New Investigation* (Yale University Press, New Haven, p. 1966), pp. 158-159.
41. L. W. Casperson, op. cit., Section 23.5, "Finding Joseph," pp. 581-589.
42. R. A. Heinlein, *Stranger in a Strange Land* (Putnam, New York, 1961).

15. ARGONAUTS IN EGYPT

"May the breeze blow softly with which we shall sail over the sea in fair weather."[1]

15.1 Introduction

Much has been written in ancient and modern times about the people and events associated with the exodus. Besides such writings, however, there may be other early sources of information that could relate to the exodus, even if those sources don't mention directly any of the familiar exodus people and events. In particular, writings from or about Egypt provide many records that could possibly be associated with the exodus.[2] Another important category of literature not usually considered includes the writings of early Greek historians. This chapter considers the possibility that the Greeks may have incorporated into their own works information from earlier Egyptian or Bible-related texts. Some of those works could include details about people and events that are otherwise not known in biblical literature today.

The principle source of information in any discussion of the exodus of the Hebrews from Egypt is, of course, the biblical book Exodus. There are, however, many other writings about exodus-

15. Argonaunts in Egypt

related happenings. Most of these writings could be referred to as Bible-related, while others might seem at first consideration to have nothing to do with the Bible or the exodus. One possibly unexpected resource is provided by the various versions of the story of Jason and the Argonauts. It has already been suggested in Section 1.5 that the *Argonautica* story as written by Apollonius Rhodius (Apollonius of Rhodes) may incorporate a creation story in a song ascribed to Orpheus that has several similarities to the creation story in Genesis. It will be suggested here that the *Argonautica* and other Greek texts may also include exodus-related components that could supplement the biblical story.

Section 15.2 provides a brief summary of who the Argonauts were and how they happened to undertake their major expedition. The geography of the Mediterranean world as understood by the ancient Greeks is considered in Section 15.3, with an emphasis on the geography of the coastal region of Egypt. The ancient understanding that the Colchians of the primary Argonaut destination were related to the Egyptians is also summarized. One of the most dramatic stories in the Old Testament concerns the escape of the Israelites from Egypt, and the biblical account of this event is reviewed in Section 15.4. Section 15.5 describes how the Argonauts were driven by a north wind onto the Mediterranean coast of or near Egypt and were carried inland by a large wave. Such a wave could have been associated with a volcanic eruption. The Argonauts were also afflicted by a multi-day period of darkness and toxic atmosphere after their landing. These effects could also have been related to a volcanic event, as considered in Section 15.6. The curious circumstance of a large horse leaping from the water after becoming free from a submerged chariot is described in Section 15.7, and this occurrence may have been adapted from some Hebrew or Egyptian recollection of the exodus.

15.2 Calm sea and prosperous voyage

As a starting point it may be helpful to consider a brief summary of the story of the Argonauts. Like the *Odyssey* and *Iliad* of Homer and other Greek writings, the *Argonautica* is structured as an epic adventure story of the struggles of heroes against villains, nature, and sometimes the gods themselves. In this story the main character is Jason, whose father Aeson, was the rightful king of Iolcos. Because of the dominance of Jason, the story is often referred to as Jason and the Argonauts. Jason's evil uncle Pelias assigned him the task of going to the faraway city of Colchis and recovering the previously stolen golden fleece. To undertake this project, Jason recruited the assistance of several of the greatest heroes of his time, some with their own unique talents. He also acquired for his purpose the ship Argo and eventually sailed off with his crew of heroes. In the course of completing his task, he fell in love with Medea, the beautiful and talented daughter of Aeetes, the king of Colchis, from whom he was attempting to recover the golden fleece.

Like most other ancient Greek adventure stories, the *Argonautica* of Apollonius Rhodius is understood to be partly a compilation of previous and sometimes very old stories and partly the introduction of new elements by the author. The purpose is in part to be as entertaining as possible and thus to bring fame and glory to the compiler/author. Apollonius wrote his story in about the middle of the third century B.C., and much

15. Argonaunts in Egypt

older versions of the Argonaut stories were already in circulation at that time. Some of his material is traceable to known earlier works, but much of it seems to have been introduced from unknown sources or from his own imagination.

Many studies have been conducted concerning the life of Apollonius and the creation of this his masterpiece, and a brief summary of such background can be found in the introductory comments to the translation by R. C. Seaton.[3] It is interesting that the first edition of this work, composed while Apollonius was still a youth in his home city of Alexandria, Egypt, was so offensive to either his general audience, or at least to some influential people, that Apollonius left (or fled from) Egypt to the island of Rhodes. While on Rhodes he revised his work, and in the revised form it received such universal acclaim that he eventually returned and was welcomed back to Egypt. Unfortunately, the first edition is lost, and it is not certain at this time what had been so offensive about it. One possibility, mentioned in Section 1.6 in connection with the Hebrew creation story, is that the original text might have included too favorable a representation of the Hebrews.

One of the most tantalizing aspects of the Greek epic stories is the possibility that they may contain references to actual historical events. In principle, these can lead to a significant increase in our understanding of ancient times. However, because of the way in which the stories were assembled, it is not always easy to tell which components might represent historical information and which are fabrications by the compiler or his sources. For perspective we quote a few comments on the story as provided by Seaton:

Adventures In Egypt

As we have it, the motive of the voyage is the command of Pelias to bring back the golden fleece, and this command is based on Pelias' desire to destroy Jason, while the divine aid given to Jason results from the intention of Hera to punish Pelias for his neglect of the honour due to her. The learning of Apollonius is not deep but it is curious; his general sentiments are not according to the Alexandrian standard, for they are simple and obvious. In the mass of material from which he had to choose the difficulty was to know what to omit, and much skill is shewn in fusing into a tolerably harmonious whole conflicting mythological and historical details. He interweaves with his narrative local legends and the founding of cities, accounts of strange customs, descriptions of works of art, such as that of Ganymede and Eros playing with knucklebones, but prosaically calls himself back to the point from these pleasing digressions by such an expression as "but this would take me too far from my song." His business is the straightforward tale and nothing else. The astonishing geography of the fourth book reminds us of the interest of the age in that subject, stimulated no doubt by the researches of Eratosthenes and others.[4]

The reason for including a discussion of the *Argonautica* in the present work is, of course, the possibility that some parts of this story might relate to the events and chronology of early Bible times. As Seaton has observed (quoted above) concerning Apollonius, "much skill is shewn in fusing into a tolerably harmonious whole conflicting mythological and historical details." Thus, in considering the subjects that he has included, one should also consider the range of historical or other material to which he had access. In this regard it is significant that he was a student of Callimachus, an important scholar and author at the

15. Argonaunts in Egypt

Library of Alexandria, the greatest library in the world at that time. Apollonius himself later became a librarian there. In that context he would have had access to vast amounts of historical material.

In order for there to be a real connection between the biblical stories and the *Argonautica*, it would have been necessary for Apollonius to have had access to a Bible, preferably in the Greek language. Israel may have regarded its historical records that evolved into the Bible as sacred, and it might not seem obvious that those records could have been included in the library's collections. However, the earliest Greek version of the Old Testament is traditionally referred to as the Septuagint, commonly abbreviated LXX:

> Just when the LXX was composed is not known, though it is probable that the Pentateuch was translated in Alexandria by the middle of the third century B.C.
>
> Traditionally the LXX was written during the reign of Ptolemy II Philadelphus (285-247 B.C.); this view is based on a work by a certain Aristeas written to a friend named Philocrates.[5]

Thus, the Pentateuch would likely have been available during Apollonius's association with the Library at Alexandria, and he would almost surely have been aware of its existence. Depending on his interests and those of his teacher Callimachus, he may even have been familiar with details of its contents. Thus it is possible that Apollonius's description of the creation or the adventures of the Argonauts could have been influenced by the biblical stories in what came to be known as the Septuagint.

15.3 Geography of Egypt

It is of interest here to focus particularly on whether any of Apollonius's writings might have been influenced by biblical stories related to the exodus. As a preliminary step, we may consider what Apollonius had to say through the characters of his story about the geography of ancient Egypt. The primary objective of the Argonauts was to recover the golden fleece from Colchis, at the eastern end of what we refer to today as the Black Sea. There seem to have been significant historical connections between Colchis of Jason's expedition and the land of Egypt. One potentially important resource for Apollonius would have been the writings of the earlier Greek historian Herodotus (c. 484-425 B.C.). Herodotus was well aware of the ancient lands of Colchis and Egypt, and he believed that the Colchians were actually Egyptians:

> For it is plain to see that the Colchians are Egyptians; and this that I say I myself noted before I heard it from others. When I began to think on this matter, I inquired of both peoples; and the Colchians remembered the Egyptians better than the Egyptians remembered the Colchians; the Egyptians said that they held the Colchians to be part of Sesostris' army.[6]

Herodotus also provides a brief record of the encounter between the Greek Argonauts and the Colchians:

> They [the Argonauts] sailed in a long ship to Aea of the Colchians and the river Phasis: and when they had done the rest of the business for which they came [recovering the golden fleece], they carried off the king's daughter Medea. When the Colchian king [Aeetes] sent a herald to

15. Argonaunts in Egypt

demand reparation for the robbery, and restitution of his daughter, the Greeks replied that as they had been refused reparation for the abduction of the Argive Io, neither would they make any to the Colchians.[7]

The exodus was of course from Egypt, and the geography of Egypt plays an important role in the exodus stories. Thus the understanding by the ancient Greeks of the geography of Egypt would be of particular interest here. Information on this geography is provided by Apollonius, as he describes how the Argonauts planned their flight from the very angry king Aeetes. In this description one learns that the "Egyptians" of Colchis kept maps showing accurate representations of seas and highways. Some of Apollonius's comments on geography are given below. Much of this material is represented as being in the words of the Argonauts' navigator Argus, son of Phrixus. Argus spoke of the different routes that the Argonauts might use in their return from Colchis to Iolcus in Greece:[8]

> And straightway Aeson's son [Jason] and the rest of the heroes bethought them of Phineus [a seer], how that he had said that their course from Aea should be different, but to all alike his meaning was dim. Then Argus [son of Phrixus] spake, and they eagerly hearkened:
> "We go to Orchomenus, whither that unerring seer, whom ye met aforetime, foretold your voyage. For there is another course, signified by those priests of the immortal gods, who have sprung from Tritonian [Egyptian] Thebes. . . . [N]or at that time was the Pelasgian land ruled by the glorious sons of Deucalion, in the days when Egypt, mother of men of an older time, was called the fertile Morning-land, and the river fair-flowing Triton [Nile River], by which all the Morning-

land is watered; and never does the rain from Zeus moisten the earth; but from the flooding of the river abundant crops spring up. From this land, it is said, a king [Sesostris, according to Herodotus[9]] made his way all round through the whole of Europe and Asia, trusting in the might and strength and courage of his people; and countless cities did he found wherever he came, whereof some are still inhabited and some not; many an age hath passed since then. But Aea [a city of Colchis] abides unshaken even now and the sons of those men whom that king settled to dwell in Aea [descendants of the Egyptians]. They preserve the writings of their fathers, graven on pillars, whereon are marked all the ways and the limits of the sea and land as ye journey on all sides round. . . ."

In this excerpt Argus, a rescued traveller from Colchis, is explaining to the Argonauts that the priests of Egyptian Thebes had long ago prepared detailed maps of Europe and Asia and left copies of them in Aea. Argus had studied those maps and (following the above excerpt) suggested a route back to Greece that would reduce the chances of their capture by angry Colchians. The king spoken of is identified by the historian Herodotus as Sesostris, but Apollonius may not have endorsed that identification.

An important implication of this excerpt is that Triton seems unmistakably to be identified as an early name for the Nile River, which is the main source of moisture and fertility in Egypt. In the words of Herodotus, the annual flooding of the Nile made agriculture in the delta region the easiest in the world:

Now, indeed, there are no men, neither in the rest of Egypt, nor in the whole world, who gain from the soil

15. Argonaunts in Egypt

with so little labour; they have not the toil of breaking up the land with the plough, nor of hoeing, nor of any other work which other men do to get them a crop; the river rises of itself, waters the fields, and then sinks back again; thereupon each man sows his field and sends swine into it to tread down the seed, and waits for the harvest; then he makes the swine to thresh his grain, and so garners it.[10]

Curiously, a different conclusion is reported in the Bible, where it is argued that farming in Canaan is much easier than it had been in Egypt:

Soon you will cross the Jordan River, and if you obey the laws and teachings I'm giving you today, you will be strong enough to conquer the land that the Lord promised your ancestors and their descendants. It's rich with milk and honey, and you will live there and enjoy it for a long time. It's better land than you had in Egypt, where you had to struggle just to water your crops. But the hills and valleys in the promised land are watered by rain from heaven, because the Lord your God keeps his eye on this land and takes care of it all year long (Deuteronomy 11:8-12).

15.4 Escape from Egypt

In the biblical exodus story, Pharaoh finally decided to free the people of Israel from their lives of slavery and let them depart from Egypt. But then he changed his mind, and this section quotes much of the story of what happened next. The following sections of this chapter emphasize interactions between the newly arrived Argonauts and the environment and activities of the escaping Israelites:

Then the Lord said to Moses, "Tell the people of Israel to turn back and encamp in front of Pi-ha-hi'roth, between Migdol and the sea, in front of Ba'al-ze'phon; you shall encamp over against it, by the sea. For pharaoh will say of the people of Israel 'They are entangled in the land; the wildernesss has shut them in.' And I will harden Pharaoh's heart, and he will pursue them and I will get glory over Pharaoh and all his host; and the Egyptians shall know that I am the Lord." And they did so.

When the king of Egypt was told that the people had fled, the mind of Pharaoh and his servants was changed toward the people, and they said, "What is this we have done, that we have let Israel go from serving us?" So he made ready his chariot and took his army with him, and took six hundred picked chariots and all the other chariots of Egypt with officers over all of them. And the Lord hardened the heart of Pharaoh king of Egypt and he pursued the people of Israel as they went forth defiantly. The Egyptians pursued them, all Pharaoh's horses and chariots and his horsemen and his army, and overtook them encamped at the sea, by Pi-ha-hi'roth, in front of Ba'al-ze'phon (Exodus 14:1-9).

The Lord said to Moses, "Why do you cry to me? Tell the people of Israel to go forward. Lift up your rod, and stretch out your hand over the sea and divide it, that the people of Israel may go on dry ground through the sea. And I will harden the hearts of the Egyptians so that they shall go in after them, and I will get glory over Pharaoh and all his host, his chariots, and his horsemen. And the Egyptians shall know that I am the Lord, when I have gotten glory over Pharaoh, his chariots, and his horsemen."

Then the angel of God who went before the host of Israel moved and went behind them; and the pillar of cloud moved from before them and stood behind them,

15. Argonaunts in Egypt

coming between the host of Egypt and the host of Israel. And there was the cloud and the darkness; and the night passed without one coming near the other all night.

Then Moses stretched out his hand over the sea; and the Lord drove the sea back by a strong east wind all night, and made the sea dry land, and the waters were divided. And the people of Israel went into the midst of the sea on dry ground, the waters being a wall to them on their right hand and on their left. The Egyptians pursued, and went in after them into the midst of the sea, all Pharaoh's horses, his chariots, and his horsemen. And in the morning watch the Lord in the pillar of fire and of cloud looked down upon the host of the Egyptians, and discomfited the host of the Egyptians, clogging their chariot wheels so that they drove heavily; and the Egyptians said, "Let us flee from before Israel; for the Lord fights for them against the Egyptians."

Then the Lord said to Moses "Stretch out your hand over the sea, that the water may come back upon the Egyptians, upon their chariots, and upon their horsemen." So Moses stretched forth his hand over the sea, and the sea returned to its wonted flow when the morning appeared; and the Egyptians fled into it, and the Lord routed the Egyptians in the midst of the sea. The waters returned and covered the chariots and the horsemen and all the host of Pharaoh that had followed them into the sea; not so much as one of them remained. But the people of Israel walked on dry ground through the sea, the waters being a wall to them on their right hand and on their left.

Thus the Lord saved Israel that day from the hand of the Egyptians; and Israel saw the Egyptians dead upon the seashore. And Israel saw the great work which the Lord did against the Egyptians, and the people feared the Lord; and they believed in the Lord and in his servant Moses (Exodus 14:15-31).

The location near Egypt at which the Israelites escaped from the armies of Pharaoh is now reasonably well understood. As described in the preceding quotations, the Israelites marched across a sea or lake at the time of an exceptionally low water level. In the biblical account the reason for the low water was a "strong east wind all night" (Exodus 14:21). Visibility was initially very poor because of the pillar of cloud and darkness, and the Egyptians were unable to locate the Israelites. When visibility had improved, the Egyptians rushed in, but by then the wind had stopped, the water was rising rapidly, chariot wheels were clogging, and the Egyptians were destroyed.

It has been an interesting question to estimate the place where the actual escape had occurred, and a possible interpretation is mentioned here. The particular sea referred to in the Bible may well have been one of the chain of lakes crossing the Isthmus of Suez: "The sea through which the Israelites passed most likely was one of the lakes in the Isthmus of Suez; most likely the Ballah Lakes, Lake Timsah, or Bitter Lakes region" . . . "The narrow area where the lesser and greater Bitter Lakes join has been proposed as a more specific point because the water can be so shallow at times as to allow the water to be forded."[11] Figure 15.1 suggests possible exodus routes that might have been followed by the Israelites as they fled from Pharaoh.[12] A strong east wind would have lowered the water level in the lesser Bitter Lake and perhaps exposed the shallows of the "Alternate Exodus Route" indicated in the figure. Initially, the Egyptians couldn't find the escaping Israelites because of the darkness and clouds between them.

15. Argonaunts in Egypt

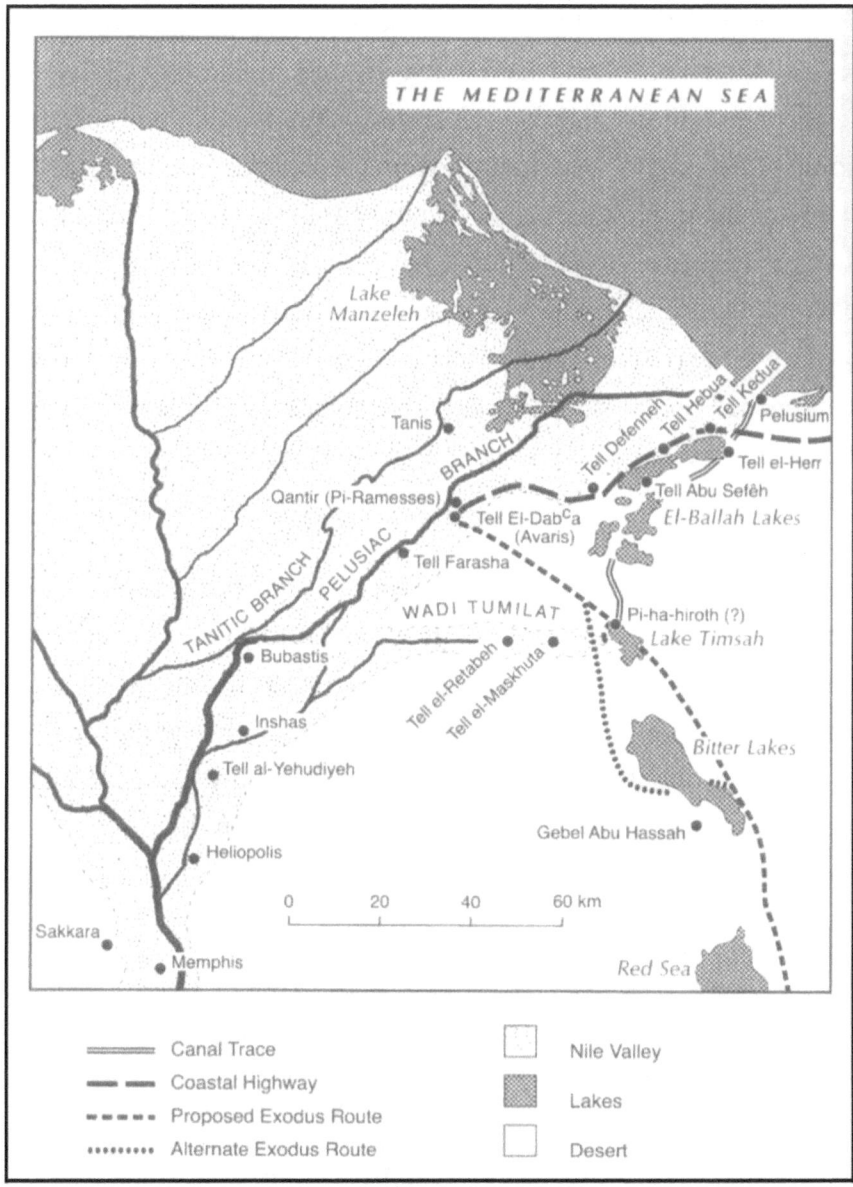

Figure 15.1. Map of the Nile Delta and North Sinai with possible exodus routes indicated. (Reproduced with permission)

During the Israelite escape, the waters of the greater and lesser Bitter Lakes may have been divided by the reduced water level. Thus, the remaining separated lakes would have offered some protection to the Israelites from the Egyptians, serving as "a wall to them on their right hand and on their left" (Exodus 14:22). It seems, however, that by the time the atmosphere had cleared, the Israelites had successfully crossed the recently dried shallows, and the wind had ceased. When the Egyptians approached them from behind, the horses and chariots of the Egyptians were trapped in the mud and returning water.

15.5 The wave

The preceding sections have introduced very briefly some of the ancient concepts relating to the geography of Egypt. The idea that maps may have been available for understanding this geography and planning travels has also been noted. With that background it is now appropriate to focus more closely on how it was that the Argonauts may have found themselves to be in Egypt. A very short version of this episode is provided by Herodotus:

> The following story is also told: – Jason (it is said) when the Argo had been built at the foot of Pelion, put therein besides a hecatomb a bronze tripod, and set forth to sail round the Peloponnesus, that he might come to Delphi. But when in his course he was off Malea, a north wind caught and carried him away to Libya [an early name for the continent of Africa]; and before he could spy land he came into the shallows of the Tritonian lake. There, while yet he could find no way out, Triton (so goes the story) appeared to him and bade Jason give him the tripod, promising so to show the shipmen the channel and send them on their way unharmed. Jason did his bidding,

15. Argonaunts in Egypt

and Triton [here a son of Poseidon] then showed them the passage out of the shallows and set the tripod in his own temple. . . .[13]

Thus, according to Herodotus, the Argonauts in the course of a voyage found themselves being driven by a powerful north wind into the shallows of a lake on the northern coast of Africa that the ancients referred to as Libya. In this excerpt the Argonauts are not said to have been traveling to or from Colchis.

A somewhat related account is recorded by Diodorus of Sicily (c. 80-20 B.C.):

> Not a few both of the ancient historians and of the later ones as well, one of whom is Timaeus, say that the Argonauts, after the seizure of the fleece, learning that the mouth of the Pontus had already been blockaded by the fleet of Aeetes, performed an amazing exploit which is worthy of mention. . . . Furthermore, when they were driven by winds to the Syrtes and had learned from Triton, who was king of Libya at that time, of the peculiar nature of the sea there, upon escaping safe out of the peril they presented him with the bronze tripod which was inscribed with ancient characters. . . .[14]

The shallows in this case are referred to as the Syrtes, considered in more detail in the following paragraphs.

Apollonius's treatment of the Argonaut story is more extensive than that of Herodotus or Diodorus. For reasons that will become obvious, we will quote and then comment on several story segments relating to the seemingly unintentional visit of the Argonauts to Egypt. As also indicated by Herodotus and Diodorus quoted above, an unexpected north wind drove the Argonauts south to Libya, and then according to Apollonius a "flood-tide" carried their ship far onto the innermost shore of a gulf:

> Now had they left behind the gulf named after the Ambracians, now with sails wide spread the land of the Curetes, and next in order the narrow islands with the Echinades, and the land of Pelops was just descried; even then a baleful blast of the north wind seized them in mid-course and swept them towards the Libyan sea nine nights and as many days, until they came far within Syrtis, wherefrom is no return for ships, when they are once forced into that gulf. For on every hand are shoals, on every hand masses of seaweed from the depths; and over them the light foam of the wave washes without noise; and there is a stretch of sand to the dim horizon; and there moveth nothing that creeps or flies. Here accordingly the flood-tide – for this tide often retreats from the land and bursts back again over the beach coming on with a rush and roar – thrust them suddenly on to the innermost shore, and but little of the keel was left in the water. And they leapt forth from the ship, and sorrow seized them when they gazed on the mist and the levels of vast land stretching far like a mist and continuous into the distance; no spot for water, no path, no steading of herdsmen did they descry afar off, but all the scene was possessed by a dead calm.[15]

According to Diodorus and Apollonius, the Argonauts were driven first by a north wind into the Syrtes or Syrtis. The word Syrtis refers to "[q]uicksands off the northern coast of Africa, considered dangerous to ships. The word Syrtis means any part of the sea that is perilous to navigation, either from storms or hidden rocks and quicksands."[16] An interaction of the Apostle Paul and others with the Syrtis is described in Acts 27:17.

15. Argonaunts in Egypt

In the version of Diodorus, the Argonauts may have been cast into the shallows of a lake ("the peculiar nature of the sea there"), while according to the longer version of Apollonius, they were thrust suddenly "on to the innermost shore" by a flood-tide. The description of the flood-tide corresponds closely to what was traditionally called a tidal wave. Tidal waves have nothing to do with ordinary tides, and they are often now called tsunamis (harbor waves), though they are also not in general caused by harbors. The ship seems to have been left aground by the wave in the shallows of a lake or gulf that opened into the Mediterranean Sea.

The reason for mentioning the encounter of the Argonauts with the large wave is the possibility that this story might in some form represent an identifiable historical event. The Argonauts may well have been washed by a wave onto an inner shore of a Tritonian Lake on the coast of Egypt. Such waves are usually caused by volcanic activity or other geophysical processes. As it happens volcanoes and earthquakes are not uncommon in the Mediterranean area, and Apollonius's indicated pattern of a withdrawal of the sea followed by a dramatic shoreward wave could be a consequence of an event such as the caldera collapse of a volcanic island or a shifting of tectonic plates. Similar effects of a much later volcanic eruption in the Mediterranean area will be considered in Section 19.7.

Before continuing, it may be appropriate to summarize some of the varied uses here of the name Triton that have been seen above in Sections 15.3 to 15.5, and others will be mentioned later. Thus, Tritonian Thebes refers to the city of Thebes in Egypt, as opposed presumably to the Thebes in Greece. The

Adventures In Egypt

"Fair-flowing Triton" is another name for the Nile River near which is located Egyptian Thebes. The Tritonian lake seems to be one of the lakes in the Nile delta through which a branch of the Nile empties into the Mediterranean Sea. The water in this lake was unusable for drinking. Triton is also the name of a son of Poseidon, though Herodotus seems to have been skeptical concerning stories about him. Triton is said by Diodorus to have been king of Libya, and he is said by Herodotus and Apollonius to have guided the Argonauts in their departure from Egypt.

It may also be of interest to consider at least briefly the origin of the name Triton that seems so closely to be identified with the Nile River in Egypt. In Greek mythology Triton is considered to be the son of the god Poseidon, who is often closely associated with the sea and other waters and who was venerated at Thebes in Greece. He was commonly represented as holding a three-pronged fishing spear or trident ("three-teeth"). Like his father Poseidon, Triton sometimes also carried a three-pronged spear. In the Argonaut stories Triton's home seems to have been near the coast of Libya (Egypt). Egypt is said from an early time to have developed and employed high-quality navigation maps, and Triton's association with the Nile River and coastal Egypt may have been encouraged by the trident-like appearance on maps of the river branches in the Nile delta.

15.6 The darkness

While the Argonauts in Apollonius's story were trying to find a solution to their stranding far from the Mediterranean shore, a strange and choking darkness overwhelmed them. The first possibility to consider is that this darkness might correspond to

15. Argonaunts in Egypt

the plague of darkness that occurred in Egypt shortly before the exodus:

> [T]he hearts of all grew numb, and pallor overspread their cheeks. And as, like lifeless spectres, men roam through a city awaiting the issue of war or of pestilence, or some mighty storm which overwhelms the countless labours of oxen, when the images of their own accord sweat and run down with blood, and bellowings are heard in temples, or when at mid-day the sun draws on night from heaven, and the stars shine clear through the mist; so at that time along the endless strand the chieftains wandered, groping their way. Then straightway dark evening came upon them; and piteously did they embrace each other and say farewell with tears, that they might, each one apart from his fellow, fall on the sand and die. And this way and that they went further to choose a resting-place; and they wrapped their heads in their cloaks and, fasting and unfed, lay down all that night and the day, awaiting a piteous death. But apart the maidens huddled together lamented beside the daughter of Aeetes [Medea]. And as when, forsaken by their mother, unfledged birds that have fallen from a cleft in the rock chirp shrilly; or when by the banks of fair-flowing Pactolus, swans raise their song, and all around the dewy meadow echoes and the river's fair stream; so these maidens, laying in the dust their golden hair, all through the night wailed their piteous lament. And there all would have parted from life without a name and unknown to mortal men, those bravest of heroes, with their task unfulfilled; . . .[17]

This toxic darkness would be compatible with the arrival of the cloud of dust and gas emitted into the atmosphere by a substantial volcanic eruption, and a discussion of the possible atmospheric consequences of volcanic eruptions has been given previously.[18] The only reasonable response of the Argonauts in their circumstance was to wrap their heads in their cloaks to avoid choking on the dust and then to lie down and wait for a piteous death. If the wave considered in the previous section was actually caused by a volcanic eruption, the subsequent arrival of a cloud of dust and gas would not have been an unlikely event. The same wind that drove their ship to the shore could at a later time have brought the noxious products of the volcano.

The previous paragraphs have introduced the idea from Herodotus, Diodorus, and Apollonius that the Argonauts were driven toward shore by a north wind near the Nile River in Egypt, where they were also cast inland by a large wave. After landing in a desolate area, according to Apollonius, the Argonauts were struck by the further misery of a multi-day period of darkness and terrifying air pollution. In the context of the exodus story, the plague that occurred last before the death of the firstborn sons in Egypt was the plague of darkness:

> Then the Lord said to Moses, "Stretch out your hand toward heaven that there may be darkness over the land of Egypt, a darkness to be felt." So Moses stretched out his hand toward heaven, and there was thick darkness in all the land of Egypt three days; they did not see one another, nor did any rise from his place for three days; but all the people of Israel had light where they dwelt (Exodus 10:21-23).

The description of the three-day plague as a "darkness to be felt" suggests more than a simple familiar lack of light, and it is not so hard to imagine in this context the presence of an opaque, choking, and perhaps foul-smelling cloud of dust and gas arising from a not-so-distant volcanic eruption. The location of such a cloud as a function of time depends in part, of course, on the direction of air currents. In the biblical story the darkness lasted three days, and for the Argonauts the duration may have been about the same. Thus, one could imagine that the Argonauts, stranded somewhere near the Mediterranean coast of Egypt, might have experienced the same period of darkness as did the Egyptians before the exodus. From the volcanic eruption records discussed in Section 8.5 and shown in Figure 8.2, this particular eruption sequence might have commenced as early as about 1442 B.C., two years before the exodus.

An account of this darkness given by Josephus is similar to the treatment of the darkness as provided by Apollonius:

> [A] thick darkness, without the least light, spread itself over the Egyptians, whereby their sight being obstructed, and their breathing hindered by the thickness of the air, they died miserably, and under a terror lest they should be swallowed up by the dark cloud. Besides this, when the darkness, after three days and as many nights, was dissipated, and when Pharaoh did not still repent and let the Hebrews go, Moses came to him and said, "How long wilt thou be disobedient to the command of God? . . ."[19]

According to Josephus, many of the Egyptians died from the suffocating "thickness of the air." A version of the exodus darkness story, including its lethality is also included in *Jasher*:

> And God sent darkness upon Egypt, that the whole land of Egypt and Pathros became dark for three days, so that a man could not see his hand when he lifted it to his mouth.

> At that time died many of the people of Israel who had rebelled against the Lord and who would not hearken to Moses and Aaron, and believed not in them that God had sent them.
>
> And who had said, we will not go forth from Egypt lest we perish with hunger in a desolate wilderness, and who would not hearken to the voice of Moses.
>
> And the Lord plagued them in the three days of darkness, and the Israelites buried them in those days, without the Egyptians knowing of them or rejoicing over them.
>
> And the darkness was very great in Egypt for three days, and any person who was standing when the darkness came, remained standing in his place, and he that was sitting, remained sitting, and he that was lying continued lying in the same state, and he that was walking remained sitting upon the ground in the same spot; and this thing happened to all the Egyptians, until the darkness had passed away.[20]

It is curious, that in the *Jasher* version of the darkness, rebellious Hebrews perished, as well perhaps as the obstinate Egyptians reported by Josephus above. The immobilizing effect of the darkness in *Jasher* is similar to that described by Apollonius.

The lethal "thickness of the air" would be compatible with an upwind volcanic eruption. Such an eruption would also be consistent with the wave that had driven the Argonauts ashore shortly before the darkness. In this interpretation the wave and darkness would have occurred at most about two years prior to the exodus. Major eruption-related events sometimes continue intermittently for a period of months or even years. The relatively complex eruption sequence at about the time of the exodus has been considered in Section 8.5 based on tree-ring evidence.

15. Argonaunts in Egypt

According to that data, the main eruption events may have commenced in about 1442 B.C. (Section 8.5, "Pre-exodus volcanic events") and again in about 1438 B.C. (Section 8.6, "Post-exodus volcanic events").

15.7 The chariot horse

An alternative possibility is that the darkness experienced by the Argonauts was that associated with the Israelites' crossing of the sea. Continuing the previous quotation from the *Argonautica*, one learns of their gradually improving circumstances:

> [B]ut as they pined in despair, the heroine-nymphs, warders of Libya, had pity on them, they who once found Athena, what time she leapt in gleaming armour from her father's head [conventional mythology of the birth of Athena], and bathed her by Trito's waters. It was noon-tide and the fiercest rays of the sun were scorching Libya; they [the heroine nymphs] stood near Aeson's son [Jason], and lightly drew the cloak from his head. And the hero cast down his eyes and looked aside, in reverence for the goddesses, and as he lay bewildered all alone they addressed him openly with gentle words:
> "Ill-starred one, why art thou so smitten with despair? We know how ye went in quest of the golden fleece; we know each toil of yours, all the mighty deeds ye wrought in your wanderings over land and sea. We are the solitary ones, goddesses of the land, speaking with human voice, the heroines, Libya's warders and daughters. Up then; be not thus afflicted in thy misery, and rouse thy comrades. And when Amphitrite [wife of Poseidon] has straightway loosed Poseidon's swift-wheeled car, then do ye pay to your mother a recompense for all her travail when she bare you so long in her womb; and so ye may return to the divine land of Achaea."

Thus they spake, and with the voice vanished at once, where they stood. But Jason sat upon the earth as he gazed around, and thus cried:

"Be gracious, noble goddesses of the desert, yet the saying about our return I understand not clearly. Surely I will gather together my comrades and tell them, if haply we can find some token of our escape, for the counsel of many is better."

He spake, and leapt to his feet, and shouted afar to his comrades, all squalid with dust, like a lion when he roars through the woodland seeking his mate; and far off in the mountains the glens tremble at the thunder of his voice; and the oxen of the field and the herdsmen shudder with fear; yet to them Jason's voice was no whit terrible – the voice of a comrade calling to his friends. And with looks downcast they gathered near, and hard by where the ship lay he made them sit down in their grief and the women with them, and addressed them and told them everything:

"Listen, friends; as I lay in my grief, three goddesses girded with goat-skins from the neck downwards round the back and waist, like maidens, stood over my head nigh at hand; and they uncovered me, drawing my cloak away with light hand, and they bade me rise up myself and go and rouse you, and pay to our mother a bounteous recompense for all her travail when she bare us so long in her womb, when Amphitrite shall have loosed Poseidon's swift-wheeled car. But I cannot fully understand concerning this divine message. They said indeed that they were heroines, Libya's warders and daughters; and all the toils that we endured aforetime by land and sea, all these they declared that they knew full well. Then I saw them no more in their place, but a mist or cloud came between and hid them from my sight."

Thus he spake, and all marvelled as they heard. Then was wrought for the Minyae the strangest of portents. From the sea to the land leapt forth a monstrous horse, of vast size, with golden mane tossing round his neck; and quickly from his limbs he shook off abundant spray and started on his course, with feet like the wind. And at once Peleus rejoiced and spake among the throng of his comrades:

"I deem that Poseidon's car has even now been loosed by the hands of his dear wife, and I divine that our mother is none else than our ship herself; for surely she bare us in her womb and groans unceasingly with grievous travailing. But with unshaken strength and untiring shoulders will we lift her up and bear her within this country of sandy wastes, where yon swift-footed steed has sped before. For he will not plunge beneath the earth; and his hoof-prints, I ween, will point us to some bay above the sea."

Thus he spake, and the fit counsel pleased all. This is the tale the Muses told; and I [Apollonius] sing obedient to the Pierides [a name of the Muses], and this report have I heard most truly; that ye, O mightiest far of the sons of kings, by your might and your valour over the desert sands of Libya raised high aloft on your shoulders the ship and all that ye brought therein, and bare her twelve days and nights alike. Yet who could tell the pain and grief which they endured in that toil? Surely they were of the blood of the immortals, such a task did they take on them, constrained by necessity. How forward and how far they bore her gladly to the waters of the Tritonian lake! How they strode in and set her down from their stalwart shoulders![21]

In Herodotus's much briefer version, quoted above, the washing ashore and the darkness episodes are not included, and the arrival from the Mediterranean Sea to the Tritonian lake is given in a single sentence: "[A]nd before he [Jason] could spy land he came into the shallows of the Tritonian lake."[22] A point of particular interest in Apollonius's more detailed account is the loosing of "Poseidon's swift-wheeled car." In the above quotations this loosing was deemed to have occurred when a "monstrous horse" leaped from the shallow gulf, shook off the water, and sped across the sandy wasteland. There are probably other interpretations, but it would seem possible that in the presence of the Argonauts a horse had become free from a recently submerged chariot ("car"). In the exodus context this picture would be very intelligible. When the Egyptian chariots and horsemen were destroyed, one horse (or more) might have been left alive but held initially by its harness to a submerged chariot:

> Then the Lord said to Moses, "Stretch out your hand over the sea, that the water may come back upon the Egyptians, upon their chariots, and upon their horsemen." So Moses stretched forth his hand over the sea, and the sea returned to its wonted flow when the morning appeared; and the Egyptians fled into it, and the Lord routed the Egyptians in the midst of the sea. The waters returned and covered the chariots and the horsemen and all the host of Pharaoh that had followed them into the sea; not so much as one of them remained. (Exodus 14:26-28)

For what it is worth, the biblical text doesn't say explicitly that the horses were all drowned, and great symbolism might have been associated with the observed escaping from the water of one large horse. Thus the horse that became free from a submerged chariot may have been a horse from Pharaoh's chariot corps that had become trapped in the waters when they had drowned Pharaoh's army at the time of the exodus. The above

15. Argonaunts in Egypt

interpretation presupposes, of course, that some observation of the aftermath of the destruction-by-water episode found its way into the records that could have been studied by Apollonius in Egypt a millennium later.

The fact that the horse was said to be monstrous and could leap from the sea to the land is reminiscent of the large breed of Urartian horses that Ezekiel referred to as "war horses" and that were reported to have leaped as far as 37 feet. This subject was considered previously in Section 6.4. Similar horses may have been developed and used by Urartian royalty (long after the exodus), and they were also exported to other countries. The leaping ability of the horse mentioned in the *Argonautica* is noted explicitly, and large horses may have been favored by the Egyptian king and his generals. A large horse might also have had a better chance of surviving the returning waters. The names Poseidon and Amphitrite could have been introduced into his story by Apollonius (at least in his second edition) to neutralize any concerns that might have been held by his anti-Hebrew readership. This would be similar to the previously-suggested possibility that a mythological creation story may have been added by Apollonius after his brief Bible-related creation account, as discussed in Section 1.5.

15.8 Conclusion

A lesser-known aspect of the story of Jason and the Argonauts is the unintentional visit of the Argonauts to Egypt (or the northeastern coast of "Libya"). That event is reported very briefly by Herodotus and Diodorus Siculus, and in more detail by Apollonius Rhodius. The present chapter has reviewed a few of the circumstances that were associated with this visit, including a powerful north wind, a huge wave, and an extended period of extreme darkness. The darkness was accompanied by a nearly unbreathable atmosphere. The wave, darkness, and toxic air could all have been caused by a volcanic eruption. Connections between the Argonauts and specific exodus-related personalities are contemplated in Chapters 16 and 17.

---

1. *Apollonius Rhodius: The Argonautica*, With an English translation by R. C. Seaton (Loeb Classical Library, Harvard University Press, Cambridge, Massachusetts, 1980), Book I, p. 33.
2. L. W. Casperson, *Patterns of Biblical Chronology* (Westbow Press, Bloomington, IN, 2012), Chapters 9-11, pp. 184-264.
3. *Apollonius Rhodius*, op. cit., "Introduction," pp. vii-xiv.
4. Ibid., pp. xii-xiii.
5. J. W. Wevers, "Septuagint," *The Interpreter's Dictionary of the Bible: An Illustrated Encyclopedia* (Abingdon Press, Nashville, 1962), Volume 4, R-Z, pp. 273-278; p. 273.
6. *Herodotus*, With an English translation by A. D. Godley, In Four Volumes (Loeb Classical Library, Harvard University Press, Cambridge, Massachusetts, 1981), Volume I, Book II, Paragraph 104, p. 391.
7. Ibid., Volume I, Book I, Paragraph 2, p. 5.
8. *Apollonius Rhodius*, op. cit., Book IV, pp. 311, 313.
9. *Herodotus*, op. cit., Volume I, Book II, Paragraphs 102-104, pp. 389, 391, 393.
10. Ibid., Volume I, Book II, Paragraph 14, p. 291.
11. J. K. Hoffmeier, *Israel in Egypt, The Evidence for the Authenticity of the Exodus Tradition* (Oxford University

Press, Inc., New York, New York, 1996), Chapter 9, "The problem of the Re(e)d Sea," pp. 199-222; p. 215, p. 212.
12. Ibid., Figure 2.
13. *Herodotus*, op. cit., Volume II, Book IV, Paragraph 179, p. 381.
14. *Diodorus Siculus in Twelve Volumes*, With an English translation by C. H. Oldfather (Loeb Classical Library, Harvard University Press, Cambridge, Massachusetts, 1979), Volume II, Book IV, 56:3-6, pp. 523, 525.
15. *Apollonius Rhodius*, op. cit., Book IV, p. 379.
16. J. E. Zimmerman, *Dictionary of Classical Mythology* (Harper and Row Publishers, New York, 1964), "Syrtes," p. 251.
17. *Apollonius Rhodius*, op. cit., Book IV, pp. 381, 383.
18. L. W. Casperson, op. cit., Chapters 21-22, pp. 523-562.
19. *The Life and Works of Flavius Josephus*, Translated by William Whiston (Holt, Rinehart and Winston, New York), *The Antiquities of the Jews*, Book II, 14:5, p. 83.
20. *The Book of Yashar*, Translated from the Hebrew and Published by Manuel Mordecai Noah (Hermon Press, New York, 1972), (cited here in the form *Jasher*), *Jasher* 80:36-41.
21. *Apollonius Rhodius*, op. cit., Book IV, pp. 383, 385, 387, 389.
22. *Herodotus*, op. cit., Volume II, Book IV, Paragraph 179, p. 381.

16. MOSES AND HERACLES

"O God, thou art my God; early will I seek thee: my soul thirsteth for thee, my flesh longeth for thee in a dry and thirsty land, where no water is" (Psalm 63:1 KJV).

16.1 Introduction

The preceding chapter began an inquiry into the possibility of relationships between Greek stories of the Argonauts and Hebrew stories of the exodus. The Argonauts were claimed by the Greeks to have landed in Egypt in ancient times, and thus they might in principle have interacted with the Hebrews. The inquiry is continued here including in particular the possibility of connections between stories of Moses and stories of Heracles. According to Apollonius, Heracles started out as one of the Argonauts on their quest to recover the golden fleece. However, part way through their voyage to Colchis, Heracles became separated from the other Argonauts and found his way to Egypt before the Argonauts were driven ashore there.

One important issue that concerned the Argonauts in Egypt, preceded by Heracles, was thirst. Several aspects of this issue are included in a quotation contained in Section 16.2. When the

16. Moses and Heracles

Argonauts were driven ashore by wind in the vicinity of Egypt, they had no obvious source of drinking water. They learned, however, that Heracles had previously obtained a flow of water by striking a rock. Drinking water was also of major concern after the exodus as the Hebrews travelled across the desert, and this situation is reviewed in Section 16.3. Moses obtained water for the Hebrews by also striking a rock. A closer parallel between the search of Heracles for water for himself and a different incident where Moses was fleeing without provisions is suggested in Section 16.4. There are enough similarities between these water-seeking stories that it seems doubtful that they could be entirely independent of each other. Considering the antiquity of writing among the Hebrews, it is possible that the Greek stories were composed and recorded at a later time and incorporated some story elements from Hebrew texts.

The significant role of serpents in both the Hebrew and Greek stories is noted in Section 16.5. The similar appearances and attitudes of Moses and Heracles are noted in Section 16.6. As represented in the stories of the *Argonautica*, Heracles would seem to have been in Egypt at about the same time as Moses and the Hebrews. Health issues of Moses and Heracles near the times of their deaths are considered and compared in Section 16.7. Another (younger) Miriam besides Moses's older sister is discussed in Section 16.8. In Section 16.9 a parallel is considered between the interaction of Heracles with Atlas and his seven daughters and the interaction of Moses with Jethro and his seven daughters. The eldest daughter of Atlas was Hespere and the eldest daughter of Jethro was Zipporah; and relationships between these various names can be contemplated. When the

Adventures In Egypt

Argonauts learned that Heracles recently had been in Egypt, they launched a semi-successful search to find him, and this subject is reviewed in Section 16.10. The manners of death of Heracles and Moses are also considered in that Section. Most of the Argonauts were eager to return to their home country, and eventually they found a way to escape from the Tritonian Lake in Egypt back into the Mediterranean Sea. Their escape is discussed in Section 16.11.

16.2 Thirsty Argonauts
Egypt is a country having very little direct rainfall. One consequence of the limited rain is that when people in ancient times lived or traveled in Egypt away from the Nile River and its branches they became very dependent for their survival on water that they brought with them or the relatively rare oases, wells, and man-made canals. Aspects of this water sensitivity would have affected not just the Egyptians, but also the Hebrews and the Argonauts if and when they were in Egypt. Some of the responses of these groups to their water shortages are considered in the paragraphs that follow.

After the Argonauts had carried their ship across the sandy waste to the sea-connected (salty) Tritonian lake, they finally had hope of a way to return home. They then could focus on a continuing desperate need – they were very thirsty. The following somewhat lengthy quotation reveals how their thirst was quenched, and several other Argonaut-related subjects are introduced for further consideration in the following sections:

> Then, like raging hounds, they rushed to search for a spring; for besides their suffering and anguish, a parching

16. Moses and Heracles

thirst lay upon them, and not in vain did they wander; but they came to the sacred plain where Ladon, the serpent of the land, till yesterday kept watch over the golden apples in the garden of Atlas; and all around the nymphs, the Hesperides, were busied, chanting their lovely song. But at that time, stricken by Heracles, he [Ladon] lay fallen by the trunk of the apple-tree; only the tip of his tail was still writhing; but from his head down his dark spine he lay lifeless; and where the arrows had left in his blood the bitter gall of the Lernaean hydra, flies withered and died over the festering wounds. And close at hand the Hesperides, their white arms flung over their golden heads, lamented shrilly; and the heroes drew near suddenly; but the maidens, at their quick approach, at once became dust and earth where they stood. Orpheus marked the divine portent, and for his comrades addressed them in prayer: "O divine ones, fair and kind, be gracious, O queens, whether ye be numbered among the heavenly goddesses, or those beneath the earth, or be called the Solitary nymphs; come, O nymphs, sacred race of Oceanus, appear manifest to our longing eyes and show us some spring of water from the rock or some sacred flow gushing from the earth, goddesses, wherewith we may quench the thirst that burns us unceasingly. And if ever again we return in our voyaging to the Achaean land, then to you among the first of goddesses with willing hearts will we bring countless gifts, libations and banquets."

So he spake, beseeching them with plaintive voice; and they from their station near pitied their pain; and lo! first of all they caused grass to spring from the earth; and above the grass rose up tall shoots; and then flourishing saplings grew standing upright far above the earth. Hespere became a poplar and Eretheis an elm, and Aegle a willow's sacred trunk. And forth from these trees their forms looked out, as clear as they were before, a marvel exceeding great, and Aegle spake with gentle words answering their longing looks:

"Surely there has come hither a mighty succour to your toils, that most accursèd man [Heracles], who robbed our guardian serpent of life and plucked the golden apples of the goddesses and is gone; and has left bitter grief for us. For yesterday came a man most fell in wanton violence, most grim in form; and his eyes flashed beneath his scowling brow; a ruthless wretch; and he was clad in the skin of a monstrous lion of raw hide, untanned; and he bare a sturdy bow of olive, and a bow, wherewith he shot and killed this monster here. So he too came, as one traversing the land on foot, parched with thirst; and he rushed wildly through this spot, searching for water, but nowhere was he like to see it. Now here stood a rock near the Tritonian lake; and of his own device, or by the prompting of some god, he smote it below with his foot; and the water gushed out in full flow. And he, leaning both his hands and chest upon the ground, drank a huge draught from the rifted rock, until, stooping like a beast of the field, he had satisfied his mighty maw."

Thus she spake; and they gladly with joyful steps ran to the spot where Aegle had pointed out to them the spring, until they reached it. And as when earth-burrowing ants gather in swarms round a narrow cleft, or when flies lighting upon a tiny drop of sweet honey cluster round with insatiate eagerness; so at that time, huddled together, the Minyae thronged about the spring from the rock. And thus with wet lips one cried to another in his delight:

"Strange! In very truth Heracles, though far away, has saved his comrades, fordone with thirst. Would that we might find him on his way as we pass through the mainland!"

So they spake, and those who were ready for this work [finding Heracles] answered, and they separated this way and that, each starting to search. For by the night winds the footsteps had been effaced where the sand was stirred. The two sons of Boreas started up, trusting in their wings; and Euphemus, relying on his swift feet, and Lynceus to cast far his piercing eyes; and with them darted off Canthus, the fifth. He was urged on by the doom of the gods and his own courage, that he might learn for certain from Heracles where he had left Polyphemus, son of Eilatus; for he was minded to question him on every point concerning his comrade. But that hero had founded a glorious city among the Mysians, and, yearning for his home-return, had passed far over the mainland in search of Argo; and in time he reached the land of the Chalybes, who dwell near the sea; there it was that his fate subdued him. And to him a monument stands under a tall poplar, just facing the sea. But that day Lynceus thought he saw Heracles all alone, far off, over measureless land, as a man at the month's beginning sees, or thinks he sees, the moon through a bank of cloud. And he returned and told his comrades that no other searcher would find Heracles on his way, and they also came back, and swift-footed Euphemus and the twin sons of Thracian Boreas, after a vain toil.[1]

16.3 Thirsty Hebrews

Thirst also has an important role in the exodus stories. In the first water-from-rock incident (Section 8.7), Moses struck the "rock at Horeb" with his rod (Exodus 17:6). Unfortunately, the word translated as rod does not have a unique meaning. Thus, one source includes among its definitions and explanations the following possibilities:

ROD. Alternately: STAFF; POLE. A stick cut from the stem or branch of a tree and used for numerous purposes.

Straight staffs, thicker at one end than at the other and of varying lengths, were the protection and support of shepherds and travelers on foot. They might also serve as poles for carrying burdens, as shafts for arrows or spears, and as instruments for inflicting punishment. A shorter staff with a knobbed end, often studded with nails or bits of flint, served the soldier and the shepherd as a weapon [i.e., a club]. Artificial rods, resembling the natural sticks, could be described by the same words – e.g., "rod of iron" (Psalm 2:9; Revelation 19:15).[2]

On the other hand, as quoted above, when Heracles was thirsty and wished to get water out of "a rock near the Tritonian lake," "he smote it below with his foot; and the water gushed out in full flow." This action by Heracles would seem to be out of character, as he is generally described in ancient stories and pictures as carrying (and often using) a large club to solve many of the problems that he might encounter in his travels. One would perhaps have expected his club also to have been his instrument of choice for obtaining water from the rock near the Tritonian lake. Early in the Argonaut story he used the club to knock out a steer to be used as a sacrifice: "[Heracles] with his club smote one steer mid-head on the brow, and falling in a heap on the spot, it sank to the ground. . . ."[3] We could imagine that Heracles wouldn't leave home without his club; and, according to Apollonius, Heracles took the club along when he joined the Argonauts on their ship Argo at the beginning of their mission: "In the middle [middle seat of the Argo] sat Ancaeus and mighty Heracles, and near him he laid his club. . . ."[4] This armored club might have been suitable for splitting rocks to obtain water (as well as splitting heads of steers), and Heracles used it later in this

16. Moses and Heracles

story for freeing a pine tree from the ground for use in replacing his broken oar:

> And quickly he laid on the ground his arrow-holding quiver together with his bow, and took off his lion's skin. And he loosened the pine from the ground with his bronze-tipped club and grasped the trunk with both hands at the bottom, relying on his strength; . . . so did Heracles lift the pine. And at the same time he took up his bow and arrows, his lion skin and club, and started on his return.[5]

However the water-yielding rock was split, the parallel here between Moses and Heracles is difficult not to notice. Perhaps in Apollonius's story, or at least in this edition, he is trying to diminish this parallel or has some other reason for having Heracles kick the rock with his foot rather than hitting it with his club (or rod) as Moses did.

It is also possible that the idea of Heracles hitting the rock with his foot came to Apollonius by a less direct route:

> As for marvels, there are none in the land [of the Scythians], save that it has rivers by far the greatest and the most numerous in the world; and over and above the rivers and the great extent of the plains there is one most wondrous thing for me to tell of: they show a footprint of Heracles by the river Tyras stamped on rock, like the mark of a man's foot, but two cubits in length.[6]

It isn't obvious why there should have been an over-size footprint of Heracles on a rock beside the river. In the present context, one can imagine that the image is a reminder of some past incident in which Heracles was already remembered to have used his foot to bring water out of a rock. Whatever the purpose of the image, it would have predated Apollonius by at least a few centuries, and via Herodotus it would have been well known. The inspiration for the image could still have been the biblical story of Moses, but if so the transformation (if that is what it was) from the rod of Moses to the foot of Heracles could have been due to some author or poet predating Herodotus.

In a second water-from-rock incident, the Hebrews became thirsty again in the wilderness:

> And the people of Israel, the whole congregation, came into the wilderness of Zin in the first month, and the people stayed in Kadesh; and Miriam died there, and was buried there.
>
> Now there was no water for the congregation; and they assembled themselves together against Moses and against Aaron. And the people contended with Moses, and said, "Would that we had died when our brethren died before the Lord! Why have you brought the assembly of the Lord into this wilderness, that we should die here, both we and our cattle? And why have you made us come up out of Egypt, to bring us to this evil place? It is no place for grain, or figs, or vines, or pomegranates; and there is no water to drink.". . .
>
> And Moses and Aaron gathered the assembly together before the rock, and he said to them, "Hear now, you rebels; shall we bring forth water for you out of this rock?" And Moses lifted up his hand and struck the rock with his rod twice; and water came forth abundantly, and the congregation drank, and their cattle. And the Lord said to Moses and Aaron, "Because you did not believe in me, to sanctify me in the eyes of the people of Israel, therefore you shall not bring this assembly into the land which I have given them." These are the waters of Mer'ibah, where the people of Israel contended with the

16. Moses and Heracles

Lord, and he showed himself holy among them (Numbers 20:1-13).

While the two biblical water-from-rock stories are closely similar, in the second account, just quoted, the Lord is represented as being dissatisfied with the performance of Moses and Aaron. One interpretation is that in the second instance Moses failed to give credit to the Lord. In this biblical context it is of interest to recall again the words of Apollonius quoted above: "Now here stood a rock near the Tritonian lake; and of his own device, or by the prompting of some god, he [Heracles] smote it below with his foot; and the water gushed out in full flow."[7] The words "of his own device, or by the prompting of some god" seem curiously reminiscent of the ambiguity implied in the biblical text of Numbers 20 quoted above. The Israelites there may have left the water-providing event with their thirst quenched but uncertain whether the credit belonged to Moses and his rod/club or to the Lord.

The story of Heracles obtaining water from a rock may be considered to have a parallel with a different Bible story. Thus it is of interest to consider what Heracles did next after achieving the flow of water: "And he, leaning both his hands and chest upon the ground, drank a huge draught from the rifted rock, until, stooping like a beast of the field, he had satisfied his mighty maw."[8] In the biblical story of the judge Gideon, his army was declared by the Lord to be too large: 'The Lord said to Gideon, "The people with you are too many for me to give the Mid'ianites into their hand, lest Israel vaunt themselves against me, saying, 'My own hand has delivered me' " ' (Judges 7:2). One of the methods of reducing the army was to sort the soldiers into two

543

categories depending on how they drank water from a stream or spring. Some of the soldiers picked up water in their hands to drink, while others knelt down to draw water into their mouths directly. The soldiers who drank directly (as Heracles also did) were sent home, while the three hundred that drank from their hands were retained in the army. Thus, had he been present, Heracles might have been sent home because of the way he drank water.

The purpose of the preceding paragraphs and quotations is, of course, to suggest that the original authors of the biblical exodus stories and the original authors of the Argonaut stories may have employed some of the same source material. Both authorships write about an important personage who, in the vicinity of Egypt obtained thirst-quenching water by striking a rock. The background of many of the stories related to Heracles and other real or imagined heroes originated so far in the misty and mythy past of the Greek-speaking world that it is often difficult to estimate their possible historical value. The first records are often assumed to have been based on oral traditions and the songs of poets. The later written records of perhaps identical events may differ widely from each other, and the authors may have been praised more for the elegance and creativity of their writings than for their historical accuracy.

In contrast, most of the early Bible-related stories seem to have originated at the times and in the places where writing was invented, and thus there would seem to be a higher likelihood that they report actual historical facts.[9] Furthermore, in the course of their evolution the written biblical stories were often considered to be sacred documents that were carefully maintained and accurately transcribed as required. However, this process was not entirely error-free as evidenced by the variations between even the earliest manuscripts. Other errors are introduced in modern times in the form of denomination- or cult-specific changes, translations, interpretations, and selections that are sometimes favored.

In short, it seems possible that the story of Moses striking a rock near Egypt with his rod/club to obtain thirst-quenching water may have been the inspiration for Apollonius's story of Heracles striking a rock with his foot for the same purpose. Other possible parallels between Moses and Heracles will be considered in following sections of this chapter. It should be recalled that while Apollonius was writing his *Argonautica* he was also serving as a librarian at the library of Alexandria in Egypt. That library could have been an unmatchable resource for records of events that had happened in ancient Egypt.

16.4 Heracles and Moses in Egypt

If Heracles were imagined to have been an actual person, one could next ask whether there are any other indications that he was ever in Egypt. One tends to think of him as an adventurer wandering about the Mediterranean world undertaking various projects or labors as fate would lead him. Then his visit to Egypt would have been just one of many incidental pauses in his exciting career. There may, however, be other interpretations of the Egyptian interlude that has just been suggested. For instance, it might be supposed that Heracles is actually a composite character representing an amalgamation of more than one real or imagined person doing many things over an extended period of time. One of those people might actually have had direct ties to Egypt, as suggested by Herodotus:

> Concerning Heracles, I heard it said that he was one of the twelve gods [of Egypt]. But I could nowhere in Egypt hear anything concerning the other Heracles, whom the Greeks know. I have indeed many proofs that the name of Heracles did not come from Hellas to Egypt, but from Egypt to Hellas (and in Hellas to those Greeks who gave the name Heracles to the son of Amphitryon); and this is the chief among them – that Amphitryon and Alcmene, the parents of this Heracles, were both by descent Egyptian; and that the Egyptians deny knowledge of the names of Poseidon and the Dioscuri, nor are these gods reckoned among the gods of Egypt. Yet had they got the name of any deity from the Greeks, it was these more than any that they were like to remember, if indeed they were already making sea voyages and the Greeks too had seafaring men, as I suppose and judge; so that the names of these gods would have been even better known to the Egyptians than the name of Heracles. Nay, Heracles is a very ancient god in Egypt; as the Egyptians themselves say, the change of the eight gods to the twelve, of whom they deem Heracles one, was made seventeen thousand years before the reign of Amasis.[10]

Thus, Herodotus seems to be saying that Heracles was born in Egypt, was given the name Heracles by the Greeks, and was eventually considered to be an Egyptian god. This is mentioned here because of the similarities of some of his activities noted in the *Argonautica* with those of Moses as recorded in Bible-related literature. Moses was, of course, born in Egypt. The name of Moses's father was Amram (Exodus 6:20) rather than Amphitryon, the Greek father of Heracles (as quoted above).

16. Moses and Heracles

Some later Hebrews may have believed that Moses was also a god (Section 16.10).

In a seemingly unrelated story, Heracles launched an attack on Ethiopia from the land of Egypt: "After this Heracles sailed up the Nile into Ethiopia, where he slew Emathion, the king of the Ethiopians, who made battle with him unprovoked, and then returned to the completion of his last Labour."[11] This account is reminiscent of an incident in which the king of Egypt ordered Moses to drive out the Ethiopians who had invaded his country.[12] The Assuan inscription reporting the conflict implies that the king of the Ethiopians was killed in this war:

> This army of his majesty overthrew those barbarians; they did [not] let live anyone among their males, according to all the command of his majesty, except one of those children of the chief of wretched Kush, who was taken away alive as a living prisoner with their people to his majesty.[13]

There is another possible interpretation for the water-finding story in the *Argonautica* and Bible-related literature. It should be noticed that in the two biblical stories mentioned thus far Moses obtained water from a rock to quench the thirst of the impatient Hebrews after the exodus. On the other hand, in the version of Apollonius, Heracles got water from a rock for the purpose of quenching his own thirst. If this difference were considered to be significant, one might look for another occasion on which Moses, like Hercules, could have been alone and very thirsty. One possibility is found when Moses fled from the king of Egypt.[14] The biblical version of the flight is the following:

Adventures In Egypt

> One day, when Moses had grown up, he went out to his people and looked on their burdens; and he saw an Egyptian beating a Hebrew, one of his people. He looked this way and that, and seeing no one he killed the Egyptian and hid him in the sand. When he went out the next day, behold, two Hebrews were struggling together; and he said to the man that did the wrong, "Why do you strike your fellow?" He answered, "Who made you a prince and a judge over us? Do you mean to kill me as you killed the Egyptian?" Then Moses was afraid, and thought, "Surely the thing is known." When Pharaoh heard of it, he sought to kill Moses.
>
> But Moses fled from Pharaoh, and stayed in the land of Mid'ian; and he sat down by a well. Now the priest of Mid'ian had seven daughters; and they came and drew water, and filled the troughs to water their father's flock (Exodus 2:11-16).

The important point here is that Moses fled from Pharaoh, presumably alone and in secret. After crossing many miles of desert he ended up in Midian beside a well. One could imagine that he had experienced hunger and thirst in the course of this journey and before he arrived at the well. A more detailed and relevant version is given by Josephus, and in that account Moses had just returned to Egypt after leading his successful military campaign against Ethiopia:

> But the Egyptians, thus saved by Moses, conceived from their very deliverance a hatred for him and thought good to pursue with greater ardour their plots upon his life, suspecting that he would take advantage of his success to revolutionize Egypt, and suggesting to the king that he should be put to death. He on his own part was harbouring thoughts of so doing, alike from envy of

16. Moses and Heracles

Moses' generalship and from fear of seeing himself abased, and so, when instigated by the hierarchy, was prepared to lend a hand in the murder of Moses. Their victim, however, informed betimes of the plot, secretly escaped, and, since the roads were guarded, directed his flight across the desert and to where he had no fear of being caught by his foes; he left without provisions, proudly confident of his powers of endurance. On reaching the town of Madian(e), situated by the Red Sea and named after one of Abraham's sons by Katura, he sat down on the brink of a well and there rested after his toil and hardships, at midday, not far from the town.[15]

Josephus states explicitly that Moses was secretly fleeing across a desert, avoiding guards and having no provisions. It would not be hard to imagine him becoming very thirsty as any water sources might have been far apart and well-guarded. Moses's circumstances would be readily compatible with the statement about Heracles in the *Argonautica*: "So he too came, as one traversing the land on foot, parched with thirst; and he rushed wildly through this spot, searching for water, but nowhere was he like to see it."[16] Thus, the description of Heracles obtaining water from a rock could in principle be an adaptation of the story of Moses on his flight from Egypt to Midian. However, while that flight story could account for his thirst, it does not include an explicit indication of whether or how Moses might have obtained water before arriving at the well in Midian.

16.5 Guardian of the garden

While searching for water, the Argonauts encountered a recently deceased serpent that had been guarding the golden apples in the garden of Atlas, as quoted more fully in Section 16.2:

> [T]hey came to the sacred plain where Ladon, the serpent of the land, till yesterday kept watch over the golden apples in the garden of Atlas; and all around the nymphs, the Hesperides, were busied, chanting their lovely song. But at that time, stricken by Heracles, he [Ladon] lay fallen by the trunk of the apple-tree; only the tip of his tail was till writhing; but from his head down his dark spine he lay lifeless; and where the arrows had left in his blood the bitter gall of the Lernaean hydra, flies withered and died over the festering wounds. And close at hand the Hesperides, their white arms flung over their golden heads, lamented shrilly. . . . [After Orpheus calmed the Hesperides, one of them spoke as follows:]
>
> "Surely there has come hither a mighty succour to your toils, that most accursèd man [Heracles], who robbed our guardian serpent of life and plucked the golden apples of the goddesses and is gone; and has left bitter grief for us. For yesterday came a man most fell in wanton violence, most grim in form; and his eyes flashed beneath his scowling brow; a ruthless wretch; and he was clad in the skin of a monstrous lion of raw hide, untanned; and he bare a sturdy bow of olive, and a bow, wherewith he shot and killed this monster here. So he too came, as one traversing the land on foot, parched with thirst; and he rushed wildly through this spot, searching for water, but nowhere was he like to see it."[17]

According to these quotations, Heracles had been searching for water when he encountered the serpent. His reaction was to kill the serpent and steal the golden apples. In principle, it could be of interest to enquire whether Moses, like Heracles, might ever have had the ability or inclination to deal with poisonous serpents. Bible-related stories show that Moses was far less intimidated by serpents than, for example, this author would be.

It may be noted first that Moses dealt successfully with poisonous serpents in his one major military campaign as leader of the Egyptian army. In that event he and his troops had surprised the Ethiopians by approaching them through an area known to be infested with poisonous serpents. To accomplish this, he had brought with him baskets full of ibises, which were highly effective at catching and dispatching the serpents.[18] Thus Moses was not only able to deal with poisonous serpents. Controlling them was an essential element in his successful military campaign.

In the burning-bush incident, Moses received God's serpent-related instructions concerning his future interactions with the Egyptians:

> "Come, I will send you to Pharaoh that you may bring forth my people, the sons of Israel, out of Egypt.". . . Then Moses answered, "But behold, they will not believe me or listen to my voice, for they will say, 'The Lord did not appear to you.'" The Lord said to him, "What is that in your hand?" He said, "A rod." And he said, "Cast it on the ground." So he cast it on the ground, and it became a serpent; and Moses fled from it. But the Lord said to Moses, "Put out your hand, and take it by the tail" – so he put out his hand and caught it, and it became a rod in his hand – "that they may believe that the Lord, the God of their fathers, the God of Abraham, the God of Isaac, and the God of Jacob, has appeared to you" (Exodus 3:10, 4:1-5).

In the actual exodus stories Moses's first encounter with Pharaoh also involved serpents:

> And the Lord said to Moses and Aaron, "When Pharaoh says to you, 'Prove yourselves by working a miracle,' then you shall say to Aaron, 'Take your rod and cast it down before Pharaoh, that it may become a serpent.' " So Moses and Aaron went to Pharaoh and did as the Lord commanded; Aaron cast down his rod before Pharaoh and his servants, and it became a serpent. Then Pharaoh summoned the wise men and the sorcerers; and they also, the magicians of Egypt, did the same by their secret arts. For every man cast down his rod, and they became serpents. But Aaron's rod swallowed up their rods" (Exodus 7:8-12).

In a post-exodus incident, the Lord punished the Hebrews by sending fiery serpents among them:

> From Mount Hor they set out by the way to the Red Sea, to go around the land of Edom; and the people became impatient on the way. And the people spoke against God and against Moses, "Why have you brought us up out of Egypt to die in the wilderness? For there is no food and no water, and we loathe this worthless food." Then the Lord sent fiery serpents among the people, and they bit the people, so that many people of Israel died. And the people came to Moses, and said, "We have sinned, for we have spoken against the Lord and against you; pray to the Lord, that he take away the serpents from us." So Moses prayed for the people. And the Lord said to Moses, "Make a fiery serpent, and set it on a pole; and every one who is bitten, when he sees it, shall live." So Moses made a bronze serpent, and set it on a pole; and if a serpent bit any man, he would look at the bronze serpent and live (Numbers 21:4-9).

16. Moses and Heracles

Here, again with God's help, Moses was able to manage a crisis involving poisonous serpents.

16.6 Like a lion
Another possible point of comparison between Moses as represented in biblical literature and Heracles as described in the *Argonautica* could be in their strength and appearance. According to Apollonius, Heracles was big and strong, and his abilities were highly respected by his colleagues on their adventure. An important step in preparing to launch their ship was planning where their equipment should be stored and where the many heroes should be seated. Heracles, together with Ancaeus, was by acclaim granted the middle seat on the Argo:

> Now when they had carefully paid heed to everything, first they distributed the benches by lot, two men occupying one seat; but the middle bench they chose for Heracles and Ancaeus apart from the other heroes, Ancaeus who dwelt in Tegea. For them alone they left the middle bench just as it was and not by lot....[19]

When the Argonauts were in trouble, Heracles was usually there to help, and interestingly he could also serve as their role model and moral compass if they were tempted to go astray. Thus, when they had become badly distracted on their voyage, it was Heracles that reminded them of their mission. Near the start of their voyage, the Argonauts landed for a time on what is today the Greek island of Lemnos in the northern Aegean Sea. At the time of their landing, the island was inhabited only by women led by the queen Hypsipyle, because these women had killed all of the males on the island. The Argonauts (including Jason) became enamored of the Lemnian women, and were in danger of never continuing on their expedition. The reproaches and sarcasm of Heracles brought them back to the original purpose of their voyage:

> Thereupon Aeson's son [Jason] started to go to the royal home of Hypsipyle; and the rest went each his way as chance took them, all but Heracles; for he of his own will was left behind by the ship and a few chosen comrades with him. And straightway the city rejoiced with dances and banquets, being filled with the steam of sacrifice; and above all the immortals they propitiated with songs and sacrifices the illustrious son of Hera and Cypris herself. And the sailing was ever delayed from one day to another; and long would they have lingered there, had not Heracles, gathering together his comrades apart from the women, thus addressed them with reproachful words:
>
> "Wretched men, does the murder of kindred keep us from our native land? Or is it in want of marriage that we have come hither from thence, in scorn of our countrywomen: Does it please us to dwell here and plough the rich soil of Lemnos? No fair renown shall we win by thus tarrying so long with stranger women; nor will some god seize and give us at our prayer a fleece that moves of itself. Let us then return each to his own; but him [Jason] leave ye to rest all day long in the embrace of Hypsipyle until he has peopled Lemnos with men-children, and so there come to him great glory."
>
> Thus did he [Heracles] chide the band; but no one dared to meet his eye or to utter a word in answer. But just as they were in the assembly they made ready their departure in all haste, and the women came running towards them, when they knew their intent.[20]

16. Moses and Heracles

Moses seems to have had a similar role in restoring the Israelites from idolatry and adultery to their mission after they strayed while journeying through the wilderness:

> When Aaron saw how happy the people were about it, he built an altar before the calf and announced, "Tomorrow there will be a feast to Jehovah!" So they were up early the next morning and began offering burnt offerings and peace offerings to the calf idol; afterwards they sat down to feast and drink at a wild party, followed by sexual immorality (Exodus 32:5-6 TLB).

> When Moses saw that the people had been committing adultery – at Aaron's encouragement, and much to the amusement of their enemies – he stood at the camp entrance and shouted, "All of you who are on the Lord's side, come over here and join me." (Exodus 32:25-26 TLB).

In a more appealing story, the Argonauts were ready to set sail from an island at dawn one morning. The winds had ceased, and any progress was to be dependent on their rowing:

> Thereupon a spirit of contention stirred each chieftain, who should be the last to leave his oar. For all around the windless air smoothed the swirling waves and lulled the sea to rest. And they, trusting in the calm, mightily drove the ship forward; and as she sped through the salt sea, not even the storm-footed steeds of Poseidon would have overtaken her. Nevertheless when the sea was stirred by violent blasts which were just rising from the rivers about evening, forspent with toil, they ceased. But Heracles by the might of his arms pulled the weary rowers along all together, and made the strong-knit timbers of the ship to quiver. But when, eager to reach the Mysian mainland,

they passed along in sight of the mouth of Rhyndacus and the great cairn of Aegaeon, a little way from Phrygia, then Heracles, as he ploughed up the furrows of the roughened surge, broke his oar in the middle. And one half he held in both his hands as he fell sideways, the other the sea swept away with its receding wave. And he sat up in silence glaring round; for his hands were unaccustomed to lie idle.[21]

Not only was Heracles big and strong like a lion. He also tended to dress and sometimes behave accordingly:

For yesterday came a man [Heracles] most fell in wanton violence, most grim in form; and his eyes flashed beneath his scowling brow; a ruthless wretch; and he was clad in the skin of a monstrous lion of raw hide, untanned; and he bare a sturdy bow of olive, and a bow, wherewith he shot and killed this monster [serpent] here. So he too came, as one traversing the land on foot, parched with thirst; and he rushed wildly through this spot, searching for water....[22]

Some of the quotations above have suggested similarities between Moses and Heracles. One area of probable difference is in manner of dress. Heracles is famous for wearing a lion skin, and it isn't suggested that Moses ever did the same. As mentioned above, it is also well known that Heracles was considered to be big and strong. It is quite possible that the same could have been said of Moses in his earlier years, and *Jasher* makes this explicit: "And the king and princes and all the fighting men [of Ethiopia] loved Moses, for he was great and worthy, his stature was like a noble lion, his face was like the sun, and his strength was like that of a lion, and he was counsellor to the king."[23] While there are some significant discrepancies between

Jasher and other texts concerning interactions with Ethiopia, there is no particular reason to question the strength and military expertise of Moses, who was a Levite of the Hebrews[24] and was raised in the royal family of Egypt. Thus, Moses may in his younger days have had the strength and bearing reminiscent of a lion, even if he didn't choose to dress himself in a lion's skin.

In a very different version of the attack on Ethiopia led by Moses, the daughter Tharbis of the Ethiopian king is said to have been completely taken with the brilliant and presumably handsome young military leader Moses:

> Tharbis, the daughter of the king of the Ethiopians, watching Moses bringing his troops close beneath the ramparts and fighting valiantly, marvelled at the ingenuity of his manœuvres and, understanding that it was to him that the Egyptians, who but now despaired of their independence, owed all their success, and through him that the Ethiopians, so boastful of their feats against them, were reduced to the last straits, fell madly in love with him; and under the mastery of this passion she sent to him the most trusty of her menials to make him an offer of marriage.[25]

This marriage may also be referred to in the Bible: "Miriam and Aaron spoke against Moses because of the Cushite woman whom he had married, for he had married a Cushite woman" (Numbers 12:1).

Quite likely, Moses was also tall, and tallness is seldom a disadvantage for leadership positions. Thus, Artapanus comments on the appearance of Moses at the time of the exodus, about sixty-seven years[26] after Tharbis "fell madly in love with him." The words of Artapanus concerning Moses are as follows: "And

Moses they say was tall and ruddy, with long white hair, and dignified: and he performed these deeds when he was about eighty-nine years old."[27] Thus, Moses was said to have been tall, and even at the time of the exodus with his long white hair he appeared "dignified."

The first king of Israel may provide another example of the advantages in antiquity of being tall:

> There was a man of Benjamin whose name was Kish, the son of Abi'el, son of Zeror, son of Beco'rath, son of Aphi'ah, a Benjaminite, a man of wealth; and he had a son whose name was Saul, a handsome young man. There was not a man among the people of Israel more handsome than he; from his shoulders upward he was taller than any of the people (1 Samuel 9:1-2).

One might like to believe that in our more sophisticated modern times a physical characteristic like tallness would have no impact on the suitability of a person to attain financially or otherwise rewarding leadership positions. It seems, however, that such a belief could be incorrect:

> Being tall comes with financial perks: In Western countries, a jump from the 25th percentile of height to the 75th – about four or five inches – is associated with a 9 to 15 percent increase in salary. Separate research shows that an extra inch of height is worth $800 more in salary each year.[28]

Elevator shoes may not have existed in ancient times, but the above results suggest that today they might sometimes be a good investment. Heracles, Moses, and Saul are all said to have been tall, and all three were leaders among their people.

16. Moses and Heracles

It may be of interest to recall that according to *Jasher*, as quoted above, the face of Moses "was like the sun." *Jasher* also suggests that Moses may have been considered a luminary even at his birth: "[S]he brought forth a son, and the whole house was filled with great light as of the light of the sun and moon at the time of their shining"[29] "She" in this quotation refers to Jochebed, the mother of Moses. Such statements from *Jasher* and other soures should probably not always be taken literally. They might better be regarded as effusive ways of saying that Moses was beautiful as a child and handsome as a young man.

Comments concerning the appearance of Moses many years later and a few months after the exodus may be of more interest:

> When Moses came down from Mount Sinai, with the two tables of the testimony in his hand as he came down from the mountain, Moses did not know that the skin of his face shone because he had been talking with God. And when Aaron and all the people of Israel saw Moses, behold, the skin of his face shone, and they were afraid to come near him. But Moses called to them; and Aaron and all the leaders of the congregation returned to him, and Moses talked with them. And afterward all the people of Israel came near, and he gave them in commandment all that the Lord had spoken with him in Mount Sinai. And when Moses had finished speaking with them, he put a veil on his face; but whenever Moses went in before the Lord to speak with him, he took the veil off, until he came out; and when he came out, and told the people of Israel what he was commanded, the people of Israel saw the face of Moses, that the skin of Moses' face shone; and Moses would put the veil upon his face again, until he went in to speak with him (Exodus 34:29-35).

Adventures In Egypt

The reported references here to the appearance of "the skin of Moses' face" could have a different interpretation, and this will be mentioned in the following section.

16.7 Humility

Despite the appearance and strength of Heracles, he tended to be humble and fair-minded in his dealings with his fellow Argonauts. Just before their departure from the beach of Pagasae, Jason addressed the Argonauts:

> "All the equipment that a ship needs – for all is in due order – lies ready for our departure. Therefore we will make no long delay in our sailing for these things' sake, when the breezes but blow fair. But, friends, – for common to all is our return to Hellas hereafter, and common to all is our path to the land of Aeetes – now therefore with ungrudging heart choose the bravest to be our leader who shall be careful for everything, to take upon him our quarrels and convenants with strangers."
>
> Thus he spake; and the young heroes turned their eyes towards bold Heracles sitting in their midst, and with one shout they all enjoined upon him to be their leader; but he, from the place where he sat, stretched forth his right hand and said:
>
> "Let no one offer this honour to me. For I will not consent, and I will forbid any other to stand up. Let the hero who brought us together, himself be the leader of the host."
>
> Thus he spake with high thoughts, and they assented, as Heracles bade; and warlike Jason himself rose up, glad at heart, and thus addressed the eager throng: . . . .[30]

Thus, Heracles argued that he did not have the best claim to the leadership role on this adventure.

16. Moses and Heracles

As the time of the exodus approached, a similar aspect of humility seems to have been shown by Moses. When he was asked to lead the Israelites out of Egypt, he offered several excuses:

> "[B]ehold, the cry of the people of Israel has come to me, and I have seen the oppression with which the Egyptians oppress them. Come, I will send you to Pharaoh that you may bring forth my people, the sons of Israel out of Egypt." But Moses said to God, "Who am I that I should go to Pharaoh, and bring the sons of Israel out of Egypt?" He said, "But I will be with you . . . (Exodus 3:9-12).

> Then Moses answered, "But behold, they will not believe me or listen to my voice, for they will say, 'The Lord did not appear to you.' " The Lord said to him, "What is that in your hand?" . . . (Exodus 4:1-2).

> But Moses said to the Lord, "Oh, my Lord, I am not eloquent, either heretofore or since thou has spoken to thy servant; but I am slow of speech and of tongue." Then the Lord said to him, "Who has made man's mouth? Who makes him dumb, or deaf, or seeing, or blind? Is it not I, the Lord? Now therefore go, and I will be with your mouth and teach you what you shall speak." But he said, "Oh, my Lord, send, I pray, some other person" (Exodus 4:10-13).

> On the day when the Lord spoke to Moses in the land of Egypt, the Lord said to Moses, "I am the Lord; tell Pharaoh king of Egypt all that I say to you." But Moses said to the Lord, "Behold, I am of uncircumcised lips; how then shall Pharaoh listen to me?" And the Lord said to Moses, "See, I make you as God to Pharaoh; and Aaron your brother shall be your prophet" (Exodus 6:28-7:1).

As Heracles considered himself not to be the best person to lead the Argonauts, so Moses was doubtful about his ability in his later years to lead the Hebrews.

In spite of his humility, Moses in Egypt had the courage to ask to see God. He was not, however, permitted fully to do so:

> Moses said, "I pray thee, show me thy glory," And he said, "I will make all my goodness pass before you, and will proclaim before you my name 'The Lord'; and I will be gracious to whom I will be gracious, and will show mercy on whom I will show mercy. But," he said, "you cannot see my face; for man shall not see me and live." And the Lord said, "Behold, there is a place by me where you shall stand upon the rock; and while my glory passes by I will put you in a cleft of the rock, and I will cover you with my hand until I have passed by; then I will take away my hand, and you shall see my back; but my face shall not be seen" (Exodus 33:17-23).

In a similar way, Heracles in Egypt wanted to see Zeus. The result, however, was about the same:

> Heracles (they say) would by all means look upon Zeus, and Zeus would not be seen by him. At last, being earnestly entreated by Heracles, Zeus contrived a device, whereby he showed himself displaying the head and wearing the fleece of a ram which he had flayed and beheaded. It is from this that the Egyptian images of Zeus have a ram's head.[31]

In short, the desire to see the face of God (or Zeus) was one more feature that Moses and Heracles had in common. This is also, of course, one more reason to consider the possibility that several of

16. Moses and Heracles

the stories of Moses and Heracles might have had a common origin.

According to the Bible, Moses is also said to have been the meekest man on earth, and meekness would not usually be considered a desirable trait in a leader:

> Miriam and Aaron spoke against Moses because of the Cushite woman whom he had married, for he had married a Cushite woman; and they said, "Has the Lord indeed spoken only through Moses? Has he not spoken through us also?" And the Lord heard it. Now the man Moses was very meek, more than all men that were on the face of the earth. And suddenly the Lord said to Moses and to Aaron and Miriam, "Come out, you three, to the tent of meeting." And the three of them came out. And the Lord came down in a pillar of cloud, and stood at the door of the tent, and called Aaron and Miriam; and they both came forward. And he said, "Hear my words: If there is a prophet among you, I the Lord make myself known to him in a vision, I speak with him in a dream. Not so with my servant Moses; he is entrusted with all my house. With him I speak mouth to mouth, clearly, and not in dark speech; and he beholds the form of the Lord. Why then were you not afraid to speak against my servant Moses?" (Numbers 12:1-8).

It should be recalled here, however, that Moses didn't always try to avoid leadership positions. Earlier in his life, as noted above, he had led the army of Egypt in the successful repulsion and destruction of an invading army from Ethiopia. Moses may have been about twenty-one years old at the start of the war against Ethiopia.[32] A short excerpt from Josephus's account is included here:

Having thus accomplished the march, he came wholly unexpected upon the Ethiopians, joined battle with them and defeated them, crushing their cherished hopes of mastering the Egyptians, and then proceeded to attack and overthrow their cities, great carnage of the Ethiopians ensuing. After tasting of this success which Moses had brought them, the Egyptian army showed such indefatigable energy that the Ethiopians were menaced with servitude and complete extirpation.[33]

A brief Greek story mentioned previously in Section 16.4 involved a visit by Heracles to Ethiopia: "After this Heracles sailed up the Nile into Ethiopia, where he slew Emathion, the king of the Ethiopians, who made battle with him unprovoked, and then returned to the completion of his last Labour."[34] Thus, both Moses and Heracles are said to have engaged in successful battles against the Ethiopians, and in both cases the king of the Ethiopians was slain. This parallel should not be overstated, as information is limited and the circumstances of the conflicts may have been quite different in these two cases. Moses led an army (with ibis auxiliaries) against an un-named Ethiopian king. The king was killed and Moses married his daughter (Tharbis, according to Josephus). On the other hand, in a very terse report Heracles (perhaps with no army) killed the Ethiopian king Emathion. No daughter of Emathion is mentioned.

The hesitancy of Moses to lead the Israelites could have been due in part to his advancing age. He was about eighty years old when he returned from Midian to Egypt,[35] and few people of that age would be enthusiastic about undertaking such a difficult and dangerous project. However, Moses's physical weaknesses (if any) might also have been due to some sort of illness. Many

16. Moses and Heracles

diseases have long existed in Egypt, and that country is sometimes believed to have been the origin of leprosy. As suggested previously, Thutmose II was probably the king of Egypt during part of the early career of Moses, and that king is likely to have died from leprosy.[36] Writings attributed to the Egyptian historian Manetho indicate that Moses was a victim of leprosy. In the words of Josephus,

> It now remains that I debate with Manetho about Moses. Now the Egyptians acknowledge him to have been a wonderful and a divine person; nay, they would willingly lay claim to him themselves, though after a most abusive and incredible manner, and pretend that he was of Heliopolis, and one of the priests of that place, and was ejected out of it among the rest, on account of his leprosy; . . .[37]

There is too little information for a definite conclusion about whether Moses himself might have had leprosy, but a few comments on this disease might not be inappropriate:

Seventy to eighty percent of all leprosy cases are of the tuberculoid type. In lepromatous (LL) leprosy, which is the second and more contagious form of the disease, the body's immune system is unable to mount a strong response to the invading organism. Hence, the organism multiplies freely in the skin. This type of leprosy is also called the multibacillary (MB) leprosy, because of the presence of large numbers of bacteria. The characteristic feature of this disease is the appearance of large nodules or lesions all over the body and face. Occasionally, the mucous membranes of the eyes, nose, and throat may be involved. Facial involvement can produce a lion-like appearance (leonine facies). This type of leprosy can lead to blindness, drastic change in voice, or mutilation of the nose. Leprosy can strike anyone; however, children seem to be more susceptible than adults. . . .Thickened nerves accompanied by weakness of muscles supplied by the affected nerve are very typical of the disease. One characteristic occurrence is a foot drop where the foot cannot be flexed upwards, affecting the ability to walk.[38]

From the above quotation it is seen that muscle weakness is typical of leprosy, and foot drop can render difficult and dangerous the otherwise simple tasks of walking and climbing. It may be recalled that Moses sat out the battle with the Amalekites, and as will be noted in Chapter 17 he may have relied on Joshua's assistance when climbing mountains. A "drastic change in voice" is common with leprosy, and Moses is quoted earlier in this section as considering himself to be "slow of speech and of tongue." Facial disfigurement is also common (leonine facies), and in the Middle Ages a victim of the disease usually wore a hood over his face.[39] As quoted in Section 16.6, "the skin of Moses' face shone; and Moses would put the veil upon his face again, until he went in to speak with him [the Lord]" (Exodus 34:35). The statement that "the skin of Moses' face shone" could be a remnant of sorts for some indication of facial disfigurement, as suggested also by his putting a veil over his face. Further consideration of Moses's possible physical limitations at the time of the exodus will be included in Chapter 17.

In a quotation earlier in this section, Miriam and Aaron attempted to usurp the leadership position that had been held by Moses. This angered the Lord, and as part of his response Miriam was afflicted with leprosy:

16. Moses and Heracles

> And the anger of the Lord was kindled against them, and he departed; and when the cloud removed from over the tent, behold, Miriam was leprous, as white as snow. And Aaron turned towards Miriam, and behold, she was leprous. And Aaron said to Moses, "Oh, my lord, do not punish us because we have done foolishly and have sinned. Let her not be as one dead, of whom the flesh is half consumed when he comes out of his mother's womb." And Moses cried to the Lord, "Heal her, O God, I beseech thee." But the Lord said to Moses, "If her father had but spit in her face, should she not be shamed seven days? Let her be shut up outside the camp seven days, and after that she may be brought in again." So Miriam was shut up outside the camp seven days; and the people did not set out on the march till Miriam was brought in again (Numbers 12:9-15).

It may be significant that Miriam is not said explicitly to have been healed, and her death at Kadesh is reported in Numbers 20:1. Aaron's death is noted in Number 20:28.

In this chapter various parallels between the lives and activities of Moses and Heracles have been considered. Another such parallel will be mentioned briefly here concerning the possible association of Moses with the disease of leprosy. One of the preliminary miracles at the time of the exodus was to include his acquiring and being healed of leprosy:

> Again, the Lord said to him, "Put your hand into your bosom." And he put his hand into his bosom; and when he took it out, behold, his hand was leprous, as white as snow. Then God said, "Put your hand back into your bosom." So he put his hand back into his bosom; and when he took it out, behold, it was restored like the rest his of flesh (Exodus 4:6-7).

Adventures In Egypt

In considering these topics one might enquire whether Heracles could ever have experienced any form of a leprosy-like condition. As it happens, it was a skin disease (ascribed to hydra poison) that led to his death as he was making a sacrifice to Cenaean Zeus: "So Hercules put it [the poison soaked tunic] on and proceeded to offer sacrifice. But no sooner was the tunic warmed than the poison of the hydra began to corrode his skin; and on that he . . . tore off the tunic, which clung to his body, so that his flesh was torn away with it."[40] This description of Heracles' deteriorating body is reminscent of *Jasher's* colorful description of the body of the leprous king identified previously as Thutmose II: "And the disorder greatly prevailed over the king, and his flesh stank like the flesh of a carcase cast upon the field in summer time, during the heat of the sun."[41]

As indicated earlier in this section, Moses was about eighty years old when he returned from Midian to Egypt, and he offered several excuses why he wasn't the best person for leading the exodus. One could imagine that in his desperation he might even have considered fleeing from Egypt. In fact, it was at about this time that Heracles/Moses arrived at Magnesian Pegasae in Greece, far to the west of Egypt. There he volunteered to join the Argonauts in their expedition to Colchis (also far from Egypt) to recover the golden fleece. Later however, before the Argonauts arrived at Colchis, Heracles was left behind and made his way back to Egypt. There Moses/Heracles led the Israelites in their exodus escape from Egypt as God had originally ordered him to do.

The story of Moses's reluctant obedience to God at the time of the exodus has a parallel in the story of Jonah's reluctant

16. Moses and Heracles

obedience concerning God's order for him to preach to the people of Nineveh:

> Now the word of the Lord came to Jonah the son of Amit'tai, saying, "Arise, go to Nin'eveh, that great city, and cry against it; for their wickedness has come up before me." But Jonah rose to flee to Tarshish from the presence of the Lord. He went down to Joppa and found a ship going to Tarshish; so he paid the fare, and went on board, to go with them to Tarshish, away from the presence of the Lord (Jonah 1:1-3).

After various adventures, including a visit to the belly of a great fish, Jonah decided that he would obey the Lord after all.

> Then the word of the Lord came to Jonah the second time, saying, "Arise, go to Nin'eveh, that great city, and proclaim to it the message that I tell you." So Jonah arose and went to Nin'eveh, according to the word of the Lord (Jonah 3:1-3).

16.8 Another Miriam

One focus of the previous section was on the meekness, modesty, and perhaps physical weakness of Moses at the time of the exodus. As noted in Section 16.6, Moses may have been about eight-nine years old at that time. It could be imagined that at his advanced age he wouldn't any longer have possessed the amazing energy and enthusiasm that he had displayed in his earlier years. He is said to have become the meekest man on earth, and as seen previously he argued strongly that he would be a poor leader: "But Moses said to the Lord, 'Oh, my Lord, I am not eloquent, either heretofore or since thou hast spoken to thy

Adventures In Egypt

servant; but I am slow of speech and of tongue' " (Exodus 4:10). Besides his age, he may have suffered from a chronic disease that affected his speech and his ability to stand and walk, as will be noted in the following chapter. His older sister Miriam (Numbers 26:59) may have been even less well, and she apparently died soon after the Israelites arrived at Kadesh: "And the people of Israel, the whole congregation, came into the wilderness of Zin in the first month, and the people stayed in Kadesh; and Miriam died there, and was buried there: (Numbers 20:1).

When she was a young girl, a sister of Moses had played an essential role in rescuing her three-month-old baby brother from Pharaoh's plan to kill all of the male children of the Israelites (Exodus 2:1-10). Thus, it would seem that she must have been several years older than her brother, who at the time of the exodus may already have been suffering from age-related and other medical issues, as discussed previously. It is of interest then that during the actual exodus event, when Moses's older sister would have been about one hundred years old, a person named Miriam was behaving like a very healthy younger person:

> Then Miriam, the prophetess, the sister of Aaron, took a timbrel in her hand; and all the women went out with her with timbrels and dancing. And Miriam sang to them: "Sing to the Lord, for he has triumphed gloriously; the horse and his rider he has thrown into the sea" (Exodus 15:20-21).

This energetic behavior is surprising considering Miriam's age, looming health problems, and soon to occur death (Numbers 20:1).

16. Moses and Heracles

A possible explanation for this seemingly youthful behavior is that there might also have been a younger person named Miriam, distinct from the sister of Aaron. One possible identification is suggested by the following text concerning some of the descendants of Judah, though not all versions agree:

> The sons of Ezrah: Jether, Mered, Epher, and Jalon. These are the sons of Bith'i-ah, the daughter of Pharaoh, whom Mered married; and she conceived and bore Miriam, Sham'mai, and Ish'bah, the father of Eshtemo'a (1 Chronicles 4:17).

It is of interest that the mother of this Miriam is said to have been a daughter of Pharaoh. Such a relationship suggests a possible identification. As a starting point, one might consider the possibility that the name of the person called Bithiah might also have been interpreted as Bathia. This possibility is of interest because, according to *Jasher*, Bathia the daughter of Pharaoh was interacting with Moses and Aaron immediately before the exodus:

> And Bathia the daughter of Pharaoh [Thutmose III] went forth with the king [Amenophis II] on that night to seek Moses and Aaron in their houses, and they found them in their houses, eating and drinking and rejoicing with all Israel. . . .
> And Moses said to her, although thou art the first born to thy mother, thou shalt not die, and no evil shall reach thee in the midst of Egypt.
> And she said, what advantage is it to me when I see the king, my brother, and all his household and subjects in this evil, whose first born perish with all the first born of Egypt?

Adventures In Egypt

> And Moses said to her, surely thy brother and his household, and subjects, the families of Egypt, would not hearken to the words of the Lord, therefore did this evil come upon them.
>
> And Pharaoh king of Egypt [Amenophis II] approached Moses and Aaron, and some of the children of Israel who were with them in that place, and he prayed to them, saying,
>
> Rise up and take your brethren, all the children of Israel who are in the land, with their sheep and oxen, and all belonging to them, they shall leave nothing remaining, only pray for me to the Lord your God.
>
> And Moses said to Pharaoh, behold though thou art thy mother's first born, yet fear not, for thou wilt not die, for the lord has commanded that thou shalt live, in order to show thee his great might and strong stretched out arm.[42]

The above excerpts and interpretations in this section, if taken together, would leave us with a princess named Bathia whose father was the Egyptian king Thutmose III, whose brother or half-brother was the exodus king Amenophis II, whose husband was from the tribe of Judah, and one of whose children was named Miriam. Although the texts are not without ambiguities, this Miriam would have been much younger than the older sister of Moses. Thus, Bathia's daughter Miriam could have been of an appropriate age at the time of the exodus to have been leading a group of Hebrew celebrants in music, dancing, and song after the exodus escape from Egypt (Exodus 15:19-21). While it is appealing to consider that there may have been two people named Miriam, it must be acknowledged that this idea appears to be contradicted in Exodus 15:20. There it is said that the dancing Miriam was the sister of Aaron. If that were so, she would have

16. Moses and Heracles

been the older sister of Moses, and she would seemingly have been approximately one hundred years old at the time of the exodus. The younger Miriam is not well known, and the reference to her being the sister of Aaron (Exodus 15:20) may be in error:[43] "It is rather puzzling, however, why Miriam is identified as the sister of Aaron, and not of Moses and Aaron. Aaron was the elder of the two, but Moses, as the genuine prophet, outranked him; therefore the omission of his name may indicate that this tradition did not consider [the younger] Miriam as the sister of Moses."

16.9 Zipporah and Hespere

The last labor of Heracles is of particular interest here. To introduce this subject it may be helpful to quote from the version of this labor that has been given by the historian Diodorus of Sicily, c. 80-20 B.C.:[44]

> The last Labour which Heracles undertook was the bringing back of the golden apples of the Hesperides, and so he again sailed to Libya. With regard to these apples there is disagreement among the writers of myths, and some say that there were golden apples in certain gardens of the Hesperides in Libya, where they were guarded without ceasing by a most formidable dragon, whereas others assert that the Hesperides possessed flocks of sheep which excelled in beauty and were therefore called for their beauty, as the poets might do, "golden apples,"[45] just as Aphroditê is called "golden" because of her loveliness. There are some, however, who say that it was because the sheep had a peculiar colour like gold that they got this designation, and that Dracon ("dragon") was the name of the shepherd of the sheep, a man who excelled in strength

of body and courage, who guarded the sheep and slew any who might dare try to carry them off. But with regard to such matters it will be every man's privilege to form such opinions as accord with his own belief. At any rate Heracles slew the guardian of the apples, and after he had duly brought them to Eurystheus and had in this wise finished his Labours he waited to receive the gift of immortality, even as Apollo had prophesied to him.

But we must not fail to mention what the myths relate about Atlas and about the race of the Hesperides. The account runs like this: In the country known as Hesperitis there were two brothers whose fame was known abroad, Hesperus and Atlas. These brothers possessed flocks of sheep which excelled in beauty and were in colour of a golden yellow, this being the reason why the poets, in speaking of these sheep as mela, called them golden mela. Now Hesperus begat a daughter named Hesperis, whom he gave in marriage to his brother and after whom the land was given the name Hesperitis; and Atlas begat by her seven daughters, who were named after their father Atlantides, and after their mother, Hesperides. . . . [T]hese Atlantides [or Hesperides] excelled in beauty. . . .

This account of Heracles' encounter with the guardian of the golden apples of the Hesperides is in part identical with the report of Apollonius quoted in Section 16.5 above. There Heracles encountered the serpent Ladon and stole the golden apples; here he encountered a dragon and stole the golden apples. On the other hand, Diodorus also provides us with an alternative version of the story. In this version the golden apples are actually sheep, and the adjective golden refers either in a poetic way to their beauty or possibly to their unusual golden color. The woolly coat of such a sheep could also be imagined as a possible interpretation of the golden fleece in the *Argonautica* story.

16. Moses and Heracles

In the "sheepish" version of the Heracles story quoted above, Atlas not only raised sheep, but he also raised seven beautiful daughters called the Hesperides. One could imagine that these daughters assisted in the raising of their father's sheep. Then the question inevitably arises whether the story of Atlas and his daughters might have a parallel in biblical literature:

> But Moses fled from Pharaoh, and stayed in the land of Mid'ian; and he sat down by a well. Now the priest of Mid'ian [Jethro, Reuel, Raguel] had seven daughters; and they came and drew water, and filled the troughs to water their father's flock. The shepherds came and drove them away; but Moses stood up and helped them, and watered their flock. When they came to their father Re'uel, he said, "How is it that you have come so soon today?" They said, "An Egyptian delivered us out of the hand of the shepherds, and even drew water for us and watered the flock." He said to his daughters, "And where is he? Why have you left the man? Call him, that he may eat bread." And Moses was content to dwell with the man, and he gave Moses his daughter Zippo'rah. She bore a son, and he called his name Gershom; for he said, "I have been a sojourner in a foreign land."
>
> In the course of those many days the king of Egypt [Thutmose III] died [1450 B.C.]. And the people of Israel groaned under their bondage, and cried out for help, and their cry under bondage came up to God. And God heard their groaning, and God remembered his covenant with Abraham, with Isaac, and with Jacob. And God saw the people of Israel, and God knew their condition.
>
> Now Moses was keeping the flock of his father-in-law, Jethro, the priest of Mid'ian; and he led his flock to the west side of the wilderness, and came to Horeb, the mountain of God. And the angel of the Lord appeared to

him in a flame of fire out of the midst of a bush. . . (Exodus 2:15-3:2).

In this brief story, Moses was fleeing for his life from the king of Egypt and ended up in the land of Midian, to the east of Egypt. There he encountered the priest/shepherd Jethro and his seven daughters. Moses eventually married Jethro's daughter Zipporah, and she bore a son he called Gershom. While leading Jethro's flock, Moses came upon the burning bush, which could be considered to mark the beginning of the exodus events.

The story of the shepherd Jethro and his seven daughters in Midian is, of course, reminiscent of Diodorus's story of the shepherd Atlas and his seven daughters. As quoted earlier in Section 16.2, Apollonius only gives proper names to three Hesperides, and the first (probably namesake and eldest) of these was called Hespere. Thus, one may be led to contemplate the possibility of some primitive Greek reconfiguration of these exodus stories in which Jethro came to be called Atlas and Zipporah became Hespere. It should be noted, however, that in other sources the location of the garden and the number and names of the Hesperides may vary.

One other possible connection may be mentioned as well. The quotation given above from Diodorus Sicilus is followed by a kidnapping story:

> And since these Atlantides excelled in beauty and chastity, Busiris the king of the Egyptians, the account says, was seized with the desire to get the maidens into his power; and consequently he dispatched pirates by sea with orders to seize the girls and deliver them into his

hands. . . . Meanwhile the pirates had seized the girls while they were playing in a certain garden and carried them off, and fleeing swiftly to their ships had sailed away with them. Heracles came upon the pirates as they were taking their meal on a certain strand, and learning from the maidens what had taken place he slew the pirates to a man and brought the girls back to Atlas their father; and in return Atlas was so grateful to Heracles for his kindly deed that he not only gladly gave him such assistance as his Labour called for, but he also instructed him quite freely in the knowledge of astrology. . . .[46]

In the biblical story quoted above, shepherds drove away the seven shepherdess daughters of Jethro. But Moses helped them and watered their flock. In the version of Diodorus quoted immediately above, pirates kidnapped the seven daughters of Atlas. But Heracles rescued them and returned them to their father.

16.10 Finding Heracles

As considered in Section 16.2 above, one of the most desperate needs of the Argonaut heroes after they washed ashore in Egypt was the obtaining of drinking water. A quotation from Apollonius summarized their situation beginning with the search for water. The then-recent presence at their location of their former colleague Heracles led to their finding water that he had caused to flow from rock. Deeply grateful, the Argonauts soon afterward became interested in catching up with Heracles himself. The last paragraph of the aforementioned quotation is repeated here. In the summary of that search, the superhuman powers of some of the Argonauts were recalled:

So they spake, and those who were ready for this work [finding Heracles] answered, and they separated this way and that, each starting to search. For by the night winds the footsteps had been effaced where the sand was stirred. The two sons of Boreas started up, trusting in their wings; and Euphemus, relying on his swift feet, and Lynceus to cast far his piercing eyes; and with them darted off Canthus, the fifth. He was urged on by the doom of the gods and his own courage, that he might learn for certain from Heracles where he had left Polyphemus, son of Eilatus; for he was minded to question him on every point concerning his comrade. But that hero had founded a glorious city among the Mysians, and, yearning for his home-return, had passed far over the mainland in search of Argo; and in time he reached the land of the Chalybes, who dwell near the sea; there it was that his fate subdued him. And to him a monument stands under a tall poplar, just facing the sea. But that day Lynceus thought he saw Heracles all alone, far off, over measureless land, as a man at the month's beginning sees, or thinks he sees, the moon through a bank of cloud. And he returned and told his comrades that no other searcher would find Heracles on his way, and they also came back, and swift-footed Euphemus and the twin sons of Thracian Boreas, after a vain toil.[47]

Like the ancient Greeks, many people today are also drawn to escapist literature (and television series) about people possessing superhuman abilities. The first superheroes mentioned in the above quotation are the sons of Boreas, who had wings with which they could fly. The names of these sons were Zetes and Calais, and their description according to Pindar includes the following: "Swiftly came they who dwell by the foot of the Pangaean mount, for with gladsome mind did their father,

16. Moses and Heracles

Boreas, lord of the winds, speedily equip Zetes and Calais, with their purple pinions heaving adown their backs"[48] It is perhaps not unreasonable to identify Zetes and Calais as superheroes, because their ability to fly is also possessed by the modern comic book superhero Warren Kenneth Worthington III, better known as Archangel. Readers who (like the author) aren't up-to-date on modern superheroes are referred to the internet for more information on Archangel and others to be mentioned below.

The next superhero mentioned in the quotation of Apollonius is "Euphemus, relying on his swift feet."[49] A more detailed description of the abilities of Euphemus is provided elsewhere by Apollonius:

> After them from Taenarus came Euphemus whom, most swift-footed of men, Europe, daughter of mighty Tityos, bare to Poseidon. He was wont to skim the swell of the grey sea, and wetted not his swift feet, but just dipping the tips of his toes was borne on the watery path.[50]

A comic book superhero analogous to Euphemus would be The Flash, and he too can run on water, as seen in a TV series.[51]

As described by Apollonius, Lynceus seems to have had telescopic vision: ". . . and Lynceus to cast far his piercing eyes."[52] Telescopic vision is also possessed by Peter Quinn as the superhero Peeper and also by several others. Elsewhere Lynceus is said even to have had what we would perhaps call X-ray vision: ". . . and Lynceus too excelled in keenest sight, if the report is true that that hero could easily direct his sight even beneath the earth."[53] A familiar comic book superhero with X-ray vision is, of course, Superman.

Adventures In Egypt

The Greek poet Pindar (c. 522-443 B.C.) was the author of one of the earliest existing accounts of the expedition of the Argonauts.[54] That account is contained in his fourth Pythian Ode, dated to c. 462 B.C. Writing at about the same time as Pindar, Herodotus includes a few mentions of the Argonauts (several referenced in these studies), but he doesn't present a continuous narrative of this subject.[55] Like Apollonius, Pindar indicates the superhuman powers of some of the Argonauts, whom he refers to as "demigods."[56]

Unfortunately, the effort to find Heracles wasn't entirely successful. Superhero Lynceus with his telescopic eyesight was apparently considered to be the most likely to find Heracles, even if Heracles was far away. Lynceus believed that he could see Heracles in the distance, but he advised against continuing the search:

> But that day Lynceus thought he saw Heracles all alone, far off, over measureless land, as a man at the month's beginning sees, or thinks he sees, the moon through a bank of cloud. And he returned and told his comrades that no other searcher would find Heracles on his way, and they also came back, and swift-footed Euphemus and the twin sons of Thracian Boreas, after a vain toil.[57]

It seems odd that after possibly seeing Heracles, the Argonauts are said to have abandoned their search for him. Perhaps in the earlier (and unacceptable) edition of Apollonius's work the Argonauts did actually catch up with Heracles, and this situation will be considered further in Chapter 17.

It may be worth noting the description, quoted above, of Lynceus thinking that "he saw Heracles all alone, far off, over

16. Moses and Heracles

measureless land, as a man at the month's beginning sees, or thinks he sees, the moon through a bank of cloud."[58] This appears to be a reference to the method employed by the Egyptians for determining the beginning date of a month in their lunar calendar system by observations of the thin lunar crescent near sunrise.[59] Partial obscuration by clouds or mist as indicated by Apollonius would cause uncertainty, varying from place to place, concerning the exact starting date of the month.

As a final topic in this chapter relating to Heracles, it may be of interest to consider his death. A preliminary consideration of this subject was included in Section 16.7. As noted there, Heracles contracted a lethal skin disease from wearing a poisoned tunic. Knowing that his death was imminent, Heracles chose to be burned on a pyre.

> But Hercules [Heracles]. . . proceeded to Mount Oeta, in the Trachinian territory, and there constructed a pyre, mounted it, and gave orders to kindle it. . . . While the pyre was burning, it is said that a cloud passed under Hercules and with a peal of thunder wafted him up to heaven. Thereafter he obtained immortality. . . .[60]

A similar description of the death of Heracles is provided by Diodorus Sicilus:

> And Heracles put on the shirt which had been anointed, and as the strength of the toxic drug began slowly to work he met with the most terrible calamity. For the arrow's barb had carried the poison of the adder [or of the Lernaean hydra], and when the shirt for this reason, as it became heated, attacked the flesh of the body, Heracles was seized with such anguish that he slew Lichas, who

had been his servant, and then, disbanding his army, returned to Trachis.

As Heracles continued to suffer more and more from his malady he dispatched Licymnius and Iolaus to Delphi to inquire of Apollo what he must do to heal the malady. ... The god gave the reply that Heracles should be taken, and with him his armour and weapons of war, unto [Mount] Oete and that they should build a huge pyre near him; what remained to be done, he said, would rest with Zeus. Now when Iolaus had carried out these orders and had withdrawn to a distance to see what would take place, Heracles, having abandoned hope for himself, ascended the pyre and asked each one who came up to him to put torch to the pyre. And when no one had the courage to obey him Philoctetes alone was prevailed upon; and he, having received in return for his compliance the gift of the bow and arrows of Heracles, lighted the pyre. And immediately lightning also fell from the heavens and the pyre was wholly consumed. After this when the companions of Iolaus came to gather up the bones of Heracles and found not a single bone anywhere, they assumed that, in accordance with the words of the oracle, he had passed from among men into the company of the gods.[61]

The essential components of this story can be summarized somewhat as follows. Heracles contracted a terrible disease affecting his skin and flesh. As this disease was unbearably painful, he built a pyre on Mount Oeta. While the pyre was burning, Heracles's bones vanished, and he was understood to have been wafted away to immortality in heaven.

This brief version has parallels in other literature including at least two in the Bible. Thus, rather than dying a normal death,

16. Moses and Heracles

Elijah is said to have been translated directly to heaven by a supernatural means, as witnessed by Elisha:

> And as they still went on and talked, behold, a chariot of fire and horses of fire separated the two of them. And Eli'jah went up by a whirlwind into heaven. And Eli'sha saw it and he cried, "My father, my father! the chariots of Israel and its horsemen!" And he saw him no more (2 Kings 2:11-12).

Of greater interest here would be the death of Moses. There is some lack of clarity on this subject. In a story that may have been written or rewritten during the reign of King Josiah (c. 637-605 B.C.), many centuries after the exodus (1440 B.C.), one finds the following information:

> And Moses went up from the plains of Moab to Mount Nebo, to the top of Pisgah, which is opposite Jericho. And the Lord showed him all the land, Gilead as far as Dan, all Naph'tali, the land of E'phraim and Manas'seh, all the land of Judah as far as the Western Sea [Mediterranean], the Negeb, and the Plain, that is, the valley of Jericho the city of palm trees, as far as Zo'ar. And the Lord said to him, "This is the land of which I swore to Abraham, to Isaac, and to Jacob, 'I will give it to your descendants.' I have let you see it with your eyes, but you shall not go over there." So Moses the servant of the Lord died there in the land of Moab, according to the word of the Lord, and he buried him in the valley in the land of Moab opposite Beth-pe'or; but no man knows the place of his burial to this day (Deuteronomy 34:1-6).

Modifications of the text of Deuteronomy are believed to have continued during the Babylonian captivity and even after the return from exile, so it isn't possible to be certain of details.

Adventures In Egypt

It is of interest that the limited biblical information concerning the death of Moses has some features in common with the stories of the death of Heracles, but the parallels are not exact. Thus, Heracles is said to have climbed to the top of a mountain, from which he was carried into immortality in heaven. Similarly, Moses is said to have climbed to the tops of the neighboring peaks Mt. Nebo and Mt. Pisgah. It isn't implied that Moses received instant immortality as Heracles and Elijah did, but it is striking that "no man knows the place of his burial to this day." Similarly, in searching for the remains of Heracles, the companions of Iolaus "found not a single bone anywhere," as quoted above.

It is possible that there existed a degree of uncertainty among the later Hebrews concerning the death of Moses:

> And after six days Jesus took with him Peter and James and John his brother, and led them up a high mountain apart. And he was transfigured before them, and his face shone like the sun, and his garments became white as light. And behold, there appeared to them Moses and Eli'jah, talking with him. And Peter said to Jesus, "Lord, it is well that we are here; if you wish, I will make three booths here, one for you and one for Moses and one for Eli'jah" (Matthew 17:1-4).

As quoted previously, Elijah was said in the Bible not to have died, and perhaps some of the Hebrews of Jesus's time may have believed that Moses, like Heracles, didn't die either:

> One poignant tradition in the rabbinical literature insisted that Moses did not die at all. . . . The same Midrashic tradition is evoked in the New Testament when the

16. Moses and Heracles

transfiguration of Jesus is witnessed by Moses and Elijah (Matthew 17:3), two biblical figures who were thought to have been granted the privilege of direct passage to heaven.[62]

16.11 Escape from the lake

In view of the many parallels between the lives of Heracles and Moses, it has been natural to imagine that some of the Greek stories about Heracles might have been adapted from earlier stories involving Moses. One further parallel may be recalled here. As noted in Section 1.6 the brief creation story sung by Orpheus had significant features in common with the biblical version. One suggested explanation for this commonality was that (according to Artapanus) Orpheus was taught by Moses. This is not so unlikely as it might seem if the Heracles who is said to have served as an Argonaut with Orpheus could actually have been Moses.

When the Argonauts in Egypt abandoned their efforts to find Heracles, their next immediate task was to find the way from the Tritonian lake back into the Mediterranean Sea. As noted in Section 15.3, the Tritonian lake seems to have been a body of water through which the river Triton (or a branch of the Nile) flowed. The fact that Heracles sought for drinking water from a rock "near the Tritonian lake" implies that the water in the lake itself was undrinkable. The easiest interpretation would seem to be that the Tritonian lake was salty because of its connection with the Mediterranean Sea. This connection is also implied by Apollonius Rhodius, who records the search by the Argonauts for the link to the sea as a possible means for their return to "the divine land of Pelops." A few excerpts concerning this adventure are included here:

> But when they had gone aboard, as the south wind blew over the sea, and they were searching for a passage to go forth from the Tritonian lake, for long they had no device, but all the day were borne on aimlessly. And as a serpent goes writhing along his crooked path when the sun's fiercest rays scorch him; and with a hiss he turns his head to this side and that, and in his fury his eyes glow like sparks of fire, until he creeps to his lair through a cleft in the rock; so Argo seeking an outlet from the lake, a fairway for ships, wandered for a long time. Then straightway Orpheus bade them bring forth from the ship Apollo's massy tripod and offer it to the gods of the land as propitiation for their return. So they went forth and set Apollo's gift on the shore; then before them stood, in the form of a youth, far-swaying Triton, . . .
>
> So he spake; and Triton stretched out his hand and showed afar the sea and the lake's deep mouth, and then addressed them: "That is the outlet to the sea, where the deep water lies unmoved and dark; on each side roll white breakers with shining crests; and the way between for your passage out is narrow. And that sea stretches away in mist to the divine land of Pelops beyond Crete. . . ."
>
> He spake with kindly counsel; and they at once went aboard, intent to come forth from the lake by the use of oars. And eagerly they sped on; meanwhile Triton took up the mighty tripod. . . . And he guided Argo on until he sped her into the sea on her course. . . . [T]hey were to cross to Crete, which rises in the sea above other islands.[63]

Thus, the Argonauts were able to re-enter the Mediterranean Sea to find their way to Crete and then back to the "divine land of

16. Moses and Heracles

Pelops." The symbolism of a serpent that "creeps to his lair through a cleft in the rock" is reminiscent of a story in Section 6.2 concerning snakes near the site of the garden of Eden: "Natural caves open at the foot of the rocks, and in the cracks of these caves large snakes may be found."

Assuming that an actual geographical place in Egypt is being described in the *Argonautica* above, the identity of the Tritonian lake would be of interest. There are several large lakes in the Nile delta and adjoining coastline, but the configuration of these lakes and their connections to the Mediterranean Sea may have evolved over the past millennia. One can, however, imagine that at the time of the Argonauts one or more of these lakes might have been too salty to assuage thirst and was also challenging to escape from in the Argo.

An early abbreviated version of the Argonauts' escape from Egypt was given by Herodotus (c. 484-430 B.C.):

> Jason (it is said) when the Argo had been built at the foot of Pelion, put therein besides a hecatomb a bronze tripod, and set forth to sail round the Peloponnesus, that he might come to Delphi. But when in his course he was off Malea, a north wind caught and carried him away to Libya [an early name for the continent of Africa]; and before he could spy land he came into the shallows of the Tritonian lake. There, while yet he could find no way out, Triton (so goes the story) appeared to him and bade Jason give him the tripod, promising so to show the shipmen the channel and send them on their way unharmed. Jason did his bidding, and Triton [here the son of Poseidon] then showed them the passage out of the shallows and set the tripod in his own temple; but first he prophesied over it, declaring the whole matter to Jason's comrades . . . .[64]

Adventures In Egypt

This version omits entirely the story of the expedition to Colchis to recover the golden fleece. That story is usually supposed to have been the main point of the adventure, but Herodotus seems to have had a different purpose. He describes how Jason gave to a person/god named Triton a valuable tripod in exchange for Triton showing Jason's comrades how to escape from the Tritonian lake and then presumably return safely to their homes. Curiously, the emphasis seems to be entirely on Jason finding a way for his comrades to leave. Thus the text can be construed to imply that Jason himself had no intention of leaving Egypt.

Herodotus also doesn't mention the interacting of the Argonauts with Talos, who has now sometimes been identified with the volcanic eruption of Thera. Prior to the story of the arrival of the Argonauts in Egypt, however, Herodotus does include one interesting observation about the island of Thera in an earlier time: "Then for seven years after this there was no rain in Thera; all their trees in the island save one were withered."[65] This brief report seems reminiscent of the massive eruption of Thera that is estimated to have occurred in 1628 B.C. After that eruption there was a seven-year famine in Canaan, Egypt, and China, as considered in Chapter 8.[66]

Another early comment suggesting the possible absence of Jason from the homeward bound Argonauts may be found in a work of Pindar (introduced in Section 16.10 above). The following words include a story of Medea on the island of Thera speaking to the crew of the Argo, presumably on their way home from Egypt:

> [T]he priestess throned beside the golden eagles of Zeus gave for them an oracle, naming Battus the coloniser of

16. Moses and Heracles

fruitful Libya, and telling how he would at once leave the holy island [Thera], and build, on a gleaming hill a city of noble chariots, and thus, in the seventeenth generation, fulfil the word spoken at Thera by Medea, which that brave daughter of Aeetes, that queen of the Colchians, breathed forth from her immortal lips, when she spake in this wise to the heroes who sailed with the warrior Jason:

"Listen ye sons of high-spirited men, ye sons of the gods! for I aver that, from this wave-washed land of Thera, the daughter of Epaphus [Libya], will, in days to come, find planted in her a root of cities that shall be fostered of men near the foundations of Zeus Ammon. . . . [stories related to the Argonauts in Egypt, Medea claiming a leadership role] . . . . The day shall come when Phoebus in his golden home shall make mention of him in his oracles, when, at a later time, he descendeth from the threshold into the Pythian shrine, telling how he shall carry many a man in his ships to the fertile precinct of the son of Cronus beside the Nile."

Verily such were the lays that Medea sang; and the god-like heroes, while they listened to her deep counsel, stirred not a whit, but bowed them down in silence.[67]

The fact that Medea is said to have spoken to the heroes who had sailed with Jason but not to Jason himself is similar to the statements by Herodotus, quoted previously in this section. Herodotus tells us that Triton explained to Jason's shipmen/comrades how to escape from the Tritonian lake, while it seems that Jason may not have been planning to accompany them. It is perhaps significant that these writings of Pindar and Herodotus may be the two oldest, and therefore probably the most historical, writings concerning the departure of the Argonauts (or some of them) from Egypt. Jason and perhaps other of the Argonauts may have decided to remain behind with

Heracles. This possibility will be considered further in Chapter 17.

16.12 Conclusion

It has been noted in this chapter that there are many similarities of plot between stories of Moses and of Heracles in Egypt. The time frame for these stories may be understood to include the important period of the Hebrew exodus from Egypt. Both Moses and Heracles experienced occasions of great thirst during their time in Egypt, and both of them obtained water by striking a rock. The similarities between these stories and several others are enough to suggest that they may have had, at least in part, a common origin. A related parallel exists between the biblical character Jethro with his seven shepherdess daughters and the Greek character Atlas and his seven shepherdess daughters. Many of the biblical texts may first have been written at nearly the times they describe. On the other hand, the Greek stories may have been adapted later in large part for esthetic rather than historical purposes. As a result the Greek versions vary substantially between their authors, but some of their actual historical content may have been influenced by Bible-related sources.

---

1. *Apollonius Rhodius: The Argonautica*, With an English translation by R. C. Seaton (Loeb Classical Library, Harvard University Press, Cambridge, Massachusetts, 1980), Book IV, pp. 389, 391, 393, 395.
2. L. E. Toombs, "Rod," *The Interpreter's Dictionary of the Bible: An Illustrated Encyclopedia*, (Abingdon Press,

Nashville, Tennessee, 1962), Volume 4, R-Z, pp. 102-103.
3. *Apollonius Rhodius*, op. cit., Book I, p. 33.
4. *Apollonius Rhodius*, op. cit., Book I, p. 39.
5. *Apollonius Rhodius*, op. cit., Book I, p. 85.
6. *Herodotus*, Volume II, With an English translation by A. D. Godley, In Four Volumes (Loeb Classical Library, Harvard University Press, Cambridge, Massachusetts, 1971), Book IV, Paragraph 82, p. 285.
7. *Apollonius Rhodius*, op. cit., Book IV, p. 393.
8. Ibid.
9. L. W. Casperson, *Patterns of Biblical Chronology* (Westbow Press, Bloomington, IN, 2012), Section 1.5, "The original authors," pp. 9-18.
10. *Herodotus*, Volume I, With an English translation by A. D. Godley, In Four Volumes (Loeb Classical Library, Harvard University Press, Cambridge, Massachusetts, 1981), Book II, Paragraph 43, p. 329.
11. *Diodorus of Sicily in Twelve Volumes*, With an English translation by C. H. Oldfather (Loeb Classical Library, Harvard University Press, Cambridge, Massachusetts, 1979), Volume 2, The Library of History of Diodorus of Sicily, Book IV, Chapter 27, p. 429.
12. L. W. Casperson, op. cit., Chapter 15, "Moses and the Ethiopians," pp. 351-385.
13. Ibid., p. 361; J. H. Breasted, *Ancient Records of Egypt: Historical Documents from the Earliest Times to the Persian Conquest*, Collected, Edited, and Translated with commentary, Volume 2, The Eighteenth Dynasty (The University of Chicago Press, Chicago, 1906), p. 50.
14. L. W. Casperson, op. cit., Chapter 16, "Exile of Moses," pp. 386-418.
15. *Josephus in Nine Volumes*, With an English translation by H. St. J. Thackeray (Loeb Classical Library, Harvard University Press, Cambridge, Massachusetts, 1978),

Volume IV, *Jewish Antiquities*, Book II, 11:1, pp. 275, 277.
16. *Apollonius Rhodius*, op. cit., Book IV, p. 393.
17. *Apollonius Rhodius*, op. cit., Book IV, pp. 389, 391, 393.
18. L. W. Casperson, op. cit., Chapter 15, "Moses and the Ethiopians," pp. 351-385.
19. *Apollonius Rhodius*, op. cit., Book I, p. 31.
20. *Apollonius Rhodius*, op. cit., Book I, p. 63.
21. *Apollonius Rhodius*, op. cit., Book I, p. 83.
22. *Apollonius Rhodius*, op. cit., Book IV, pp. 391, 393.
23. *The Book of Yashar*, Translated from the Hebrew and Published by Manuel Mordecai Noah (Hermon Press, New York, 1972), (cited here and below in the form *Jasher*), *Jasher* 72:24.
24. L. W. Casperson, op. cit., Section 16.6, "Levite origins," pp. 409-416.
25. *Josephus in Nine Volumes*, op. cit., Volume IV, *Jewish Antiquities*, Book II, 10:2, p. 275.
26. L. W. Casperson, op. cit., Section 16.5, "Chronology of the exile," pp. 403-409; Table 16.1, p. 408.
27. L. W. Casperson, op. cit., "Appendix A: Artapanus," pp. 622-629; Paragraph 40, p. 629.
28. *The Week*, Volume 15, Issue 721, p. 34 (29 May 2015); based on the article by J. Pinsker, "The financial perks of being tall," *The Atlantic*, (18 May 2015).
29. *Jasher* 68:4.
30. *Apollonius Rhodius*, op. cit., Book I, pp. 25, 27.
31. Herodotus, Volume I, op. cit., Book II, Paragraph 42, p. 327.
32. L. W. Casperson, op. cit., Section 16.5, "Chronology of the exile," pp. 403-409; Table 16.1, p. 408.
33. The full account may be found in L. W. Casperson, op. cit., Section 15.2C, "Josephus," pp. 354-356; or see *Josephus in Nine Volumes*, op. cit., Volume IV, *Jewish Antiquities*, Book II, 10:1-2, pp. 269, 271, 273, 275.

34. Diodorus of Sicily, op. cit., Book IV, Chapter 27, p. 429.
35. L. W. Casperson, op. cit., Section 16.5, "Chronology of the exile," Table 16.1, p. 408.
36. L. W. Casperson, op. cit., Section 17.2, "Sickness of the king," pp. 420-427.
37. L. W. Casperson, op. cit., "Appendix B: Manetho," Chapter 31, p. 644.
38. "Leprosy," *Gale Encyclopedia of Medicine*, Third Edition (The Gale Group, Inc., 2006). Web. 28 February 2017.
39. C. Creighton, "Leprosy," *The Encyclopædia Britannica: A Dictionary of Arts, Sciences, and General Literature*, Volume 14 (The Werner Company, Chicago, 1896), pp. 468-470.
40. *Apollodorus, The Library*, With an English translation by Sir James George Frazer (The Loeb Classical Library, G. P. Putnam's Sons, New York, 1921), Volume II, 7:7, p. 269.
41. L. W. Casperson, op. cit., Section 17.3, "Death of the king," pp. 427-432; p. 432; *Jasher* 76:57.
42. *Jasher* 80:49,52-57.
43. D. M. Beegle, *Moses, The Servant of Yahweh* (William B. Eerdmans Publishing Company, Grand Rapids Michigan, 1972), p.162.
44. *Diodorus of Sicily*, op. cit., Book IV, Chapters 26-27, pp. 427, 429.
45. The word translated here as "apple" also means "sheep." See *Diodorus of Sicily*, op. cit., Book IV, Chapter 26, p. 427, note 1.
46. *Diodorus of Sicily*, op. cit., Book IV, Chapter 27, pp. 429, 431.
47. *Apollonius Rhodius*, op. cit., Book IV, pp. 393, 395.
48. *The Odes of Pindar: Including the Principal Fragments*, With an introduction and an English translation by John Sandys (The Loeb Classical Library, The Macmillan Company, New York, 1915) *The Pythian Odes*, pp. 151-

311; "Pythian IV: For Arcesilas of Cyrene," pp. 196-231; p. 219.
49. *Apollonius Rhodius*, op. cit., Book IV, p. 395.
50. *Apollonius Rhodius*, op. cit., Book I, p. 15.
51. *The Flash*, Season 1, Episode 5, "Plastique" (11 November 2014). Web. 28 February 2017.
52. *Apollonius Rhodius*, op. cit., Book IV, p. 395.
53. *Apollonius Rhodius*, op. cit., Book I, p. 13.
54. "Pindar," Wikipedia. Web. 29 March 2017.
55. *Herodotus*, With an English translation by A. D. Godley, In Four Volumes (Loeb Classical Library, Harvard University Press, Cambridge, Massachusetts, 1981), for example, Volume II, Book IV, pp. 345, 381.
56. *The Odes of Pindar: Including the Principal Fragments*, op.cit., pp. 217, 219, 221.
57. *Apollonius Rhodius*, op. cit., Book IV, p. 395.
58. Ibid.
59. L. W. Casperson, op. cit., Sections 6.2-6.3, pp. 112-123.
60. *Apollodorus, The Library*, op. cit., Volume II, 7:7, pp. 269, 271.
61. *Diodorus of Sicily*, op. cit., Book IV, Chapter 38, pp. 465, 467.
62. J. Kirsch, *Moses, A Life* (Ballantine Books, New York, 1998), p. 350.
63. *Apollonius Rhodius*, op. cit., Book IV, pp. 399, 401, 403, 405.
64. *Herodotus*, Volume II, op. cit., Book IV, Paragraph 179, p. 381.
65. Ibid., Book IV, Paragraph 151, p. 353.
66. See also L. W. Casperson, op. cit., Section 22.5, "Famine duration," pp. 557-560; p. 560.
67. *The Odes of Pindar: Including the Principal Fragments*, op.cit., pp. 199, 201, 203, 205.

17. JOSHUA AND JASON

"And Moses said to Joshua, 'Choose for us men, and go out, fight with Am'alek; tomorrow I will stand on the top of the hill with the rod of God in my hand' " (Exodus 17:9).

17.1 Introduction

Chapters 15 and 16 introduced the idea of the possible presence of real or fictional Mycenaean (c. 1600-1100 B.C.) Greek Argonauts in Egypt at about the same location as the exodus. Such a presence could have impacted any written accounts maintained by the Egyptians, Greeks, and Hebrews. Bible-related literature discussed in those earlier chapters describes the activities of the Hebrews in Egypt mostly prior to their departure at the time of the exodus. On the other hand, Greek writings, exemplified especially by the *Argonautica* of Apollonius Rhodius, describe the adventures of the Argonauts following their involuntary landing in the vicinity of the Nile River. Egyptian texts relating to the exodus may have existed as well, and references by the early historians Artapanus, Manetho, and Cheremon support this conclusion.[1] Not only are the settings for these exodus-related and Argonaut-related stories about the same, but the plots also have much in common. If the events in the exodus and Argonaut stories did not actually occur simultaneously, they may have been close enough in time for them to have become entangled. Thus, for example, it appears possible that some of the activities ascribed to Heracles may have been adapted from earlier stories concerning Moses. For the most part, the previous discussions here ended with the exodus itself, with little consideration of what happened afterward. Specifically, the discussion of the very important person Joshua has been left for the present chapter.

Joshua is found to have appeared abruptly, and several aspects of his arrival are considered in Section 17.2. Initially, most of Joshua's interactions with the Hebrews, at least as recorded in the Bible, seem to have occurred through the person of Moses, and examples of such interactions are reviewed in Section 17.3. There is very little information about the family of Joshua beyond the oft-repeated statement that he was the son of Nun. This situation is reviewed in Section 17.4, and the possibility is suggested that Joshua may not originally have had a conventional biblical family tree. The Hebrew ancestry of Joshua as indicated in 1 Chronicles 7:20-27 also reveals a tragedy that struck the family of Ephraim. Some of the details of this event are reviewed in Section 17.5. Joshua's first recorded activity was his organizing of an army of Hebrews and leading them to victory in a battle against the Amalekites. Some of the long-term consequences of this battle are considered in Section 17.6. The possibility of Joshua being the same person as Jason (of the Argonauts) is contemplated in Section 17.7. The ending of the *Argonautica* is reviewed in Section 17.8, including a possible

17. Joshua and Jason

encounter by the Argonauts with an erupting volcano as they returned to their original departure site.

17.2 Arrival of Joshua

One of the most striking features of the exodus stories is the sudden appearance of the warrior Joshua among the Israelites. Joshua isn't even mentioned in the Bible until events of a few weeks after the exodus. The first use of his name is in Exodus 17:9, which is quoted at the top of this chapter. It may be worth noting that at the appearance of such an important person as Joshua, his origin and ancestry are not clearly indicated. It is evident from the outset that Joshua is going to be significant. He is immediately given responsibility by Moses for organizing an army of Hebrew refugees from Egypt and leading them in battle against the Amalekites. It appears that Moses may have known a lot more about what proved to be the extraordinary organizational and fighting ability of Joshua than would seem to be suggested by his nondescript introduction. In any case, the Hebrew army under Joshua seems to have been entirely successful in its first battle. This conflict occurred at Rephidim after Moses struck the rock with his rod and obtained water for the people to drink, as quoted in Section 8.6. That previous quotation is followed immediately with the words:

> Then came Am'alek and fought with Israel at Reph'idim. And Moses said to Joshua, "Choose for us men, and go out, fight with Am'alek; tomorrow I will stand on the top of the hill with the rod of God in my hand." So Joshua did as Moses told him, and fought with Am'alek; and Moses, Aaron, and Hur went up to the top

of the hill. Whenever Moses held up his hand, Israel prevailed; and whenever he lowered his hand, Am'alek prevailed. But Moses' hands grew weary; so they took a stone and put it under him, and he sat upon it, and Aaron and Hur held up his hands, one on one side, and the other on the other side; so his hands were steady until the going down of the sun. And Joshua mowed down Am'alek and his people with the edge of the sword (Exodus 17:8-13).

The organization of this and preceding texts suggests that the attack by the Amalekites occurred shortly after the Israelites obtained drinking water for themselves and their cattle: "Then came Amalek and fought with Israel. . ." (Exodus 17:8). If the Israelites had been experiencing an unexpected drought at Rephidim, the Amalekites at that location would likely have been thirsty as well. It is perhaps natural to assume that the Amalekites may have attacked the Israelites to obtain some (or all) of this new water resource for themselves.

It was suggested in Section 16.7 that Moses may not have been as strong at the time of the exodus as he was in his youth. A further indication here of his possible weakness was his need for physical support soon after the exodus. Thus, Moses's stated role in this battle was to sit on a rock atop a hill, watching the fighting while Aaron and Hur supported his hands.

The Hebrews arrived at Rephidim at some time later than one month after the exodus (Exodus 16:1), and they left Rephidim at some time before two and one half months after the exodus (Exodus 19:1-2). While the Hebrews were at Rephidim they were attacked by the Amalekites as mentioned above, and though the Amalekites were defeated they would be considered enemies by the Hebrews for several centuries. Even after the extinction of the

17. Joshua and Jason

actual Amalekites, this group would be employed as a concept and a justification for attacks by the Hebrews on other peoples, as discussed in Section 17.6 below.

An important aspect of the above quotation is that the name Joshua appears here in the biblical record for the first time. There is as yet no indication of Joshua's ancestry, or where he came from, or why Moses chose him to organize the Hebrew army, or why he was chosen to lead that army in battle, or why he later came to replace Moses as overall leader of the Hebrews. Joshua's essential contributions in the battle are well recognized, and he seems to have been given much of the credit for the victory: "And Joshua mowed down Am'alek and his people with the edge of the sword" (Exodus 17:13).

17.3 Joshua and Moses
After the battle with the Amalekites, Moses may have continued to depend on Joshua as his commander-in-chief. Also, from this first battle Moses seems to have been grooming Joshua as his successor: "And the Lord said to Moses, 'Write this as a memorial in a book and recite it in the ears of Joshua, that I will utterly blot out the remembrance of Am'alek from under heaven' " (Exodus 17:14). In addition, Joshua seems to have served Moses as his personal assistant and perhaps bodyguard wherever Moses went:

> The Lord said to Moses, "Come up to me on the mountain, and wait there; and I will give you the tables of stone, with the law and the commandments, which I have written for their instruction." So Moses rose with his ser-

vant Joshua, and Moses went up into the mountain of God" (Exodus 24:12-13).

And Moses turned, and went down from the mountain with the two tables of the testimony in his hands, tables that were written on both sides; on the one side and on the other were they written. And the tables were the work of God, and the writing was the writing of God, graven upon the tables. When Joshua heard the noise of the people as they shouted, he said to Moses, "There is a noise of war in the camp" (Exodus 32:15-17).

Now Moses used to take the tent and pitch it outside the camp, far off from the camp; and he called it the tent of meeting. And every one who sought the Lord would go out to the tent of meeting, which was outside the camp. Whenever Moses went out to the tent, all the people rose up, and every man stood at his tent door, and looked after Moses, until he had gone into the tent. When Moses entered the tent, the pillar of cloud would descend and stand at the door of the tent, and the Lord would speak with Moses. And when all the people saw the pillar of cloud standing at the door of the tent, all the people would rise up and worship, every man at his tent door. Thus the Lord used to speak to Moses face to face, as a man speaks to his friend. When Moses turned again into the camp, his servant Joshua the son of Nun, a young man, did not depart from the tent (Exodus 33:7-11).

And a young man ran and told Moses, "Eldad and Medad are prophesying in the camp." And Joshua the son of Nun, the minister of Moses, one of his chosen men, said, "My lord Moses, forbid them." But Moses said to him, "Are you jealous for my sake? Would that all the Lord's people were prophets, that the Lord would put his spirit upon them! And Moses and the elders of Israel returned to the camp (Numbers 11:27-30).

17. Joshua and Jason

> The Lord said to Moses, "Send men to spy out the land of Canaan, which I give to the people of Israel; from each tribe of their fathers shall you send a man, every one a leader among them." So Moses sent them from the wilderness of Paran, according to the command of the Lord, all of them men who were heads of the people of Israel. And these were their names: . . . ; from the tribe of E'phraim, Hoshe'a the son of Nun; . . . These were the names of the men whom Moses sent to spy out the land. And Moses called Hoshe'a the son of Nun Joshua (Numbers 13:1-4, 8, 16).

The Bible provides us with only very limited personal information about Joshua. The texts quoted above are among the first to introduce the Bible reader to this important warrior. He initially had shown up unexpectedly as the Hebrew leader in the battle against the Amalekites. There he was called simply Joshua, and with no further introduction he was asked to organize the Hebrew refugees into an effective fighting force to oppose the Amalekites (Exodus 17:9). Remarkably, he was successful. Later Joshua is referred to as the "servant" of Moses, and he accompanied Moses when he climbed the "mountain of God" to receive the law and the commandments of the Lord (Exodus 24:13). Consistent with his experience and responsibility as a warrior, Joshua was very concerned about the "noise of war in the camp" (Exodus 32:17). He also accompanied Moses as his servant when Moses talked with the Lord in the tent of meeting (Exodus 33:11), and Joshua seems to have served as the caretaker of the tent.

In the story of the tent, Joshua acquired an ancestry of sorts, as beginning there he is referred to as "Joshua the son of Nun" (Exodus 33:11). He is also given an approximate age, as he is said to have been "a young man." It seems likely also that Joshua was the "young man" who ran from the camp to the tent of meeting to warn Moses of "prophesying in the camp," as quoted above (Numbers 11:27). Instead of being concerned about the prophesying, however, Moses reprimanded Joshua: "Would that all the Lord's people were prophets, that the Lord would put his spirit upon them!" This story is reminiscent of the New Testament story in which Jesus was warned by one of his disciples that some unknown or unidentified person was casting out demons in Jesus's name (Mark 9:38-40, Luke 9:49-50). Similar to the situation with Moses and Joshua, Jesus reminded the disciples of the following maxim: "For he that is not against us is for us."

Not surprisingly, Joshua was chosen as one of the spies to investigate the land of Canaan (Numbers 13:16). He was said to be representing the tribe of Ephraim, and he had proven himself to be an extremely able warrior. Thus, his presence would have increased the chances of a successful mission. Moses may at that time have officially assigned to Joshua his already indicated name and parentage: "And Moses called Hoshea the son of Nun Joshua" (Numbers 13:16). Joshua himself later sent spies to Jericho, presumably a standard military strategy preceding an invasion (Joshua 2).

17.4 Genealogy of Joshua

It may be recalled that in the Bible Moses had a well-defined family tree back to Levi son of Jacob:

17. Joshua and Jason

These are the names of the sons of Levi according to their generations: Gershon, Kohath, and Merar'i, the years of the life of Levi being a hundred and thirty-seven years. The sons of Gershon: Libni and Shim'e-i by their families. The sons of Kohath: Amram, Izhar, Hebron, and Uz'ziel, the years of the life of Kohath being a hundred and thirty-three years. The sons of Merar'i: Mahli and Mushi. These are the families of the Levites according to their generations. Amram took to wife Joch'ebed his father's sister and she bore him Aaron and Moses, the years of the life of Amram being one hundred and thirty-seven years (Exodus 6:16-20).

While the indicated ages of some of these patriarchs at the times of their deaths seem high by modern standards, the number of generations between Levi and Moses is not necessarily unreasonable. These data will be considered further below.

Joshua's genealogy is more troublesome than that of Moses. In an apparently much later writing, Joshua was given a lengthy (but somewhat confusing) ancestry back to Ephraim, grandson of Jacob:

The sons of E'phraim: Shuthe'lah, and Bered his son, Tahath his son, Ele-a'dah his son, Tahath his son, Zabad his son, Shuthe'lah his son, and Ezer and E'le-ad, whom the men of Gath who were born in the land slew, because they came down to raid their cattle. And E'phraim their father mourned many days, and his brothers came to comfort him. And E'phraim went in to his wife, and she conceived and bore a son; and he called his name Beri'ah, because evil had befallen his house. His daughter was She'erah, who built both Lower and Upper Beth-hor'on, and Uz'zen-she'erah. Rephah was his son, Resheph his son, Telah his son, Tahan his son, Ladan his son,

Ammi'hud his son, Eli'shama his son, Nun his son, Joshua his son (1 Chronicles 7:20-27).

In commenting on this genealogy, the following has been said: "However, the Ephraimite genealogy here is quite confused (perhaps containing three rescensions of the earlier list in Num. 26:35-37)...."[2]

The date of the writing of 1 Chronicles has also been difficult to establish. It will be sufficient here to quote a summary of this subject:

> It has been maintained that the first form of the Chronicler's work was produced in the early postexilic period. If one author is postulated, the date proposed must be later than the latest event mentioned. This will depend on the chronology of Ezra and Nehemiah and on such matters as the evaluation of lists of high priests (especially in Nehemiah 12) which may show generations down to a particular moment. But such evidence is inconclusive. If the late date for Ezra is assumed (398 B.C.), then a date around 350 B.C. is possible. There do not appear to be any allusions to the fall of the Persian Empire (331 B.C.) and to Greek rule, but this too is uncertain. The history of Judah during the period from 400 to 200 B.C. is too little known to be of substantial use in dating documents.[3]

Many commentators would probably agree that 1 Chronicles was written in about 350 B.C. or later. If this is correct then Joshua's first substantial biblical ancestry dates from about one thousand years after his death. In this circumstance all details of that genealogy should be approached with caution.

17. Joshua and Jason

A tentative interpretation of the above genealogical information from 1 Chronicles concerning the ancestries of Moses and Joshua is listed more formally in Table 17.1 for the principal names from Jacob to the time of the exodus. Besides other uncertainties in the interpretation of the text quoted above (1 Chronicles 7:20-27), the proper placement of Sheerah in the table is not clear. She is "[e]ither the daughter of Ephraim and the sister of Beriah [as shown in the table] or the daughter of Beriah the son of Ephraim."[4]

Table 17.1. Genealogies from Jacob to Moses (Exodus 6:16-20) and Joshua at the time of the exodus. This doubtful genealogy from Ephraim to Joshua is a tentative interpretation of 1 Chronicles 7:20-27.

```
                        Jacob
                          |
          ┌───────────────┴─────┐
         Levi                 Joseph
          |                     |
        Kohath               Ephraim
          |                     |
        Amram   Shuthelah  Ezer  Elead   Beriah   Sheerah (d.)
          |       |                        |
        Moses   Bered                    Rephah
                  |                        |
                Tahath                   Resheph
                  |                        |
                Eleadah                   Telah
                  |                        |
                Tahath                   Tahan
                  |                        |
                Zabad                    Ladan
                  |                        |
                Shuthelah               Ammihud
                                           |
                                        Elishama
                                           |
                                          Nun
                                           |
                                        Joshua
```

605

Adventures In Egypt

As discussed previously,[5] the period of time covered by these genealogies between when Jacob and his family entered Egypt and the exodus is probably about one hundred eighty-five years. If Kohath was a young son of Levi at the time of the entry into Egypt, then in the genealogy of Moses it could be inferred that Kohath and Amram may each have been about fifty years old at the births of their sons.[6] Similarly, the biblical chronology from Joseph through his son Manasseh to the time of the exodus is readily compatible with the genealogy of Moses for the same period.[7] Thus, for the genealogy through Manasseh a generation length would appear to have had the reasonable value of about forty years.

On the other hand, with the genealogy shown in Table 17.1 based on 1 Chronicles 7:20-27, a generation length in the Ephraim-Joshua line would have averaged roughly seventeen years. This result is unreasonable and suggests that, if the overall time interval of one hundred eighty-five years is about right, there must be one or more substantial errors in Joshua's family tree or its interpretation as shown in the table. In this context a brief alternative listing of the descendants of Ephraim could also be of interest:

> These are the sons of E'phraim according to their families: of Shuthe'lah, the family of the Shuthe'lahites; of Becher, the family of the Bech'erites; of Tahan, the family of the Ta'hanites. And these are the sons of Shuthe'lah: of Eran, the family of the E'ranites. These are the families of the sons of E'phraim according to their number, thirty-two thousand five hundred (Numbers 26:35-37).

17. Joshua and Jason

It is not clear how one could reconcile or interpret the genealogical information that has just been indicated. The reader may, of course, address these discrepancies as he or she wishes, but it will be assumed here that the sequence shown in the table is probably incorrect.

The only other suggestion that Joshua might be descended from Ephraim is his indicated participation as a representative of that tribe among the spies sent to investigate Canaan (Numbers 13:1-16). However, given Joshua's auspicious introduction as leader of the Hebrews in their battle with the Amalekites, one might have expected him to have been the leader/protector of the spying mission rather than only one of its members. Regarding one of the spy narratives, it has been stated that "it is not impossible that only Caleb is original here."[8] Thus, in spite of the spy reports and the genealogies, Joshua may not actually have been a descendant of Ephraim. The late and mysterious appearance of such a key figure in the exodus story, the later assignment to him of a father, and the still later placement of him and his father in an unreasonable family tree combine to suggest that essentially nothing is known with certainty about the background of Joshua. Thus, he may not even have been a conventional Hebrew. This consideration will continue in Section 17.7. First, however, it may be of interest to consider a different aspect of the family tree discussed above.

17.5 Trouble for Ephraim

Besides the genealogical implications of 1 Chronicles 7:20-27, as quoted in Section 17.4, those verses also report in the briefest possible way a disaster that struck the family of Ephraim. For convenience the relevant verses are quoted again here in an abbreviated form:

> The sons of E'phraim: . . . Ezer and E'le-ad, whom the men of Gath who were born in the land slew, because they came down to raid their cattle. And E'phraim their father mourned many days, and his brothers came to comfort him. And E'phraim went in to his wife, and she conceived and bore a son; and he called his name Beri'ah, because evil had befallen his house (1 Chronicles 7:20-23).

According to this story, Ezer and Elead, two of the sons of Ephraim, entered the vicinity of Gath for the purpose of stealing cattle. This raid was unsuccessful and Ezer and Elead were both killed by the inhabitants of Gath.

It should be noted that Gath was located near the coastal plain of Canaan and later became one of the chief cities of the Philistines, together with Gaza, Ashkelon, Ashdod, and Ekron. On the other hand, Ephraim was the second son of Joseph, and the biblical stories indicate clearly that he and his family lived in Egypt: "And to Joseph in the land of Egypt were born Manas'seh and E'phraim, whom As'enath, the daughter of Poti'phera the priest of On, bore to him" (Genesis 46:20). These sons also spent their childhood in Egypt: "And Jacob lived in the land of Egypt seventeen years" (Genesis 47:28). On his death-bed in Egypt Jacob said the following to Joseph: "And now your two sons, who were born to you in the land of Egypt before I came to you in Egypt, are mine; E'phraim and Manas'seh shall be mine. . ." (Genesis 48:5). From Table 13.1, Ephraim and Manasseh were each at least nineteen years old (seventeen plus two) and at most twenty-six years old (seventeen plus nine) in Egypt when Jacob

17. Joshua and Jason

died there. Thus, it would have been many years after Jacob's death when sons of Ephraim might have invaded Gath.

The Israelites are said to have migrated to Egypt because of the severe famine in Canaan. That famine may have been a consequence of a major volcanic eruption on the island of Thera.[9] On the other hand, the people known in the Bible as Philistines are said to have migrated to Canaan from Caphtor: "Did I not bring up Israel from the land of Egypt, and the Philistines from Caphtor and the Syrians from Kir?" (Amos 9:7). Thus, it is of interest to consider both the likely location of Caphtor and also the reason why the so-called Philistines or other "people of the sea" might have departed from that location:

> In Egyptian sources a place name Keftiu (kftyw or kftiw) is found in texts from 2200 to 1200 B.C. It is commonly accepted among Egyptologists that keftiu is the Egyptian form of Kaptara/Caphtor. On the basis of geographical, historical, and literary considerations it is clear that the island of Crete, with which Egypt had commercial relations after ca. 2200, is meant by this term.[10]

As discussed previously, an eruption of the Thera volcano in about 1628 B.C. probably caused the famine that led to the migration of Jacob and his family from Canaan to Egypt,[11] and a smaller eruption may have been associated with the famine that caused the earlier journey of Abram and his household to Egypt, perhaps in about 1728 B.C. as discussed in Section 8.2.[12] These eruptions or others in about the same locations could have caused substantial changes to civilizations that existed in Mediterranean areas near or downwind of the events. Many of the people that

Adventures In Egypt

would have been impacted by such eruptions would have had no way to escape from the subsequent storms, toxic vapors, and destruction of food sources. On the other hand, members of sea-based civilizations ("people of the sea") would have had boats that could have carried many of them to Crete, Canaan, and other more hospitable environments. Thus, events similar to those that caused the temporary migrations of early Hebrews to Egypt could have been responsible for migrations of other societies as well. There are many Bible-related references to the Philistines beginning with stories of their interactions with Abraham and Isaac (Genesis 21:22-34; 26:1-33, Jubilees 24:18-33). That the Philistines were originally from Caphtor is mentioned explicitly in Jubilees 24:30. One of the earlier events in the three-stage Thera eruption sequence (Section 8.2) might have brought the Philistines to Canaan. In any case they are said to have been present there when Abram returned from Egypt in about 1722 B.C. (Section 13.2).

If the famine that led to the migration of the Israelites to Egypt was indeed caused by a volcanic eruption, then it follows that with the easing of the famine the Israelites might have been contemplating a return to their ancestral homeland in Canaan. Such a return could explain the reported presence of the children of Ephraim back in Canaan. Unfortunately, their time of return is reported to have violated instructions of the Lord (see quotations below), and the Ephraimites seem not to have been adequately prepared to deal with the inhabitants of the land.

A more detailed report of the conflict between the children of Ephraim and the men of Gath is provided by *Jasher*. A few verses from that source are quoted here:

17. Joshua and Jason

> And these men [of Gath] were engaged in battle with the children of Ephraim, and the Lord delivered the children of Ephraim into the hands of the Philistines. And they smote all the children of Ephraim, all who had gone forth from Egypt, none were remaining but ten men who had run away from the engagement. For this evil was from the Lord against the children of Ephraim, for they transgressed the word of the Lord in going forth from Egypt, before the period had arrived which the Lord in the days of old had appointed to Israel.
>
> And of the Philistines also there fell a great many . . , and their brethren carried them and buried them in their cities. And the slain of the children of Ephraim remained forsaken in the valley of Gath for many days and years, and were not brought to burial, and the valley was filled with men's bones.
>
> And the men who had escaped from the battle came to Egypt, and told all the children of Israel all that had befallen them. And their father Ephraim mourned over them for many days, and his brethren came to console him. And he came unto his wife and she bare a son, and he called his name Beriah, for she was unfortunate in his house.[13]

The implication of these verses is that it was not wrong for the children of Ephraim to wish to return to their ancestral homes in Canaan. The problem is that they didn't wait for the time that the Lord had appointed for their return, i.e., the time of the exodus.

As quoted above, the Philistines who were killed were buried in their cities, and evidence of Bronze Age tombs has been found in Gath.[14] It is perhaps of interest that in *Jasher*'s version the slain of the children of Ephraim were not buried for many years. As a result, the valley of Gath was said to be filled with men's bones. This circumstance brings to mind a different story about a valley filled with bones:

> The hand of the Lord was upon me, and he brought me out by the spirit of the Lord, and set me down in the midst of the valley; it was full of bones. And he led me round among them; and behold, there were very many upon the valley; and lo, they were very dry. . . . Then he said to me, "Son of man, these bones are the whole house of Israel. Behold, they say, 'Our bones are dried up, and our hope is lost; we are clean cut off.' Therefore prophesy, and say to them, Thus says the Lord God: Behold, I will open your graves, and raise you from your graves, O my people; and I will bring you home into the land of Israel (Ezekiel 37:1-2, 11-12).

This story continues with the assurance that the people whose bones these were could be brought back to life, and in a similar way the discouraged Israelites in captivity would be restored to their homes in the land of Israel.

As a final point, it may be noticed that in *Jasher's* version of the destruction of the children of Ephraim, as quoted above, Ephraim's brothers "came to console him." It is usually understood that Joseph and his wife Asenath had two sons, Manasseh and Ephraim (Genesis 41:50-52). Thus, we might expect that Ephraim could only have had one brother rather than brothers (plural) as in *Jasher*. From another point of view, however, all of Joseph's brothers could perhaps be considered as brothers of Ephraim. On his deathbed Jacob spoke to Joseph:

> And Jacob said to Joseph, "God Almighty appeared to me at Luz in the land of Canaan and blessed me, and said to me 'Behold, I will make you fruitful, and multiply you,

17. Joshua and Jason

and I will make of you a company of peoples, and will give this land to your descendants after you for an everlasting possession.' And now your two sons, who were born to you in the land of Egypt before I came to you in Egypt, are mine; E'phraim and Manas'seh shall be mine, as Reuben and Simeon are. And the offspring born to you after them shall be yours; they shall be called by the name of their brothers in their inheritance (Genesis 48:3-6).

Thus, Ephraim and Manasseh would seem each to have been given a full share of the inheritance of Jacob among his natural children. This may have represented a double portion for Joseph, while Jacob's oldest son Reuben was deprived of his birthright:

The sons of Reuben the first-born of Israel (for he was the first-born; but because he polluted his father's couch, his birthright was given to the sons of Joseph the son of Israel, so that he is not enrolled in the genealogy according to the birthright; though Judah became strong among his brothers and a prince was from him, yet the birthright belonged to Joseph (1 Chronicles 5:1-2).

Jacob also suggested the possibility of Joseph having other sons after Manasseh and Ephraim: "And the offspring born to you [Joseph] after them [Manasseh and Ephraim] shall be yours; they shall be called by the name of their brothers in their inheritance" (Genesis 48:6). Perhaps these additional sons and daughters, if any, would not be given a separate inheritance as Manasseh and Ephraim were but rather would be counted among their families. As noted above, Ephraim is said in *Jasher* to have been consoled by his brothers following the deaths of his children at the hands of the men of Gath. In view of the elevated status of

Ephraim and Manasseh considered above, the other "brothers" could in principle have included other sons of Jacob.

The idea of the sons of Ephraim being killed for attempting to steal the cattle or sheep of the men of Gath has a parallel in the story of the Argonauts in Egypt (or Libya). One of the Argonauts named Canthus attempted to steal sheep to help provide sustenance for his fellow Argonauts. Unfortunately for Canthus, the shepherd was able to defend himself:

> But thee, Canthus, the fates of death seized in Libya. On pasturing flocks didst thou light; and there followed a shepherd who, in defence of his own sheep, while thou wert leading them off to thy comrades in their need, slew thee by the cast of a stone; for he was no weakling, Caphaurus, . . . who on that day in defending his sheep slew Canthus. But he escaped not the chieftains' avenging hands, when they learned the deed he had done. And the Minyae [Argonauts], when they knew it, afterwards took up the corpse [of Canthus] and buried it in the earth, mourning; and the sheep they took with them.[15]

It is possible that, as with other stories mentioned here, the Bible-related account of the deaths of the sons of Ephraim could have provided part of the inspiration for the Argonaut story of the death of Canthus. On the other hand, it was probably not a particularly uncommon occurrence for hungry adventurers like the sons of Ephraim or the Argonauts to sometimes have to depend for survival on animals or other food items that they could steal from native populations.

17. Joshua and Jason

17.6 The Amalekite question

Because of the importance of the Israelite conflict with the Amalekites, it may be worthwhile to consider some implications of this event. The war with Amalek did not end with the first battle:

> And the Lord said to Moses, "Write this as a memorial in a book and recite it in the ears of Joshua, that I will utterly blot out the remembrance of Am'alek from under heaven." And Moses built an altar and called the name of it, The Lord is my banner, saying, "A hand upon the banner of the Lord! The Lord will have war with Am'alek from generation to generation" (Exodus 17:14-16).

These instructions and subsequent events were taken as justification for the final extermination of the Amalekites and occupation of their lands during the reign of Hezekiah, more than seven hundred years after the exodus:

> And some of them, five hundred men of the Simeonites, went to Mount Se'ir, having as their leaders Pelati'ah, Neari'ah, Rephai'ah, and Uz'ziel, the sons of Ishi; and they destroyed the remnant of the Amal'ekites that had escaped, and they have dwelt there to this day (1 Chronicles 4:42-43).

In spite of the above, Jewish hatred of any people that they declare to be Amalekites has continued into more modern times:

> In 1839 the British missionary Joseph Wolff, who was active in both Palestine and Yemen, found it "remarkable that the Armenians, who are detested by the Jews as the supposed descendants of the Amalekites, are the only

Christian church who have interested themselves for the protection and conversion of the Jew." Similarly in their 1842 account of their extensive missionary efforts among Jews in both Europe and the Middle East, the Scottish missionaries Bonar and McCheyne suggested that "the peculiar hatred which the Jews bear toward the Armenians may arise from a charge often brought against them, namely that Haman was an Armenian, and that the Armenians were the Amalekites of the Bible, attributing this to the fact that Armenians were the first nation to adopt Christianity in 301 AD."[16]

The Jews seem to have shown little sympathy during the Armenian genocide. Regarding their attitude, the Israeli historian Yair Auron said the following: "A slight grimace on their lips, a short heartfelt sigh, and nothing more. The Armenians are not Jews, and according to folk tradition the Armenians are nothing more than Amaleks! Amaleks? We would give them help? To whom? To Amaleks? Heaven forbid!"[17] As mentioned in Section 4.3, about one and a half million Armenians were killed by Ottoman Turks in the years preceding, during, and after World War I in the event referred to as the Armenian Genocide. The genocide of Armenian Christians was of great interest to Adolph Hitler, and this event may have served in many ways as a model and inspiration for his genocidal policies toward the Jews and others.[18]

If the Armenians could have been declared by the Jews to be Amalekites, others could as well. Thus, in recent years it sometimes has been taught in schools in Israel that the Palestinians are Amalekites:

17. Joshua and Jason

In February 1980, Rabbi Yisrael Hess, the former campus rabbi of Bar-Ilan University, published an article in the student bulletin Bat Kol, the title of which, 'The Genocide Commandment in the Torah' (in Hebrew, 'Mitzvat Hagenocide Batorah') leaves no place for ambiguity. The article ends with the following: 'The day is not far when we shall all be called to this holy war, this commandment of the annihilation of Amalek.' Hess quotes the biblical commandment according to which he believes Israel, in the tradition of Joshua from biblical times, should act: 'Go and strike down Amalek; put him under the ban with all that he possesses. Do not spare him, but kill man and woman, baby and suckling, ox and sheep, camel and donkey.'. . . Clearly, for Hess, Amalek is synonymous with the Palestinian Arabs, who have a conflict with Israeli Jews, and they must be 'annihilated', including women, children and infants.[19]

It would seem that in a strategy of expansion one might first define an imagined enemy or potential victim to be an Amalakite, and then one can claim justification in following the ancient biblical imperative to kill all Amalekites.

On the other hand, it has also been suggested that the actual historical Amalekites could more easily be interpreted as the victims in their encounter with the Hebrews than as the aggressors:

> . . . The Amalekites could well be regarded as the archetypal victims in the Pentateuch, in that divine instructions to dispose of this people are given on more than one occasion. No doubt other peoples were to be eliminated from the 'Promised Land', but no single group acquired quite the same status as them. They also symbolize a further classic device: the rhetorical move – very familiar from the twentieth century – of portraying

the victim as aggressor in order to justify his/her elimination. This resonates, surely, with the experiences of Jews in Germany and Poland, Blacks in north America and South Africa, Asians in Africa, Muslims in Bosnia, Armenians in Turkey, Christians in Pakistan, Algerians in France, Pakistanis in Britain, and innumerable others – all labelled as a threat and a danger to the dominant culture and then victimized as a punishment. The biblical evidence is unambiguous: thus Exodus 17:14-16 (after Moses' defeat of the Amalekites), reports that

"The Lord said to Moses, 'Write this as a reminder in a book and recite it in the hearing of Joshua: I will utterly blot out the remembrance of Amalek from under heaven'. And Moses built an altar and called it, The Lord is my banner. He said, 'A hand upon the banner of the Lord! The Lord will have war with Amalek from generation to generation.' "

The more frequently quoted passage, in Deuteronomy 25:17-19, is where the denunciation of the victim as the aggressor is most explicitly found:

"Remember what Amalek did to you on your journey out of Egypt, how he attacked you on the way, when you were faint and weary, and struck down all who lagged behind you; he did not fear God. Therefore when the Lord your God has given you rest from all your enemies on every hand, in the land that the Lord your God is giving you as an inheritance to possess, you shall blot out the remembrance of Amalek from under heaven; do not forget."

Almost every commentary on this passage (including a range of Christian websites which would try the patience of a saint) (1) assumes that the Bible's account is factual, and (2) reads that account as an unproblematic description of unprovoked attack. But surely even in the Bible's own terms the story is of a vast army of people making incursions into a territory which is not theirs and no doubt making huge demands on the economic resources of a region not famous for agricultural surpluses. The Amalekites' defence of their home territory is hardly surprising: how odd, then, that we should so readily accept the Deuteronomists' claim that the victim is the aggressor.[20]

An economic resource of particular interest to both the Israelites and the Amalekites would have been drinking water for people and cattle, as discussed above in Section 17.2.

The date and circumstances of the writing of Deuteronomy are, as with Chronicles considered previously, not entirely understood. It is, however, common to associate the authorship and advancement of this work with the era of King Josiah,[21] c. 637/36 - 605 B.C.[22] This timing would be about eight hundred years after the exodus-related events that are referred to, and historical accuracy should not be expected. The political independence of the kingdom of Judah had been ended when Sennacherib forced Hezekiah to start paying tribute to Assyria in about 701 B.C.:[23]

> If Josiah, after this catastrophe, wished to make himself politically independent, he had to fall back upon the old method of procedure, conscription of the free peasants; for the building of an effective mercenary force was much too expensive for the empty coffers of the state. Actually it can be proved by a series of statements from the historical work of Deuteronomy and Chronicles that Josiah in his striving for political expansion did go back to this ancient form of military organization.[24] Since

Deuteronomy must be connected anyway with the events under Josiah, it is natural to connect the warlike spirit of Deuteronomy, which emerged so abruptly, with this reorganization. Up until that time politics and waging war had been a concern of the king and his officials, officers, and mercenaries. With the calling out of the militia, forces suddenly came into a central position which had been excluded for centuries: old traditions from the holy wars, as they were waged in the days before statehood, came alive again and were adapted in a makeshift fashion to the demands of the new age.

Thus, Deuteronomy may have been created and promoted, not primarily as a historical document, but as a motivational instrument to bring the people back onto a war footing. The commandments to kill all Amalekites (after they were said already to be extinct) could have been part of an effort to re-ignite the religious fervor of the ancient holy wars.

One could hope that these Old Testament commands of doubtful origin would never again be a sufficient basis for the self-serving destruction of innocent people. Replacement commands with more authority have since been given:

> You have heard that it was said, 'You shall love your neighbor and hate your enemy.' But I say to you, love your enemies, bless those who curse you, do good to those who hate you, and pray for those who spitefully use you and persecute you, that you may be sons of your Father in heaven (Matthew 5:43-45 NKJV).

17.7 Joshua and Jason

Considerations in the previous chapter have concerned possible relationships between the life and times of Heracles, as

17. Joshua and Jason

represented by Apollonius Rhodius and others, and of Moses as represented in the Bible. A somewhat similar possibility of relationships between Joshua of Bible-related texts and Jason of the *Argonautica* will be considered here. As is well known, Joshua in the Bible was a younger contemporary of Moses, who lived after Moses's death to lead the Israelites into Canaan. Similarly, Jason was a younger contemporary of Heracles. In the same way that we have considered parallels between the activities of Heracles and Moses, it could seem reasonable to look for connections between the lives of Jason and Joshua.

A. Names

As a starting point, it is worthwhile to consider directly the names Joshua and Jason. The name Joshua appears in various forms in the Bible, depending in part on the translation but including at least Hoshea (Numbers 13:8, 16; Deuteronomy 32:44 RSV), Jehoshua (Numbers 13:16 KJV), Jehoshuah (I Chronicles 7:27 KJV), Oshea (Numbers 13:8, 16 KJV), Jesus (Acts 7:45 KJV; Hebrews 4:8 KJV), Jeshua (Nehemiah 8:17 RSV).[25] The differences between these names are not necessarily significant, and in one text we find the following words: "And Moses called Hoshea the son of Nun Joshua" (Numbers 13:16). The name Joshua is often understood to mean Jehovah is salvation.

On the other hand, the name Jason is of course Greek, and its meaning is understood to be "healer," perhaps not so different from the interpretation of the name Joshua or Jesus as "savior": "[Y]ou shall call his name Jesus, for he will save his people from their sins" (Matthew 1:21). In modern usage the names Jason and

Adventures In Egypt

Joshua are considered variants of each other. In earlier times the similarity of the appearance and the interpretation of the names Joshua and Jason, may also have led to a degree of interchangeability between them. Sometimes for their own safety or political advantage Jews of the past sought to disguise or disown their Jewish ancestry by changing their names to Greek names and adopting the Greek way of living. Thus, Jason is said to be a "Greek name assumed by Jews who bore the Hebrew name Joshua."[26] An example of a person called Jesus changing his name to the Greek Jason (and his brother Onias changing his name to the Greek Menelaus) is provided by Josephus:

> About this time, upon the death of Onias the high priest, they gave the high priesthood to Jesus his brother. . . . But this Jesus, who was the brother of Onias, was deprived of the high priesthood by the king, who was angry with him, and gave it to his younger brother, whose name also was Onias; for Simon had these three sons, to each of which the priesthood came. . . . This Jesus changed his name to Jason, but Onias was called Menelaus. Now as the former high priest, Jesus, raised a sedition against Menelaus, who was ordained after him, the multitude were divided between them both. And the sons of Tobias took the part of Menelaus, but the greater part of the people assisted Jason; and by that means Menelaus and the sons of Tobias were distressed, and retired to Antiochus, and informed him that they were desirous to leave the laws of their country, and the Jewish way of living according to them, and to follow the king's laws, and the Grecian way of living. . . . Accordingly, they left off all the customs that belonged to their own country, and imitated the practices of the other nations.[27]

17. Joshua and Jason

In the same way that a Jew named Joshua might have changed his name to Jason, one could imagine that a Greek named Jason who chose to live among Israelites might have changed his name to Joshua.

B. Did Jason become Joshua?

The above comments lead back to the question of who exactly was the abruptly appearing person named Joshua, and some readers might already be wincing at the inevitable suggestion: Maybe the similarity between the names Joshua and Jason could be seen as a hint that Joshua of the exodus stories and Jason of the Argonaut stories were the same person.

After washing ashore near Egypt, Jason helped direct the transport of the Argo to an escape route involving the Tritonian lake at the mouth of the Triton (Nile) River in Egypt. In their desperate search for water the Argonauts came upon the place where Heracles had obtained drinking water by striking a rock. This is reminiscent to the modern reader of how Moses in the same geographical area obtained drinking water for the Israelites by striking a rock. This similarity suggests at a minimum that Apollonius, with access to the best collection of historical records in Egypt, may have adapted dramatic stories from Egyptian and Bible-related histories to enrich his own epic drama. If not done with sufficient care, however, this enrichment based on Jewish history could have made the resulting work offensive to some Greek readers and thus also could have resulted in Apollonius's flight to Rhodes.

After satisfying their physical thirsts, the Argonauts sent out searchers to find Heracles (as perhaps represented in the Bible by

Adventures In Egypt

Moses), their greatest hero and former comrade. After tentatively sighting him in the distance, they inexplicably abandoned that quest and instead continued on their journey back to Greece. An alternative ending to the story could have been that Jason himself, and perhaps other of the Argonauts, actually caught up with Heracles/Moses and joined him in his own quest. As noted in Section 16.11, Herodotus described the efforts by Jason to ensure that his shipmen could find their way safely out of the Tritonian lake and back to their homes among the Greeks. The curious wording employed by Herodotus may be understood to suggest that Jason himself chose to stay in Egypt rather than returning home with his comrades.

It seems that the warrior Jason (as Joshua) may have preferred to rejoin his faithful friend Heracles (Moses to us) in leading the Hebrews toward their promised land. Concerning this friendship, Jason and Heracles had each endorsed the other to be the leader of the Argonaut expedition as discussed in Section 16.7. However, including an ending to the story in which established Greek heroes were closely associated with the Hebrews might not have been satisfactory to Greek or Egyptian readers of the Argonaut story at the time in which it was written.

C. Who was Nun?

One possible problem in identifying Joshua with Jason is that, once Joshua is said to be a son of Nun, that name is used commonly in the Bible whenever Joshua is mentioned. It could be worth considering that Joshua may in some sense have been a son of Nun, but Nun may not have been a Hebrew. If Nun was not a Hebrew name, the next possibility to consider is that it could be an Egyptian name. In fact, the name Nun would not necessarily represent the actual name of the earthly father of a young Hebrew (or other nationality) man named Joshua (or Jason). No person named Nun is mentioned in the Bible besides the father of Joshua, whose late and implausible biblical ancestry considered in Section 17.4 was given only once. It also may be remarked that any family such as a wife or children that Joshua may have had seem not to be suggested in the Bible.

In the present context it may be noted that Nun is also the name of the oldest Egyptian god. His name meant primeval waters or seas (or chaotic waters, or stormy waters), and he was the father of Re the sun god.[28] Thus, the name or appellation "son of Nun" could even have been something of a nickname meaning the "son of stormy waters" or perhaps more simply the "son of the sea." This name would be especially appropriate if Joshua's origin had something to do with the sea and his human father was unknown to a biblical author familiar with Egypt. After Jason and the Argonauts washed ashore in Egypt, as discussed in Section 15.4, Jason's actual human ancestry would likely have been unknown and irrelevant to the Hebrews. In the absence of an ancestry and in view of his apparent origin, it might have occurred to them to refer to him as Joshua the "son of the sea."

D. Age of Joshua

Besides the similarity and sometimes interchangeability of the names Joshua and Jason, it may be of interest to consider other similarities between the appearances and activities of these men as reported in biblical sources and the *Argonautica*. One parallel concerns the age of Joshua at his appearance in the biblical record. After Joshua's arrival on the scene, the wilderness wandering of the Hebrews lasted about forty years; and Joshua had a substantial career after that, leading the Hebrews in their initial conquest of much of Canaan. Therefore, he could not have been very old at his first appearance. This is confirmed in the exodus account: "Thus the Lord used to speak to Moses face to face, as a man speaks to his friend. When Moses turned again into the camp, his servant Joshua the son of Nun, a young man, did not depart from the tent" (Exodus 33:11). The key words for this purpose are that Joshua was said to be "a young man," and this conforms well to the rest of the biblical story. As suggested in Section 17.3, Joshua may also have been the "young man" mentioned in Numbers 11:27-30.

The corresponding question about Jason concerns his age at the time of the Argonauts' expedition. Conveniently, Apollonius provides us with approximate information on this subject:

> Now the Doliones and [their king] Cyzicus himself all came together to meet them [the Argonauts] with friendliness, and when they knew of the quest and their lineage welcomed them with hospitality, and persuaded them to row further and to fasten their ship's hawsers at the city harbour. . . . As with Jason, the soft down was just blooming on his [the king's] chin . . . .[29].

In short, King Cyzicus and Jason seem both to have been teenagers at a time near the beginning of the Argonauts' adventures. It isn't clear how long these adventures continued, but Jason would seem likely still to have been a "young man" at his arrival in Egypt.

17. Joshua and Jason

If Jason, like Joshua, was a young man at the time of the exodus, then he might have satisfied the criterion that only those less than twenty years old at that time could enter the promised land:

> And the Lord's anger was kindled on that day, and he swore, saying, "Surely none of the men who came up out of Egypt, from twenty years old and upward, shall see the land which I swore to give to Abraham, to Isaac, and to Jacob, because they have not wholly followed me; none except Caleb the son of Jephun'neh the Ken'izzite and Joshua the son of Nun, for they have wholly followed the Lord . . . (Numbers 32:10-12).

It is curious that at the same time this entrance criterion was established (which Joshua may have met) a waiver of the criterion was created for both Joshua and Caleb.

It might also be noted that if Jason was a typical Greek warrior he would probably not initially have been fluent in either the Hebrew or the Egyptian language. In that circumstance it would be of interest to consider how he might have communicated with the exodus refugees from Egypt. It may be noted that almost all of Joshua's initial verbal communication seems to have been with Moses himself. Several of their conversations were quoted above in Section 17.3. Moses as a prince of Egypt is likely to have received a broad and excellent education. There are also many indications that he served as a priest of Osiris in Heliopolis.[30] Moses's ancestor Abram is said to have lived with those priests during his visit to Egypt in a time of famine.[31] About a century later, Abraham's great grandson Joseph married the daughter of the priest of Heliopolis: "And

Pharaoh called Joseph's name Zaph'enath-pane'ah; and he gave him in marriage As'enath, the daughter of Poti'phera priest of On [Heliopolis]" (Genesis 41:45). It is clear from the Bible and other texts that Moses himself was well-trained in astronomy and its use in maintaining the calendar system of Egypt. Moses's design for the tabernacle and its orientation facing east may have been modeled on the prayer-houses of Heliopolis.[32]

Moses probably received training in many other fields.[33] He is said to have been a scribe in Egypt, and the first books of the Bible have always been understood to owe much to the reading and writing ability of Moses. He could translate hieroglyphics and was probably fluent in several of the languages of his time. He had received legal training, and many early biblical laws have close parallels in other ancient law codes. Moses also seems likely to have received medical training, and studies in military strategy and warfare were probably requirements for a potential king.

Given his broad education, Moses would almost certainly have been able to communicate well with an Argonaut. If his activities were in any way to be associated with those of the person known as Heracles, Moses would already have been well-acquainted with Jason and his colleagues before their arrival in Egypt.

E. Style of fighting

Among Joshua's especially relevant talents were his abilities as a warrior and leader. This is, of course, entirely consistent with Jason's role as leader of the Argonauts. It is tempting to go even further and compare what little is known about the fighting styles

17. Joshua and Jason

of Joshua and Jason. As noted in Section 17.2, Joshua's approach to battle was very direct. His victory was achieved in one day "until the going down of the sun. And Joshua mowed down Am'alek and his people with the edge of the sword" (Exodus 17:12-13). The image suggested there is of Joshua mowing down enemy soldiers with his sword as a farmer might cut his crops with a sickle until sunset. Thus, it would be of interest to see if any information is available on Jason's approach to battle. Fortunately, Apollonius provides an answer to this inquiry. In one of the more dramatic episodes of the *Argonautica*, the "earthborn men" were fully armed supernatural soldiers who sprang from the ground and attempted to destroy Jason, fighting alone in a ferocious one-day battle:

> And as when abundant snow has fallen on the earth and the storm blasts have dispersed the wintry clouds under the murky night, and all the host of the stars appear shining through the gloom; so did those warriors shine springing up above the earth. . . . And even as a fiery star leaps from heaven, trailing a furrow of light, a portent to men, whoever see it darting with a gleam through the dusky sky; in such wise did Aeson's son [Jason] rush upon the earthborn men, and he drew from the sheath his bare sword, and smote here and there, mowing them down, many on the belly and side, half risen to the air – and some that had risen as far as the shoulders – and some just standing upright, and others even now rushing to battle. And as when a fight is stirred up concerning boundaries, and a husbandman, in fear lest they should ravage his fields, seizes in his hand a curved sickle, newly sharpened, and hastily cuts the unripe crop, and waits not for it to be parched in due season by the beams of the sun; so at that time did Jason cut down the crop of the

Earthborn; and the furrows were filled with blood, as the channels of a spring with water. And they fell, some on their faces biting the rough clod of earth with their teeth, some on their backs, and others on their hands and sides, like to sea-monsters to behold. And many, smitten before raising their feet from the earth, bowed down as far to the ground as they had risen to the air, and rested there with the damp of death on their brows. . . . And the day died, and Jason's contest was ended.[34]

It is striking that in both the story of Joshua and the Amalekites and the story of Jason and the earthborn men the same symbolism is employed of the enemy being mowed down "until the going down of the sun" (Joshua) or until "the day died" (Jason).

The above quotation from the *Argonautica* suggests that Apollonius or his sources may have had personal experience in astronomy, including the observation of meteors or "shooting stars." Thus, the first sentences make an analogy between the appearance of the shining earthborn men emerging from the ground and shining stars seen in a murky night sky of winter. Then, as a shooting star leaves a trail of light across the sky, Jason left a trail of bodies as he mowed down the earthborn men. A photograph of the trail of the vaporizing Chelyabinsk meteor as it crossed a winter sky in Russia in 2013 may be found in the reference for Table 9.1.

F. Royalty of Joshua

In the *Argonautica* story Aeson had been the rightful king of Iolcus in Thessaly. However, Aeson had been overthrown but not killed by his aggressive half-brother Pelias. Many years later Aeson's son Jason was a strong young man and was determined

17. Joshua and Jason

to gain the throne that was rightfully his. Pelias told Jason that he could have the throne only if he would recover the long-stolen golden fleece.

To the extent that Jason might correspond to Joshua, the warrior Joshua should also have exhibited kingly attributes and ambitions. His military skills and leadership abilities, as already discussed, would certainly be compatible with him having or obtaining a role as king. The kingly aspects of Joshua's career have also been noted previously:

> The figure of Joshua has also been invested with a certain royal aura. He is described as having the "spirit" (Numbers 27:18; Deuteronomy 34:9; cf. 1 Samuel 10:10; 16:13; Isaiah 11:2). He is called the "servant of Yahweh" (Joshua 24:29). He calls the people together for the makng of a covenant (Joshua 24:1; cf. 2 Samuel 5:3). His name is said to have been changed from Hoshea to Joshua (Numbers 13:16), recalling the common royal practice of assuming a regnal name (see King). His division of the land is parallel to the king's erection of administrative districts (1 Kings 4:7-19). Likewise, Joshua is called upon to decide claims (cf. Joshua 14:6-15); 17:4, 14-18), a function of the judge which also fell to the king (cf. Isaiah 11:3b-4). The priestly glossators have heightened this atmosphere by providing Joshua with a priestly associate, Eleazar.[35]

Thus, the rightful heir to the throne of Iolcus may have preferred instead to earn and accept a leadership position among the Hebrews. The potential of such a position may have been suggested by Jason's friend and former shipmate Heracles, assuming that he had returned to Egypt to resume his quest as Moses to lead his people the Hebrews to freedom. Moses was much older than Jason and was probably anxious to arrange for his own replacement by a person that he personally knew would be an outstanding leader.

17.8 End of the *Argonautica*
The actual ending of the *Argonautica* included in the final version by Apollonius occupies only a few pages at the end of Book IV of his work. This conclusion may seem to the reader of that work to be something of a letdown, compared to the detail and drama of the earlier part of the story. The Argonauts boarded their ship Argo in Egypt and, as indicated in Section 16.11, found a way to get from the Tritonian lake into the Mediterranean Sea. In the version by Herodotus, Jason is mentioned but may not have been planning to return to Greece. The geography is not difficult to visualize. Thus, the Tritonian lake could be imagined as an ancient parallel to one of the lakes that even today occur in the Nile Delta and drain into the Mediterranean Sea.

When the Argonauts arrived near Crete, probably without Jason, they encountered Talos, the giant man of bronze, who prevented them from landing by throwing rocks at them:

> And Talos, the man of bronze, as he broke off rocks from the hard cliff, stayed them from fastening hawsers to the shore, when they came to the roadstead of Dicte's haven. He was of the stock of bronze, of the men sprung from ash-trees, the last left among the sons of the gods; and the son of Cronos gave him to Europa to be the warder of Crete and to stride round the island thrice a day with his feet of bronze. Now in all the rest of his body and limbs was he fashioned of bronze and invulnerable; but beneath the sinew by his ankle was a blood-red vein; and

this, with its issues of life and death, was covered by a thin skin. So the heroes, though outworn with toil, quickly backed their ship from the land in sore dismay. And now far from Crete would they have been borne in wretched plight, distressed both by thirst and pain, had not Medea addressed them as they turned away.[36]

After the Argonauts left Egypt, it seems that Medea had in effect become their leader. She volunteered to subdue Talos by means of her sorcery, and her efforts were successful:

So Talos, for all his frame of bronze, yielded the victory to the might of Medea the sorceress. And as he was heaving massy rocks to stay them from reaching the haven, he grazed his ankle on a pointed crag; and the ichor gushed forth like melted lead; and not long thereafter did he stand towering on the jutting cliff. But even as some huge pine, high up on the mountains, which woodmen have left half hewn through by their sharp axes when they returned from the forest – at first it shivers in the wind by night, then at last snaps at the stump and crashes down; so Talos for a while stood on his tireless feet, swaying to and fro, then at last, all strengthless, fell with a mighty thud. For that night there in Crete the heroes lay; then, just as dawn was growing bright, they built a shrine to Minoan Athena, and drew water and went aboard, so that first of all they might by rowing pass beyond Salmone's height.[37]

The story of Talos might seem at a first to be clearly and entirely mythical. However, several aspects of the description of Talos could possibly be interpreted in terms of volcanic phenomena. According to Apollonius, Talos was encountered as the Argonauts were preparing to land on what is said to have been the island of Crete. Talos was reported to have been a huge man of bronze who threw rocks at their ship. An erupting volcano on an island could, of course, be said to be throwing rocks. An early discussion of the interpretation of Talos in terms of the eruption of the island of Thera was given by Luce:[38]

> To return to Talos: what can one make of this bronze warder who hurls rocks at ships trying to sail to Crete? Is he simply a figure of folk-tale and imagination? Or can he be rationalized? Talos has been explained as the Minoan sun-god, and the all-seeing sun always makes a good watchman. But other features, such as the stone-throwing, the ankle vein, and the collapse and death of the giant are not at all appropriate to a solar myth. A quite different and very ingenious explanation was advanced by J. Schoo.[39] Schoo suggested that the figure of Talos embodies an early Greek memory of the Thera volcano. Thera 'guards' the northern approaches to Crete which would have been used by the early Mycenaean sailors. His frame of 'unbreakable bronze' represents the wall of the newly formed crater on the mountain peak of Thera as it then was. The rocks which he throws are the 'bombs' shot from the vent of the volcano. His 'heel' is a subsidiary volcano on the coast of the island, like Cape Kolumbo or Cape Mavrorachidi. He collapses and becomes quiescent when all his ichor has flowed out like 'molten lead' – a reminiscence of the cooling off of lava streams after the end of an eruption. Finally like the Cyclops (another stone-thrower), he is left with a great blind eye when the caldera has been formed.

Another volcanic phenomenon could, at a time of poor visibility and vivid imagination, be thought to resemble a giant man of bronze. Under some eruption conditions, one of the more conspicuous effects is in the form of a rising plug or spine of

17. Joshua and Jason

viscous lava extruding at the summit crater. Depending on the rate at which the spine is forced upward from a crater, it might still contain molten lava in its interior. For higher rates of rise, lava could begin to leak out at the base of the spine, leaving nothing to prevent the spine's fall to one side. These circumstances would account in some measure for the description of Talos by Apollonius.

One of the better-documented eruptions illustrating several of the phenonena just described was the 1902 eruption of Mt. Pelée on the island of Martinique in the West Indies. A few brief sentences on this eruption and a subsequent eruption are reproduced here:

> On April 25, [1902,] the mountain emitted a large cloud containing rocks and ashes from its top, where the Étang Sec caldera was located. The ejected material did not cause a significant amount of damage.

> The main eruption:
> On May 8, 1902, Ascension Day, a volcanic eruption destroyed the town of Saint-Pierre, about 6.4 kilometres (4.0 mi) south of the summit. [The eruption killed about 30,000 people.]

> Beginning in October 1902, a dramatic volcanic spine grew from the crater floor in the Étang Sec crater, reaching a maximum width of about 100 to 150 m (300 to 500 ft) and a height of about 300 m (1,000 ft). Called the "Needle of Pelée" or "Pelée's Tower", this extraordinary volcanic feature rose up to 15 m (50 ft) a day, and became twice the height of the Washington Monument and more or less the same volume as the Great Pyramid of Egypt. It became unstable and collapsed into a pile of rubble in March 1903, after 5 months of growth.

On September 16, 1929, Mount Pelée began to erupt again. This time, there was no hesitation on the part of authorities and the danger area was immediately evacuated. Although there were pyroclastic flows, the activity was not as violent as the 1902 activity. It culminated in another "spine" or lava plug, albeit smaller than the 1902 plug, being emplaced at the summit. The activity ended in late 1932.[40]

The rock throwing by Talos, as reported by Apollonius, may have been suggested by a fairly ordinary effect often occurring in volcanic eruptions and reported explicitly for the Mt. Pelée eruption of 1902. An earlier eruption of Mt. Nemrud near the land of Eden was mentioned previously in Section 6.7 and provides a more detailed and colorful description of this and other eruption effects:

> In 1441 [A.D.] a great sign took place, for the mountain called Nemrud, which lies between Kelath and Bitlis, suddenly began to rumble like heavy thunder. This set the whole land into terror and consternation, for one saw that the mountain was rent asunder to the breadth of a city; and from out of this cleft flames arose, shrouded in dense, whirling smoke, of so evil a stench that men fell ill by reason of the deadly smell. Red-hot stones glowed in the terrible flames, and boulders of enormous size were hurled aloft with peals of thunder. Even in other provinces men saw all this distinctly.[41]

The hurling aloft of "boulders of enormous size" would certainly justify the response of the Argonauts in their ship, as quoted previously: "So the heroes, though outworn with toil, quickly backed their ship from the land in sore dismay."

17. Joshua and Jason

Less common, but prominent after the main eruption of Mt. Pelée, was the gradual growth of a huge lava spine 1,000 feet in height. This monolith would be far larger than necessary to serve the ancient mythologists as a model for the giant man of bronze in Apollonius's description. The inside of this tower was so large that it did not cool quickly: "At night the sides of this magnificent monolith [were] marked by traces of red incandescent cracks from the still hot lava in its interior. . . . It finally became unstable and collapsed into a pile of rubble in March 1903."[42] As quoted above, a smaller lava spine was created during the subsequent eruption of Mt. Pelée beginning in 1929.

Assuming that the giant man of bronze in the story of Talos was based on a lava spine similar to those of Mt. Pelée, it might have collapsed in a slightly different way. According to Apollonius, "he [Talos] grazed his ankle on a pointed crag; and the ichor gushed forth like melted lead." This statement seems to be a way of representing that very hot and fluid lava began gushing from the foot (or base) of the Talos monolith. Having thus lost its (or his) footing, Talos could have toppled to one side in somewhat the manner of a large pine tree as described by Apollonius earlier in this section.

While eruptions on Crete have occurred in the past, the above paragraphs have suggested that the report of the encounter of the Argonauts with Talos on their return journey could be a representation of an eruption of Thera north of Crete. The departure of many of the Argonauts westward from Egypt would have coincided closely with the exodus of the Hebrews eastward. Thus, the eruption activity on Thera associated with the story of Talos could also have been responsible for the unusual atmospheric conditions experienced by the escaping Hebrews.

Adventures In Egypt

It may be worth mentioning that the homeward voyage of the Argonauts is also said to have included another darkness episode, reminiscent of the one that followed the wave-borne landing of the Argonauts near Egypt and the darkness plague before the exodus. The darkness experienced by the Argonauts after their Talos encounter is described as follows:

> But straightway as they sped over the wide Cretan sea night scared them, that night which they name the Pall of Darkness; the stars pierced not that fatal night nor the beams of the moon, but black chaos descended from heaven, or haply some other darkness came, rising from the nethermost depths. And the heroes, whether they drifted in Hades or on the waters, knew not one whit; but they committed their return to the sea in helpless doubt whither it was bearing them.[43]

Like the Argonauts, the Hebrews also experienced a brief period of darkness after the exodus: "Then the angel of God who went before the host of Israel moved and went behind them; and the pillar of cloud moved from before them and stood behind them, coming between the host of Egypt and host of Israel. And there was the cloud and the darkness; and the night passed without one coming near the other all night" (Exodus 14:19-20). As in earlier incidents, this darkness may have been related to volcanic activity in the Mediterranean region. A post-exodus eruption of Thera could, as in earlier eruptions associated with migrations by Abram and Jacob, have been followed in the vicinity of Egypt by a period of famine. That famine may be reflected in the severe water shortage at Rephidim as reported in Exodus 7:1-14. The shortage has also been suggested as contributing to the resistance

17. Joshua and Jason

by the Amalekites to the invading Hebrews considered in Section 17.2 above (Exodus 17:8-16).

The Argonauts did survive the final phase of their voyage, but the anticipated excitement and tension of their arrival home with the golden fleece and the nondestruction of their leader seem to vanish within the uninformative final sentence of the book:

> For now have I come to the glorious end of your [the Argonauts'] toils; for no adventure befell you as ye came home from Aegina, and no tempest of winds opposed you; but quietly did ye skirt the Cecropian land and Aulis inside of Euboea and the Opuntian cities of the Locrians, and gladly did ye step forth upon the beach of Pagasae.[44]

The lack of drama or a proper ending to the story would be compatible with the idea that an earlier version, whether by Apollonius himself or by one of his sources, might have ended differently. This also leaves open the possibility, suggested above, that perhaps Jason/Joshua and some of the other Argonauts may have joined Heracles/Moses and, contrary to the original plan, never returned to Greece at all. This possibility was suggested previously in Section 16.11 following quotations from the early accounts by Herodotus and Pindar. It may be noted that Jason is mentioned briefly by Apollonius (at least in his later version of the story) as being among the Argonauts on their voyage home from Egypt. However, his alleged role in that version is extremely minor; and it seems possible that, if he didn't actually do so, he might as well have stayed near Egypt with Heracles/Moses:

Adventures In Egypt

Jason held Medea's hand.
"[A]nd Aeson's son [Jason] took her hand in his and guided her way along the thwarts."[45]

Jason cried in his distress.
"But Jason raised his hands and cried to Phoebus with mighty voice, calling on him to save them; and the tears ran down in his distress. . . ."[46]

Jason pondered and interpreted a prophecy.
"Jason pondered a prophecy of the Far-Darter [Apollo] and lifted up his voice and said. . . ."[47]

There is not a word in the conclusion of the *Argonautica* about the achievement of the primary goal of the entire mission – Jason's recovery of the golden fleece. Neither is anything said about Jason's being given the throne of Iolcus. This silence could be a consequence of Jason's determination to stay with Heracles/Moses rather than returning to his home.

17.9 Attributes of Jason/Joshua

Jason of the Argonaut stories had several exceptional physical and character traits, and some of those are indicated here:

1. Humble – After being the primary person involved in organizing the Argonaut expedition, Jason would have been content to have Heracles as the actual expedition leader.[48]

2. Physically powerful – In spite of his youth, Jason was a mighty warrior. He single-handedly defeated a supernatural army of earth-born men, as well as other opponents.[49]

3. Faithful to Medea – After escaping from Colchis, Jason wed Medea as he had promised.[50]

4. Faithful to Heracles – After the Argonauts' unintentional arrival in Egypt, Jason (as Joshua) became a participant in the exodus events of the Israelites. In that role he faithfully assisted Heracles (as Moses) in every way that he could.

5. Effective leader – Eventually, Joshua is said to have taken over the leadership role of the aging Moses and led the Hebrews in their conquest of Canaan. Joshua was already an experienced leader, and he successfully carried out the plans that had been developed by Moses.

Medea, on the other hand, seems to have preferred to leave behind Egypt and her husband rather than participate in the exodus. Thus, she is said to have left Egypt and travelled by ship with many of the Argonauts toward Greece. While there are legends to the contrary, there seem to be no biblical indications that Joshua ever married again.

Jason was initially, of course, a Greek-speaking Argonaut; and it could be of interest to see how he may or may not have adapted to the religious concepts of the Hebrews. Table 17.2 includes a list of quotation segments from the biblical book of Joshua. These quotations are meant to represent statements made by Joshua to the Hebrews as they together set about conquering Canaan. Some of the words in the quotations are written here in boldface for the purpose of indicating whether or not Joshua considered himself to be a God-worshiping Hebrew as might ordinarily be assumed.

Table 17.2. Statements by Joshua concerning the relationship of God to Joshua and to the Hebrews.

Verse in Joshua	Bible Text	Whose God?
1: 11	the land which the Lord **your** God gives you	people
1: 13	The Lord **your** God is providing you a place	people
1: 15	land which the Lord **your** God is giving them	people
3: 3	the covenant of the Lord **your** God	people
3: 9	hear the words of the Lord **your** God	people
4: 5	Pass on before the ark of the Lord **your** God	people
4: 23	For the Lord **your** God dried up the waters	people
4: 23	as the Lord **your** God did to the Red Sea	people
4: 24	that you may fear the Lord **your** God	people
8: 7	the Lord **your** God will give it into your hand	people
10: 19	the Lord **your** God has given them into your hand	people
18: 3	the Lord, the God of **your** fathers, has given you	people
18: 6	here before the Lord **our** God	people/Joshua
22: 3	the charge of the Lord **your** God	people
22: 4	the Lord **your** God has given rest	people
22: 5	love the Lord **your** God	people
23: 3	all that the Lord **your** God has done	people
23: 5	The Lord **your** God will push them back	people
23: 8	cleave to the Lord **your** God as you have done	people
23: 10	it is the Lord **your** God who fights for you	people
23: 13	land which the Lord **your** God has given you	people
23: 14	the Lord **your** God promised concerning you	people
23: 15	the Lord **your** God promised concerning you	people
28: 16	the covenant of the Lord **your** God	people
24: 15	as for me and my house, we will serve the Lord	Joshua

The striking implication of the table is that Joshua did not usually seem to include himself among the Hebrews. Thus, he often spoke to the people about "**your**" God but almost never about "**our**" God. Thus, one could imagine that Joshua may have been involved in the exodus and conquest stories mainly because of his devotion to Moses, rather than for his otherwise perhaps nonexistent Hebrew background and worship experience. Only in Joshua 18:6 does the leader Joshua use the form "**our**" God rather than the form "**your**" God or its equivalent. The uniqueness of this example could cause one to suspect that it represents an error in either the original writing or in a later transcription. Thus, Joshua's personal religious beliefs, if any, could initially have been different from those of the Hebrews that he was called to lead.

The above (speculative) interpretation of Joshua's worship practices may have changed dramatically near the end of his leadership career and of his life. At that time Joshua made a beautiful and powerful declaration that he intended to serve the Lord whether anyone else did or not:

> Now therefore fear the Lord, and serve him in sincerity and in faithfulness; put away the gods which **your** fathers served beyond the River [Euphrates], and in Egypt, and serve the Lord. And if you be unwilling to serve the Lord, choose this day whom you will serve, whether the gods **your** fathers served in the region beyond the River, or the gods of the Amorites in whose land you dwell; but as for me and my house, we will serve the Lord (Joshua 24:14-15).

17.10 Conclusion

The above discussions have entertained the idea that some of the wondrous stories about Jason and the Argonauts written by Apollonius Rhodius in his *Argonautica* might have taken part of their inspiration from Bible-related stories of the exodus era. The imaginative reader may wish to consider other parallels between the biblical and ancient Greek stories. For example, one could wonder whether in the many silent years between Moses's flight from Egypt and his participation in the exodus he may have travelled more widely in the ancient world. His activities at that time could be imagined even to have contributed to the Heracles literature. Given his royal upbringing and education in Egypt, Moses would likely have been able to communicate directly with Jason and the other Argonauts. Similarly, after the Argonauts washed ashore in Egypt, Jason's activities may have contributed to the Joshua literature in the Bible.

---

1. L. W. Casperson, *Patterns of Biblical Chronology* (Westbow Press, Bloomington, IN, 2012), Chapter 9, "Late Exodus Stories," pp. 184-220.
2. R. L. Hicks, "Shuthelah," in *The Interpreter's Dictionary of the Bible, An Illustrated Encyclopedia* (Abingdon Press, Nashville, 1962), Volume 4, R-Z, p. 342.
3. P. R. Ackroyd, "Chronicles, I and II," in The *Interpreter's Dictionary of the Bible: An Illustrated Encyclopedia* (Abingdon Press, Nashville, 1976), Supplementary Volume, pp. 156-158; p. 157.
4. "Sheerah," in *The Interpreter's Dictionary of the Bible, An Illustrated Encyclopedia* (Abingdon Press, Nashville, 1962), Volume 4, R-Z, p. 317.
5. L. W. Casperson, op. cit., Chapter 19, "From Abraham to the Exodus," pp. 472-488.
6. Ibid., pp. 487-488.
7. Ibid., Section 19.4, "Generations," pp. 483-488. See especially Figure 19.1, "Hebrews having genealogies that are continuous from their entry into Canaan until the exodus," p. 486.
8. E. M. Good, "Joshua son of Nun," in *The Interpreter's Dictionary of the Bible: An Illustrated Encyclopedia*

(Abingdon Press, Nashville, 1962), Volume 2, E-J, pp. 995-996; p. 996.
9. L. W. Casperson, op. cit., Chapter 21, "Cause of the Famine," pp. 523-542; and Chapter 22, "Duration of the Famine, pp. 543-562.
10. J. C. Greenfield, "Caphtor," in *The Interpreter's Dictionary of the Bible, An Illustrated Encyclopedia* (Abingdon Press, Nashville, 1962), Volume 1, A-D, p. 534.
11. L. W. Casperson, op. cit., Section 21.3, "The Thera Eruption," pp. 526-532.
12. See also L. W. Casperson, op. cit., Section 24.4, "Abram and the Famine," pp. 605-609.
13. *The Book of Yashar*, Translated from the Hebrew and Published by Manuel Mordecai Noah (Hermon Press, New York, 1972), (cited here in the form *Jasher*), *Jasher* 75:15-22, p. 227.
14. "Gath (city)," Wikipedia. Web. 6 April 2017.
15. *Apollonius Rhodius: The Argonautica*, With an English translation by R. C. Seaton (Loeb Classical Library, Harvard University Press, Cambridge, Massachusetts, 1980), Book IV, pp. 395, 397.
16. E. Horowitz, *Reckless Rites: Purim and the Legacy of Jewish Violence* (Princeton University Press, Princeton, New Jersey, 2006), pp. 10-11.
17. Y. Auron, *The Banality of Indifference: Zionism and the Armenian Genocide*, (Transaction Publishers, London, 2002), p.126.
18. H. Travis, "Did the Armenian genocide inspire Hitler? Turkey, past and future," *The Middle East Quarterly*, Winter 2013, pp. 27-35.
19. N. Masalha, *Imperial Israel and the Palestinians: the Politics of Expansion* (Pluto Press, Sterling, VA, 2000), pp. 130-131.

20. A. G. Hunter, "(De)nominating Amalek: Racist stereotyping in the Bible and the justification of discrimination," in *Sanctified Aggression: Legacies of biblical and post-biblical vocabularies of violence*, Edited by Jonneke Bekkenkamp and Yvonne Sherwood (T & T Clark International, New York, NY, 2003), pp. 92-108; pp. 98-99.
21. P. L. Hammer, "Deuteronomy," *The Interpreter's Dictionary of the Bible, An illustrated encyclopedia*, (Abingdon Press, Nashville, 1962), Volume 1, A-D, pp. 831-838.
22. A possible chronology for the reign of Josiah is considered in L. W. Casperson, op. cit., Table 3.4, "Chronology of the late southern kingdom," p. 55.
23. G. von Rad, "Deuteronomy," in *The Interpreter's Dictionary of the Bible: An Illustrated Encyclopedia* (Abingdon Press, Nashville, 1962), Volume 1, A-D, pp. 831-838; p. 836.
24. See E. Junge, *Der Wiederaufbau des Heerwesens des Reiches Juda unter Josia* (Kohlhammer, Stuttgart, 1937).
25. E. M. Good, op. cit., pp. 995-996.
26. "Jason (2)," *International Standard Bible Encyclopedia*," General Editor James Orr, 1915.
27. F. Josephus, *The Life and Works of Flavius Josephus*, Translated by William Whiston (Holt, Rinehart and Winston, New York), *The Antiquities of the Jews*, Book XII, 5:1, pp. 361-362.
28. "Nun," *Encyclopædia Brittanica Online* (Encyclopædia Britannica Inc., 2016).
29. *Apollonius Rhodius: The Argonautica*, op. cit., Book I, pp. 69-71.
30. L. W. Casperson, op. cit., Section 14.4, "Moses as a priest of Osiris," pp. 339-344.
31. Ibid., Section 4.6, "Abram in Canaan and Egypt," pp. 610-618; p. 613.

32. Ibid., Section 6.4, "Ancient observatories," pp. 123-129.
33. Ibid., Section 14.5, "Other youthful studies," pp. 345-348.
34. *Apollonius Rhodius*, op. cit., Book III, pp. 287-289.
35. E. M. Good, op. cit., p. 996.
36. *Apollonius Rhodius*, op. cit., Book IV, p. 407.
37. Ibid., Book IV, p. 409.
38. J. V. Luce, *Lost Atlantis: New Light on an Old Legend* (McGraw-Hill Book Company, New York, 1969), p. 149.
39. J. Schoo, *Vulkanische und seismische Aktivität des ägäischen Meeresbeckens im Spiegel der griechische Mythologie*, Mnemosyne, Third Series, Vol. 4, Fasc. 4 (1937), pp. 257-94.
40. "Mount Pelée," Wikipedia. Web. 15 May 2015.
41. F. Oswald, *A Treatise on the Geology of Armenia*, Thesis accepted by the University of London for the Degree of Doctor of Science (Published by the author at Iona, Beeston, Notts, 1906), p. 154.
42. "Mount Pelée Eruption (1900)," *How Volcanoes Work*, San Diego State University. Wikipedia. Web. 15 May 2015.
43. *Apollonius Rhodius*, op. cit., Book IV, pp. 409, 411.
44. Ibid., Book IV, p. 415.
45. Ibid., Book IV, p. 407.
46. Ibid., Book IV, p. 411.
47. Ibid., Book IV, p. 413.
48. Ibid., Book I, pp. 25, 27.
49. Ibid., Book III, pp. 285, 287, 289.
50. Ibid., Book IV, pp. 373, 375, 377.

PART 4
CHRONOLOGY IN CONTEXT

18. BEFORE THE BEGINNING

"Jesus said to them, 'Truly, truly, I say to you, before Abraham was, I am' " (John 8:58).

18.1 Introduction

As indicated in Chapter 1, there are many Bible-related stories concerning the origin of the universe and its first inhabitants. These stories appear to have originated at different times and in different places, and they also vary in their chronologies of the creation events. Of particular interest in this study is, of course, the biblical version of creation. Perhaps not surprisingly, this version has been suggested to have close parallels in cultures that developed in the geographical areas that at one time or another were associated with the ancient Hebrews. The biblical creation stories have been accepted as literal history by their advocates for many centuries and up to the present.

One purpose of this chapter is to review the most basic approaches that might be taken in obtaining estimates for the age of the universe (or world or earth). For a Bible-related study the obvious answer is that the creation of the universe occurred when the Bible says it did – a few thousand years B.C., depending on

18. Before the Beginning

texts and translations. On the other hand, many scientists might say that, contrary to the Bible, the universe was created millions of years ago. Similarly, many Bible readers might anticipate that the universe as we know it will come to an end "soon," while many scientists would anticipate that the universe in about its present form will be around for many more millions of years. Some of the arguments for and against these varying interpretations and conclusions will be introduced in this chapter and will continue into the two following chapters.

The most essential element of the Bible-related stories may be the concept that the universe was created by God. Beyond this basic principle, the beliefs of God-fearing people concerning creation exhibit significant variations. Thus, agreement wouldn't always be obtained on when the universe was created, and the dates that might be inferred vary among sources. Disagreement might also arise on how long creation took, or whether the seven-day creation duration indicated in the Bible should be understood to correspond literally to one week, given that in other texts the biblical seven-day week may have been introduced many centuries later. How to interpret the data discovered by modern geologists is also not always agreed upon. The belief that the universe was created by God is, of course, in fundamental disagreement with the atheist notion that God does not exist. The present chapter is intended to provide a brief review of a few of the past and current concepts regarding the origin of the universe that have been held by those who accept or deny the existence of God.

The chronologies that have been adopted in at least a qualitative way by many modern atheists are reviewed briefly in

Section 18.2. These conclusions are in fairly direct disagreement with the conclusions reached in most religious studies. A historically important example of such a conflict concerns the structure of the solar system, and this issue is summarized in Section 18.3. Various Bible-related concepts on the date of the origin of the earth are reviewed in Section 18.4, and the date of its biblically-projected demise is considered in Section 18.5. One of the most explicit New Testament statements concerning the end of the universe is associated with the Apostle Peter as summarized in Section 18.6. The credentials of Peter for having better than average insight into the future are reviewed in Section 18.7, and some ancient writings by and about Peter are considered in Section 18.8.

18.2 Chronologies of atheists

The relative brevity of human lifetimes and perhaps the busyness of human schedules tend to foster the idea that any fundamental changes in the world around us are either non-existent or are too slow to be of any practical consequences. Thus, there may be minor ongoing geophysical or climatological changes, and perhaps even changes in the larger universe, but the condition of the earth in the more distant past or remote future may seem to be of little relevance to our lives. The world we inherited is essentially the same as the one we will experience throughout our lifetimes and will also be the world we leave to our descendants.

In spite of the prevalence of such seemingly self-centered points of view, there have always been some people who have had the time and the inclination to take a serious interest in the history of the universe. Many religions offer explanations of the origin and fate of the universe, and philosophers and scientists have also been drawn to such subjects. The religious interpretations vary widely, and the scientific investigations tend to be extremely complex. The difficulty of pursuing questions of this sort is, of course, that most of the major events are remote from our practical experience and relevant data may be difficult to find and interpret.

For conventional Bible-related approaches to any possible creation events, only a limited amount of data is available. The biblical stories themselves have come down to us in several different versions. While the general forms of these versions have much in common, the actual chronological details vary significantly from one version to another. Concerning the "standard" biblical date of creation, the most that can be said with confidence is that it was a few thousand years B.C.

The non-biblical approaches of scientists to the subject of creation are much more complicated and have varied significantly with time. Information on this subject is readily available, and a few internet quotes are given here. One of the early concepts was that the universe and the laws of nature have always been pretty much as they appear today. This general understanding has often been referred to as uniformitarianism:

> Uniformitarianism is the scientific observation that the same natural laws and processes that operate in the universe now have always operated in the universe in the past and apply everywhere in the universe. . . . The planet [earth] has almost always looked and behaved as it does now. Change is continuous, but leads nowhere. The earth is in balance: a dynamic steady state.[1]

The basic concepts of uniformitarianism were introduced and developed beginning in about the eighteenth century, and they remained fairly dominant until the mid-twentieth century:

> From around 1850 to 1980, most geologists endorsed uniformitarianism ("The present is the key to the past") and gradualism (geologic change occurs slowly over long periods of time) and rejected the idea that cataclysmic events such as earthquakes, volcanic eruptions, or floods of vastly greater power than those observed at the present time, played any significant role in the formation of the earth's surface. Instead they believed that the earth had been shaped by the long term action of forces such as volcanism, earthquakes, erosion, and sedimentation, that could still be observed in action today.[2]

At the same time that uniformitarianism was dominant, the seemingly contrary theory called catastrophism was also developing among a growing minority of geologists:

> Catastrophism is the theory that the Earth has been affected in the past by sudden, short-lived, violent events, possibly world-wide in scope. . . . Since the early disputes, a more inclusive and integrated view of geologic events has developed, in which the scientific consensus accepts that there were some catastrophic events in the geologic past, but these were explicable as extreme examples of natural processes which can occur. . . .
> Neocatastrophism is the explanation of sudden extinctions in the paleontological record by high magnitude, low frequency events (such as asteroid impacts, super-volcanic eruptions, supernova gamma ray bursts, etc.), as opposed to the more prevalent geomorphological thought which emphasizes low magnitude, high frequency events. . . .

18. Before the Beginning

Today most geologists combine catastrophist and uniformitarianist standpoints, taking the view that Earth's history is a slow, gradual story punctuated by occasional natural catastrophic events that have affected Earth and its inhabitants.[3]

For the chronologist having a serious interest in the history and future of the earth and the universe as a whole, it is not sufficient to have a general understanding of the types of changes that have occurred. It is also necessary to assign at least approximate dates to the principle events of the past and the anticipated future. For many centuries the pursuit of an accurate chronology of historical and geological events of the past has been undertaken by both religious and nonreligious investigators. A few approaches to the attainment of such chronologies are indicated in the following sections.

18.3 Science and religion

Many Christian researchers over the centuries have been interested in the chronology of the world beginning with the creation events reported in Genesis. One of the most influential chronologies of relatively recent times was that developed in the seventeenth century by James Ussher, the Archbishop of Armagh (Church of Ireland). Ussher's work was based to a large extent on the Masoretic version of the Old Testament, and he inferred that the starting date of creation was in 4004 B.C.[4] Using other versions of the Bible and other interpretations of the biblical texts, the dates associated with the life of Adam could be closer to the middle of the third millenium B.C., as considered in Section 12.5 of this study.

Besides chronology, Christian researchers have been interested in other aspects of the biblical texts. Sometimes individual researchers may have found themselves in disagreement with their own churches on questions of scientific facts or theological interpretations. An important example from the past is provided by the experiences of the Italian physicist Galileo Galilei. Galileo was born in 1564 and died in 1642, and he is often considered to have been the father of experimental science. Among other achievements, he built a telescope that was sufficiently powerful to observe the orbits of the moons of the planet Jupiter and the phases of the planet Venus. His studies in astronomy led him to subscribe to the Copernican theory and to conclude that the earth itself was a planet orbiting the sun. Unfortunately, this conclusion disagreed with the position of the Catholic Church that the earth was at the center of the universe. This irreconcilable difference led to Galileo's spending the last years of his life under house arrest.

One might have imagined that these discordant interpretations of the structure of the solar system would have been resolved in Galileo's favor several centuries ago, but that was not the case:

> On 31 October 1992, 350 years after Galileo's death, Pope John Paul II gave an address on behalf of the Catholic Church in which he admitted that errors had been made by the theological advisors in the case of Galileo. He declared the Galileo case closed, but he did not admit that the Church was wrong to convict Galileo on a charge of heresy because of his belief that the Earth rotates round the sun.[5]

18. Before the Beginning

Two of the numbered paragraphs from an English translation of the text of Pope John Paul II's address are mostly reproduced below. These paragraphs relate especially to the case of Galileo:[6]

> 10. From the beginning of the Age of Enlightenment down to our own day, the Galileo case has been a sort of "myth", in which the image fabricated out of the events was quite far removed from reality. In this perspective, the Galileo case was the symbol of the Church's supposed rejection of scientific progress, or of "dogmatic" obscurantism opposed to the free search for truth. This myth has played a considerable cultural role. It has helped to anchor a number of scientists of good faith in the idea that there was an incompatibility between the spirit of science and its rules of research on the one hand and the Christian faith on the other. A tragic mutual incomprehension has been interpreted as the reflection of a fundamental opposition between science and faith. The clarifications furnished by recent historical studies enable us to state that this sad misunderstanding now belongs to the past.

> 12. [A] lesson which we can draw is that the different branches of knowledge call for different methods. Thanks to his intuition as a brilliant physicist and by relying on different arguments, Galileo, who practically invented the experimental method, understood why only the sun could function as the centre of the world, as it was then known, that is to say, as a planetary system. The error of the theologians of the time, when they maintained the centrality of the earth, was to think that our understanding of the physical world's structure was, in some way, imposed by the literal sense of Sacred Scripture. . . . In fact, the Bible does not concern itself with the details of the physical world, the understanding of which is the competence of human experience and reasoning. There

exist two realms of knowledge, one which has its source in Revelation and one which reason can discover by its own power. To the latter belong especially the experimental sciences and philosophy. The distinction between the two realms of knowledge ought not to be understood as opposition. The two realms are not altogether foreign to each other, they have points of contact. The methodologies proper to each make it possible to bring out different aspects of reality.

Not all readers will be in full agreement with the above statements, and it is good to be aware of the limitations of all sources of human knowledge. Christians tend to believe that God is in control of the universe, and thus no amount of "human experience and reasoning" will ever allow the exact prediction of the future. On the other hand, with Revelation, as identified by the pope, biblical truths have reached most of us as written and transcribed by imperfect human hands, and thus no amount of study (or wishful thinking) will remove all errors and uncertainties. Concerning the movements of the sun and earth, "different aspects of reality" continue even today:

> 1 In 4 Americans Thinks The Sun Goes
> Around The Earth, Survey Says
>
> A quarter of Americans surveyed could not correctly answer that the Earth revolves around the sun and not the other way around, according to a report out Friday from the National Science Foundation. The survey of 2,200 people in the United States was conducted by the NSF in 2012 and released on Friday at an annual meeting of the American Association for the Advancement of Science meeting in Chicago. To the question "Does the Earth go

around the Sun, or does the Sun go around the Earth," 26 percent of those surveyed answered incorrectly.[7]

Giordano Bruno (1548-1600) was a near contemporary of Galileo, and Bruno also agreed with and extended the Copernican model of the universe. While Galileo was condemned to lifetime house arrest for his "heresies," Bruno was less fortunate. "On 20 January 1600, [not-so-clement] Pope Clement VIII declared Bruno a heretic and the Inquisition issued a sentence of death." On 17 February 1600, Bruno "was hung upside down naked before he was finally burned at the stake."[8]

18.4 The origin of the earth

As mentioned above, the conflict between Galileo and the Catholic Church concerned the question of whether the earth orbits the sun or the sun orbits the earth. In Galileo's time it was often understood to be implied by the Bible that the sun orbits about the earth, though the actual biblical basis for this conclusion may not always seem compelling today. For example, one biblical argument used for the movement of the sun was the following: "The sun rises and the sun goes down, and hastens to the place where it rises" (Ecclesiastes 1:5). We might commonly say that the sun rises in the east and sets in the west. Use of such words would usually not be construed today to suggest our belief in the actual orbiting of the sun about the earth.

Similar questions concern the origin of the universe. The belief of fundamentalist Christian researchers concerning this origin might be that the universe was created by God at some specific time in the past, though the precise date and process of this creation may not be determinable. Genesis addresses this

Chronology In Context

subject, but discussions presuppose that the writers would have had knowledge of the relevant facts. There is no evidence that the authors of Genesis did their first writing until long after the creation of the universe. As a result, the creation dates that can be determined from Bible-related data are subjective and depend on details of the texts employed. As discussed in Chapter 12, the dates determined solely on the basis of such interpretations tend to be in the range of a few thousand years B.C., depending on biblical versions and assumptions relating to generation lengths.

The biblical dating methods just referred to are still accepted by many Christians and were, for example, the basis for the specific chronology developed many years ago by Archbishop Ussher, as mentioned in Section 18.3. As is also well known, the dates based solely on biblical information disagree dramatically with the results of most scientific studies of the history of the earth. Thus, data obtained from research in geology, biology, and other scientific disciplines consistently indicate that the earth must be at least many millions of years in age. This seeming discrepancy between the biblical accounts and modern scientific methods is, of course, similar to the controversy in which Galileo found himself embroiled. It is important to consider how such a significant disparity might be resolved. One obvious possibility for a Christian is to observe that data obtained by scientists might not actually have required the immense periods of time that are usually claimed. As told the author many years ago, God might have created the earth just as we find it today, in the blink of an eye and complete with dinosaur bones and all.

Another possible resolution of the disagreement just mentioned is that a chronology based only on versions of Genesis

18. Before the Beginning

might be disregarding other more relevant biblical information. The Bible is fairly explicit that time intervals may be expressed very differently as they relate to the activities of God as opposed to the activities of men. An illustration of this ambiguity is evident in Psalm 90, which is attributed to Moses:

A Prayer of Moses, the man of God.

1. Lord, thou hast been our dwelling place in all generations.
2. Before the mountains were brought forth, or ever thou hadst formed the earth and the world, from everlasting to everlasting thou art God.

3. Thou turnest man back to the dust, and sayest, "Turn back, O children of men!"
4. For a thousand years in thy sight are but as yesterday when it is past, or as a watch in the night.

5. Thou dost sweep men away; they are like a dream, like grass which is renewed in the morning:
6. in the morning it flourishes and is renewed; in the evening it fades and withers (Psalm 90:1-6, RSV).

It is also, of course, Moses who is often credited with being the author (or compiler, or editor) of Genesis, and that book is sometimes accepted as the main source of our biblical information on the chronology of creation. In this context, verse 4 above is of particular interest. It is stated that a thousand years are for God as only a moment in time. Thus, the few human lifetimes included in Genesis may be negligible compared to the lifetimes of the mountains and the earth. The above verses do not, like Genesis, imply that the universe is only a few thousand years old. On the contrary, God is said to have been active in the universe long ages before the earth was formed. It is also not stated that the earth was created in a day. As suggested in Section 1.2, the much later post-exodus seven-day-week may have been imposed on an earlier creation account by a subsequent editor. The messages of Psalm 90 form the basis for the familiar church hymn "Oh God, Our Help in Ages Past," published by Isaac Watts in 1719.

18.5 The end of the earth

Just as the date and circumstances of the creation of the earth have been a source of particular interest and sometimes contention between Christians and non-Christian scientists, so too have been the expected date and circumstances of the end of the earth. The Bible seems to provide us with a substantial amount of information on this subject, and one might hope that this information would give us clear and specific answers. The idea that the earth will end someday, followed by a judgment, is one of the oft-repeated messages of the Bible.

The most popular biblical source for information on the end times of the earth is the book Revelation. That book reports the visions brought by an angel to a person named John concerning future events. Revelation holds a somewhat unique place in biblical scripture. It was the most controversial of New Testament books, and it was the last book to be included in the New Testament canon. Even its authorship is not agreed upon, and it wasn't widely accepted until several centuries A.D. Some Christian churches do not include it at all.[9]

There are a variety of opinions concerning the identity of this person John. Some authors have decided that John who wrote Revelation was the same person as the author of the gospel of John and the three epistles of John, but others consider this conclusion to be unlikely. John of Revelation doesn't make this claim, and differences of style and thematic content may support the idea that the author of Revelation is distinct from the author of these other so-called Johannine works. Also, the author of Revelation "mentions the twelve apostles as figures from the past (Revelation 21:14) and does not refer to himself as one of them. The traditional identification of the John of the book of Revelation with the apostle of the same name is thus questionable."[10] The author of Revelation is now often identified by the more specific name John of Patmos, referring to the indicated island location of his writing (Revelation 1:9).

One of the persistent claims of Revelation is that Jesus will be returning "soon" to judge the inhabitants of the earth, and then the earth as we know it will come to an end. That this second coming should have been soon after the writing of his book was emphasized repeatedly by John of Patmos. However, John was writing in the first or second century A.D., and now in the twenty-first century this coming has not yet occurred. Such a seeming discrepancy merits at least a few tentative comments. To introduce this subject it may be helpful to quote as an example part of a brief devotional:

Coming Soon!

A "COMING SOON!" announcement often precedes future events in entertainment and sports, or the launch of the latest technology. The goal is to create anticipation and excitement for what is going to happen, even though it may be months away.

While reading the book of Revelation, I was impressed with the "coming soon" sense of immediacy permeating the entire book. Rather than saying, "Someday, in the far distant future, Jesus Christ is going to return to earth," the text is filled with phrases like "things which must shortly take place" (1:1) and "the time is near" (v.3). Three times in the final chapter, the Lord says, "I am coming quickly" (Rev. 22:7,12,20). Other versions translate this phrase as, "I'm coming soon," "I'm coming speedily," and "I'm on My way!"

How can this be – since 2,000 years have elapsed since these words were written? "Quickly" doesn't seem appropriate for our experience of time.

Rather than focusing on a date for His return, the Lord is urging us to set our hearts on His promise that will be fulfilled. We are called to live for Him in this present age "looking for the blessed hope and glorious appearing of our great God and Savior Jesus Christ" (Titus 2:13).

Live as if Christ is coming back today.[11]

Thus, John of Patmos repeatedly told his readers that Jesus would return soon, and he ended his book with what appears to be a plea for that to happen: "Even so, come, Lord Jesus!" (Revelation 22:20). The author of the above devotional message makes the point that Jesus did not come soon, but that nevertheless we should live for him in the present, being confident of his promised return at some time in the future. This reasonable plan does, however, leave open the question of why Jesus failed to return soon as promised in the message left us by John. The most obvious possibility is that John's prophecy was in error on at least that subject. If John's frequently repeated

18. Before the Beginning

prophecy concerning the time of the return of Jesus should be regarded as untrue, it then follows that John of Patmos, whoever he was, should perhaps not be considered a true prophet.

The Bible recognizes clearly the possibility of false prophets, and a blunt Old Testament statement on this reality is the following:

> But the prophet who presumes to speak a word in my name which I have not commanded him to speak, or who speaks in the name of other gods, that same prophet shall die. And if you say in your heart, "How may we know the word which the Lord has not spoken?" – when a prophet speaks in the name of the Lord, if the word does not come to pass or come true, that is a word which the Lord has not spoken; the prophet has spoken it presumptuously, you need not be afraid of him (Deuteronomy 18:20-22).

Revelation is a popular book and probably serves as the principal resource used by many Bible readers in attempting to discover information about end times. Some people might believe that Revelation is the only biblical source for such information. However, that is not the case.

There are several statements in the Bible relating either to the end of the earth or the end of human life on the earth. Some examples from the Old Testament are given here first:

> For a fire is kindled by my anger, and it burns to the depths of Sheol, devours the earth and its increase, and sets on fire the foundations of the mountains (Deuteronomy 32:22).

> The nations rage, the kingdoms totter; he utters his voice, the earth melts (Psalm 46:6).

> Fire goes before him, and burns up his adversaries round about. His lightnings lighten the world; the earth sees and trembles. The mountains melt like wax before the Lord, before the Lord of all the earth (Psalm 97:3-5).

> The mountains quake before him, the hills melt; the earth is laid waste before him, the world and all that dwell therein (Nahum 1:5).

> "Therefore wait for me," says the Lord, "for the day when I arise as a witness. For my decision is to gather nations, to assemble kingdoms, to pour out upon them my indignation, all the heat of my anger; for in the fire of my jealous wrath all the earth shall be consumed" (Zephaniah 3:8).

In most of the examples above, the earth either melts or is burned up. Isaiah also includes the idea that after the destruction, God will create new heavens and a new earth:

> "For behold, I create new heavens and a new earth; and the former things shall not be remembered or come into mind" (Isaiah 65:17).

> "For as the new heavens and the new earth which I will make shall remain before me, says the Lord; so shall your descendants and your name remain" (Isaiah 66:22).

Based on these verses one can reasonably conclude that several of the Old Testament prophets predicted that the earth would be melted or burned by fire. It is also of interest to know whether there are New Testament references to the return of Jesus and the fate of the earth besides the prophesies in Revelation. As it happens, the New Testament contains many such references, and several of those will be considered in the following section.

18. Before the Beginning

18.6 A man called Peter

Additional texts concerning the return of Jesus are included in the following devotional excerpt:

> Waiting
>
> . . .
>
> Christ's second coming is also the theme of several New Testament passages. As Christ ascended into heaven, the angels told His disciples that Christ "will come back in the same way" they saw Him go (Acts 1:11). Jesus said His return would be unannounced and could occur at any moment; therefore we are to "Be on guard! Be alert!" (Mark 13:33-37). The early Christians believed that Jesus's return was "almost here" (Romans 13:11-14). The apostle James encouraged believers to "be patient and stand firm, because the Lord's coming is near" (James 5:8). The anticipation that Jesus could come any moment led some Christians in Thessalonica to become idle, quitting their jobs and waiting for Him to return. But Paul told them to get back to work and live meaningful lives (2 Thessalonians 3:11-13).
>
> "While we [patiently] wait for the blessed hope" (Titus 2:13) – that wonderful day of Jesus's return – we can ask the Spirit to help us to live "holy and godly lives . . . spotless, blameless and at peace with him" (2 Peter 3:11, 14).[12]

Some of the most significant New Testament statements concerning the end of the earth and the promise of the Lord's return are contained in 2 Peter:

> This is now the second letter that I have written to you, beloved, and in both of them I have aroused your sincere mind by way of reminder; that you should remember the predictions of the holy prophets and the

commandment of the Lord and Savior through your apostles. First of all you must understand this, that scoffers will come in the last days with scoffing, following their own passions and saying, "Where is the promise of his coming? For ever since the fathers fell asleep, all things have continued as they were from the beginning of creation." They deliberately ignore this fact, that by the word of God heavens existed long ago, and an earth formed out of water and by means of water, through which the world that then existed was deluged with water and perished. But by the same word the heavens and earth that now exist have been stored up for fire, being kept until the day of judgment and destruction of ungodly men.

But do not ignore this one fact, beloved, that with the Lord one day is as a thousand years, and a thousand years as one day. The Lord is not slow about his promise as some count slowness, but is forbearing toward you, not wishing that any should perish, but that all should reach repentance. But the day of the Lord will come like a thief, and then the heavens will pass away with a loud noise, and the elements will be dissolved with fire, and the earth and the works that are upon it will be burned up.

Since all these things are thus to be dissolved, what sort of persons ought you to be in lives of holiness and godliness, waiting for and hastening the coming of the day of God, because of which the heavens will be kindled and dissolved, and the elements will melt with fire! But according to his promise we wait for new heavens and a new earth in which righteousness dwells (2 Peter 3:1-13).

Peter is understood to have died as a martyr in about 65-67 A.D., and that this book was actually written by him is accepted by many readers. Others consider that it was probably written by a person with a close awareness of Peter and his teachings and who

18. Before the Beginning

was publishing Peter's teachings in his honor. This latter possibility will be considered in Section 18.8 below.

The teachings concerning end times are far more brief in 2 Peter than in the complex apocalyptic writings of Revelation. On the other hand, Revelation isn't quite so clear on what will cause the end of the earth. After lengthy discussions in Revelation on future plagues and wars, the present earth is last heard of in a few sentences summarizing the flight of the earth and sky, the resurrection of the dead, the judgment and punishment, and the arrival of a new heaven and earth:

> Then I saw a great white throne and him who sat upon it; from his presence earth and sky fled away, and no place was found for them. And I saw the dead, great and small, standing before the throne, and books were opened. Also another book was opened, which is the book of life. And the dead were judged by what was written in the books, by what they had done. And the sea gave up the dead in it, Death and Hades gave up the dead in them, and all were judged by what they had done. Then Death and Hades were thrown into the lake of fire. This is the second death, the lake of fire; and if any one's name was not found written in the book of life, he was thrown into the lake of fire.
> Then I saw a new heaven and a new earth; for the first heaven and the first earth had passed away, and the sea was no more... (Revelation 20:11 - 21:1).

These verses from Revelation inform us simply that earth and sky will flee away. In contrast to Revelation, 2 Peter is very specific on what will happen to the earth someday. As quoted more extensively above one finds the following:

Chronology In Context

> [Y]ou should remember the predictions of the holy prophets (2 Peter 3:2).
>
> But by the same word the heavens and earth that now exist have been stored up for fire, being kept until the day of judgment and destruction of ungodly men (2 Peter 3:7).
>
> But the day of the Lord will come like a thief, and then the heavens will pass away with a loud noise, and the elements will be dissolved with fire, and the earth and the works that are upon it will be burned up.
>   Since all these things are thus to be dissolved, what sort of persons ought you to be in lives of holiness and godliness, waiting for and hastening the coming of the day of God, because of which the heavens will be kindled and dissolved, and the elements will melt with fire! (2 Peter 3:10-12).

It seems clear in these verses that the reader is being asked to look back to the writings of the Old Testament prophets. There the reader will learn that the heavens and the earth will be destroyed by fire. The description of this event is compatible with the Old Testament prophecies cited previously in Section 18.5, in which it is also stated or implied that the earth will be destroyed by fire.

In these chronological studies the time of the destruction of the earth is at least as important as its mechanism. On the subject of chronology, Revelation and 2 Peter differ greatly. As considered previously, Revelation states repeatedly that Jesus would be coming soon. However, he did not do so, and it is of interest to consider more closely what Peter (or a writer representing him) had to say on this same subject of chronology. For clarity it may be helpful to repeat more selectively a few

18. Before the Beginning

other sentences from the text of Peter quoted previously in this section:

> This is now the second letter that I have written to you, beloved, and in both of them I have aroused your sincere mind by way of reminder; that you should remember the predictions of the holy prophets and the commandment of the Lord and Savior through your apostles. First of all you must understand this, that scoffers will come in the last days with scoffing, following their own passions and saying, "Where is the promise of his coming? For ever since the fathers fell asleep, all things have continued as they were from the beginning of creation" (2 Peter 3:1-4)

> But do not ignore this one fact, beloved, that with the Lord one day is as a thousand years, and a thousand years as one day. The Lord is not slow about his promise as some count slowness, but is forbearing toward you, not wishing that any should perish, but that all should reach repentance (2 Peter 3:8-9).

There are several important points in these texts, such as the following:

1. The reader is asked to "remember the predictions of the holy prophets." In contrast to false prophets, the predictions of the holy prophets would by definition have been accurate. Thus, this invitation may be understood to call attention to those whom the writer might already recognize as false prophets, perhaps including John of Patmos.

2. The reader is also asked to remember the "commandment of the Lord and Savior through your apostles." This could be a non-

subtle reminder that Peter, whom the writer is representing, was an apostle, whereas other people who were prophesying may not have been apostles.

3. The statement on scoffers probably represents a situation with which the early Christians were already familiar and which continues today. In spite of (false) prophecies concerning the immediacy of Christ's return, he has not yet returned. The statement that "with the Lord one day is as a thousand years" may be telling us that his return might not be for a very long time (many thousands of years). It would seem that Peter was well aware of the prayer of Moses as considered above in Section 18.4.

4. If the gospels of Mark and others are to be understood literally on this subject, no one including Jesus could even know when his return is to be: "And then they will see the Son of man coming in clouds with great power and glory. . . . But of that day or that hour no one knows, not even the angels in heaven, nor the Son, but only the Father" (Mark 13:26,32; similarly Matthew 24:30,36; Luke 21:27,28; Acts 1:7).

18.7 The credentials of Peter

The clarity of the prophecies of Peter and their reasonable consistency with several earlier prophecies suggest the question of why Peter might have had greater insight into the future than some other writers known to us. In fact Peter held a very special place among the disciples, and he seems often to be the leading human spokesperson in Jesus's initially small human entourage.

18. Before the Beginning

Thus it is possible that more information about the future may have been revealed to Peter than to the others. Some of the events that make him seem special are listed here with supporting information from Matthew and other sources. These events are listed numerically somewhat in the order that they appear in Matthew. For further insight one should read the full biblical texts including the versions given in the other gospels and Acts.

1. Peter and his brother Andrew were the first disciples chosen by Jesus: "As he [Jesus] walked by the Sea of Galilee, he saw two brothers, Simon who is called Peter and Andrew his brother, casting a net into the sea; for they were fishermen. And he said to them, 'Follow me, and I will make you fishers of men.' Immediately they left their nets and followed him" (Matthew 4:18-20).

2. Jesus visited Peter's house and healed Peter's mother-in-law who had been sick with a fever: "And when Jesus entered Peter's house, he saw his mother-in-law lying sick with a fever; he touched her hand, and the fever left her, and she rose and served him" (Matthew 8:14-15). This was not the first healing by Jesus, but he seems not to have routinely visited the homes of the disciples or carried out healings on their behalf.

3. Peter retained his status as the first of Jesus's twelve disciples:

> And he called to him his twelve disciples and gave them authority over unclean spirits, to cast them out, and to heal every disease and every infirmity. The names of the twelve apostles are these: first, Simon, who is called Peter, and Andrew his brother; James the son of

Zeb'edee, and John his brother; Philip and Bartholomew; Thomas and Matthew the tax collector; James the son of Alphaeus, and Thaddaeus; Simon the Cananaean, and Judas Iscariot, who betrayed him (Matthew 10:1-4).

This could simply recall that Peter was appointed chronologically first of the twelve, but there are indications that he could also have been considered the greatest leader among them.

4. Peter seems always to have been the boldest and bravest of the twelve disciples. One example will be mentioned here, and others will be suggested later. This example is presented as another devotional excerpt:

Walking on Water

. . .
Do you avoid taking risks at all costs? Many of us are reluctant to step out of our comfort zones in case we fail, get hurt, or look stupid. But if we allow that fear to bind us, we'll end up afraid to do anything.

The story of Peter's water-walking adventure and why it supposedly failed is a popular choice for preachers (Matthew 14:22-33). But I don't think I've ever heard any of them discuss the behavior of the rest of the disciples. In my opinion, Peter was a success. He felt the fear but responded to the call of Jesus anyway. Maybe it was those who never tried at all who failed.

Jesus risked everything for us. What are we prepared to risk for Him?[13]

5. Peter was inquisitive.

And he [Jesus] called the people to him and said to them, "Hear and understand: not what goes into the mouth defiles a man, but what comes out of the mouth, this defiles a man.". . . But Peter said to him "Explain the parable to us." And he said, "Are you also still without understanding? Do you not see that whatever goes into the mouth passes into the stomach, and so passes on? But what comes out of the mouth proceeds from the heart, and this defiles a man. For out of the heart come evil thoughts, murder, adultery, fornication, theft, false witness, slander. These are what defile a man; but to eat with unwashed hands does not defile a man" (Matthew 15:10,15-20).

6. Peter may have understood better than the other disciples who Jesus really was, and he was not afraid to declare his witness:

> Now when Jesus came into the district of Caesare'a Philip'pi, he asked his disciples, "Who do men say that the Son of man is?" And they said, "Some say John the Baptist, others say Eli'jah, and others Jeremiah or one of the prophets." He said to them, "But who do you say that I am?" Simon Peter replied, "You are the Christ, the son of the living God." And Jesus answered him, "Blessed are you, Simon Bar-Jona! For flesh and blood has not revealed this to you, but my Father who is in heaven. And I tell you, you are Peter, and on this rock I will build my church, and the powers of death shall not prevail against it. I will give you the keys of the kingdom of heaven, and whatever you bind on earth shall be bound in heaven, and whatever you loose on earth shall be loosed in heaven" (Matthew 16:13-19).

7. Peter may sometimes have tried too hard to protect Jesus. It could have stung him to be reprimanded for his concern about what was at least in part the purpose of Jesus's mission on earth:

> From that time Jesus began to show his disciples that he must go to Jerusalem and suffer many things from the elders and chief priests and scribes, and be killed, and on the third day be raised. And Peter took him and began to

rebuke him, saying, "God forbid, Lord! This shall never happen to you." But he turned and said to Peter, "Get behind me, Satan! You are a hindrance to me; for you are not on the side of God, but of men" (Matthew 16:21-23).

8. Peter, James, and John, the brother of James, were present at the transfiguration of Jesus. Of the first four disciples, Andrew, the brother of Jesus, was absent. Peter wanted to do something there to honor Jesus, Moses, and Elijah, but he wasn't given the opportunity: "And Peter said to Jesus, 'Lord, it is well that we are here; if you wish, I will make three booths here, one for you and one for Moses and one for Eli'jah' " (Matthew 17:4). It was sometimes believed that Moses, like Elijah, never died an earthly death (Section 16.10). The arrival of a bright cloud and a voice from within it interrupted Peter's suggested plan.

9. Peter was sometimes treated as the spokesperson for Jesus and his disciples. When Peter was asked by the tax collectors whether Jesus paid taxes, fisherman Peter was the one to obtain the tax money from the mouth of a fish (Matthew 17:24-27).

10. Peter sought to learn from Jesus: "Then Peter came up and said to him, 'Lord, how often shall my brother sin against me, and I forgive him? As many as seven times?' Jesus said to him, 'I do not say to you seven times, but seventy times seven' " (Matthew 18:21-22). Considering event number "8." above and other occasions, one can wonder if there might sometimes have been difficulties between Peter and his brother. Andrew was mentioned and present less often than, for example, James or John. Perhaps Peter, as represented in these verses, considered that Andrew had sinned against him, and that conflict may have been keeping Andrew away.

18. Before the Beginning

11. In his boldness Peter could sometimes make inappropriate suggestions ("8." above) and ask what some might consider to be inappropriate questions: "Then Peter said in reply, 'Lo, we have left everything and followed you. What then shall we have?'" (Matthew 19:27). Jesus's answer included in part the following encouraging promise: "And every one who has left houses or brothers or sisters or father or mother or children or lands, for my name's sake, will receive a hundred fold, and inherit eternal life" (Matthew 19:29). Curiously, the word "wife" is missing from this list of family members, and Peter's wife is known sometimes to have accompanied him on his journeys (Section 18.8 below).

12. When Jesus told the disciples that they would fall away, Peter made a promise that he couldn't keep:

> And when they had sung a hymn, they went out to the Mount of Olives. Then Jesus said to them, "You will all fall away because of me this night; for it is written, 'I will strike the shepherd, and the sheep of the flock will be scattered.' But after I am raised up, I will go before you to Galilee." Peter declared to him, "Though they all fall away because of you, I will never fall away." Jesus said to him, "Truly, I say to you, this very night, before the cock crows, you will deny me three times." Peter said to him, "Even if I must die with you I will not deny you." And so said all the disciples (Matthew 26:30-35).

It is clear again here that Peter seemed to be the inspiration and leader among the disciples. He was the first to declare that he would never fall away, "And so said all the disciples." It isn't related to the point of this paragraph, and it may not be so well acknowledged that, like many modern worshippers, Jesus and his disciples sang hymns.

13. When Jesus went to pray in Gethsemane he took with him his special disciples Peter, James, and John: "Then Jesus went with them to a place called Gethsem'ane, and he said to his disciples, 'Sit here, while I go yonder and pray.' And taking with him Peter and the two sons of Zeb'edee, he began to be sorrowful and troubled" (Matthew 26:36-37). As usual, Peter is named first, and in this case the personal names of James and John are not used. Peter's brother Andrew seems again to have been absent.

14. The disciples didn't stay awake while Jesus was praying: "And he came to the disciples and found them sleeping; and he said to Peter, 'So, could you not watch with me one hour? Watch and pray that you may not enter into temptation; the spirit indeed is willing, but the flesh is weak' " (Matthew 26:40-41). It is striking that the three disciples he had taken with him had all fallen asleep, but only Peter was reprimanded by name. The special interest that Jesus seems to have had in Peter may suggest that he was sometimes held to a different standard from the others.

15. Unlike the other disciples, Peter was willing and able to fight for Jesus. A crowd of soldiers directed by Judas came to Gethsemane to arrest Jesus and lead him to the high priest. At least one of the disciples (John said it was Peter) was armed with a sword, and he used that sword to strike and cut off the ear of Malchus, a slave of the high priest (John 18:10). Jesus then ordered Peter to put away his sword. Most of the disciples

18. Before the Beginning

forsook Jesus and fled, but Peter and John followed as he was led away (Matthew 26:47-56, Mark14:43-50, Luke 22:47-53, John 18:1-11). As a possible point of interest, it may be mentioned that Luke, "the beloved physician" (Colossians 4:14), informs us that Jesus had healed the slave's severed ear (Luke 22:51).

16. Those who had seized Jesus led him to Caiaphas the high priest, and Peter followed at a distance. Peter entered the courtyard of the high priest in an effort to monitor the proceedings. While there he three times denied knowing Jesus. Afterward he regretted this denial and "wept bitterly" (Matthew 26:57-75). Peter's action is often represented as a major failure by him. It was certainly regretted by Peter himself, but it doesn't seem so bad when considered in the context of Jesus's arrest and crucifixion. At the arrest Peter was the only disciple who is said to have fought for his Lord against "a great crowd with swords and clubs, from the chief priests and the elders of the people" (Matthew 26:47). He apparently ended his resistance only because he was ordered to do so by Jesus (see number "15." above).

The fact that Peter had injured the slave of the high priest would have made his presence in the high priest's courtyard especially dangerous for him, and while there he was recognized by a relative of Malchus (John 18:26). The only other disciple that seems to have been present was John, who acknowledged that he was an acquaintance of the high priest and seemed able to come and go as he pleased (John 18:15). As will be considered further in the following section, acquaintanceship with the high priest would seem to be a strange credential in which to take pride for a disciple of the person the high priest is determined to destroy. Instead of focusing only on Peter's weakness in denying a knowledge of Jesus, one might also note Peter's bravery at Gethsemane and his devotion in entering the hazardous courtyard.

A somewhat similar indication of John's curious relationship to the high priest may be suggested by John's presence at the crucifixion of Jesus. No other disciples are said to have been present at the cross, and the others may have kept their distance out of concern for their lives: "And all his acquaintances and the women who had followed him from Galilee stood at a distance and saw these things" (Luke 23:49).

17. The specialness of Peter among the disciples can also be seen in other ways and in other texts. There are about three times as many mentions by name of the disciple Peter in the synoptic gospels (Matthew, Mark, and Luke) as there are of either of the next two most commonly mentioned disciples James and John, the sons of Zebedee. Furthermore, James and John are usually mentioned only as being present, while Peter is usually at the center of the action.

18. The book of Acts also includes many amazing stories about the activities of Peter after the ascension of Jesus. From the beginning of this book and after the ascension, Peter had a leading role in guiding the new Christian church. One of his first tasks was to find a replacement apostle for Judas who had betrayed Jesus and then had died:

18. Before the Beginning

> In those days Peter stood up among the brethren (the company of persons was in all about a hundred and twenty), and said, "Brethren, the scripture had to be fulfilled, which the Holy spirit spoke beforehand by the mouth of David, concerning Judas who was guide to those who arrested Jesus. . . . For it is written in the book of Psalms,
>
> 'Let his habitation become desolate,
> and let there be no one to live in it'; and
> 'His office let another take.'
>
> So one of the men who have accompanied us during all the time that the Lord Jesus went in and out among us, beginning from the baptism of John [the Baptist] until the day when he was taken up from us – one of these men must become with us a witness to his resurrection." And they put forward two, Joseph called Barsab'bas, who was surnamed Justus, and Matthi'as. . . . And they cast lots for them, and the lot fell on Matthi'as; and he was enrolled with the eleven apostles (Acts 1:15-26).

Thus, according to Acts the membership of the church seems to have increased from just the eleven "apostles whom he had chosen" (Acts 1:2) at the ascension to about one hundred and twenty when Peter addressed the "brethren." Through the leadership of Peter the deceased Judas was replaced by Matthias, restoring the number of apostles to twelve.

19. On the day of Pentecost, Peter taught the people with a powerful and persuasive sermon. A few excerpts from that sermon and the people's response are included here:

> "Men of Israel, hear these words: Jesus of Nazareth, a man attested to you by God with mighty works and

> wonders and signs which God did through him in your midst, as you yourselves know – this Jesus, delivered up according to the definite plan and foreknowledge of God, you crucified and killed by the hands of lawless men. But God raised him up, having loosed the pangs of death, because it was not possible for him to be held by it. . . . Let all the house of Israel therefore know assuredly that God has made him both Lord and Christ, this Jesus whom you crucified."
>
> Now when they heard this they were cut to the heart, and said to Peter and the rest of the apostles, "Brethren what shall we do?" And Peter said to them, "Repent, and be baptized every one of you in the name of Jesus Christ for the forgiveness of your sins; and you shall receive the gift of the Holy Spirit. For the promise is to you and to your children and to all that are far off, every one whom the Lord our God calls to him." And he testified with many other words and exhorted them, saying, "Save yourselves from this crooked generation." So those who received his word were baptized, and there were added that day about three thousand souls. And they devoted themselves to the apostles' teaching and fellowship, to the breaking of bread and the prayers (Acts 2:22-24, 36-42).

As a result of Peter's sermon, the membership in the new church was increased by "about three thousand souls."

20. As will be mentioned further in "21." below, one of Peter's most dramatic gifts was that of healing:

> Now Peter and John were going up to the temple at the hour of prayer, the ninth hour. And a man lame from birth was being carried, whom they laid daily at that gate of the temple which is called Beautiful to ask alms of those who entered the temple. Seeing Peter and John about to go into the temple, he asked for alms. And Peter

18. Before the Beginning

directed his gaze at him, with John, and said, "Look at us." And he fixed his attention upon them, expecting to receive something from them. But Peter said, "I have no silver and gold, but I give you what I have; in the name of Jesus Christ of Nazareth, walk." And he took him by the right hand and raised him up; and immediately his feet and ankles were made strong. And leaping up he stood and walked and entered the temple with them, walking and leaping and praising God. And all the people saw him walking and praising God, and recognized him as the one who sat for alms at the Beautiful Gate of the temple; and they were filled with wonder and amazement at what had happened to him (Acts 3:1-10).

21. In consideration of prophecies and end times, it is significant that Peter's interest in this subject is already indicated in Acts. He preached an important public sermon in Jerusalem following his healing of the man who was lame from birth. The sermon included the following statements:

> And now, brethren, I know that you acted in ignorance, as did also your rulers. But what God foretold by the mouth of all the prophets, that his Christ should suffer, he thus fulfilled. Repent therefore, and turn again, that your sins may be blotted out, that times of refreshing may come from the presence of the Lord, and that he may send the Christ appointed for you, Jesus, whom heaven must receive until the time for establishing all that God spoke by the mouth of his holy prophets from of old (Acts 3:17-21).

While not very detailed, this quotation from Peter refers to the return of Jesus and the predictions of the "holy prophets," as discussed more explicitly in 2 Peter 3 (Section 18.6).

The sermon, like its predecessor mentioned above in "19.", was very persuasive:

> And as they [Peter accompanied by John] were speaking to the people, the priests and the captain of the temple and the Sad'ducees came upon them, annoyed because they were teaching the people and proclaiming in Jesus the resurrection from the dead. And they arrested them and put them in custody until the morrow, for it was already evening. But many of those who heard the word believed; and the number of the men came to about five thousand (Acts 4:1-4).

Few modern preachers could claim to have inspired such revivals, or to have done so under such threats of persecution.

22. Besides the ability to heal, Peter could also have a role in the punishing of people on God's behalf. Such was the case in the frightening deaths of Ananias and his wife Sapphira (Acts 5:1-11). These people had lied about their donations to meet the needs of the newly growing company of believers.

23. A summary of the amazing healing program of the apostles, including especially Peter, is the following:

> Now many signs and wonders were done among the people by the hands of the apostles. And they were all together in Solomon's Portico. None of the rest dared join them, but the people held them in high honor. And more than ever believers were added to the Lord, multitudes both of men and women, so that they even carried out the sick into the streets, and laid them on beds and pallets, that as Peter came by at least his shadow might fall on some of them. The people also gathered from the towns

around Jerusalem, bringing the sick and those afflicted with unclean spirits, and they were all healed (Acts 5:12-16).

In short, Peter seems to have held a central position in an extraordinarily successful ministry of preaching and healing, even while the priests and other officials were doing everything in their power to shut down his program. Several other of his activities and achievements are recorded in Acts, but those indicated above should suggest the scope and success of Peter's ministry. His message of salvation and the return of Jesus may well have been carried forward on his behalf even following his martyrdom, as perhaps represented in 2 Peter and mentioned in Section 18.6 above.

18.8 Writings by and about Peter

It is commonly believed that the Apostle Peter had help from others in the actual writing of the several works with which he is associated. One of his most important assistants seems to have been John Mark, author of the Gospel of Mark:[14]

> There is much evidence that St. Peter chose St. Mark as his secretary or amanuensis.
> "Peter's claim to literary fame rests more firmly on his relation to the Gospel of Mark. Papias of Hierapolis recorded the fact that 'Mark, the interpreter of Peter, wrote down carefully what he remembered, both the sayings and the deeds of Christ, but not in chronological order, for he did not hear the Lord and he did not accompany him. At a later time, however, he did accompany Peter, who adapted his instruction to the needs [of his hearers], but not with the object of making a

connected series of discourses of our Lord. So Mark made no mistake in writing the individual discourses in the order in which he recalled them.'

"On this authority it is believed that Mark served as translator for Peter when he preached in Rome. As Peter told and retold his experiences with Jesus, Mark interpreted them again and again to Christian groups. This frequent repetition gave Mark an almost verbatim memory of Peter's recollections. After the death of Peter, Mark, realizing the value of Peter's first-hand account, recorded what he remembered so clearly in the document we know as the first of the Gospel records. Matthew and Luke obviously used Mark's Gospel in the writing of their lives of Jesus. In this manner Peter became the source for our earliest Gospel and thus to a large extent supplied the material for the first written record of our Lord. If this reconstruction of events is accurate, Mark's Gospel can be considered Peter's personal remembrance of his life with Jesus. As such it remains one of Peter's greatest contributions to the Christian church."[15]

The epistles of Peter were probably also written by his disciples after his martyrdom in Rome. The use of interpreters by Paul and Peter has been commented on by Jerome:

> Now, Saint Paul had a perfect knowledge of the holy Scriptures, he was naturally eloquent, and he possessed the gift of speaking in tongues, as he prides himself in the Lord, saying, "I praise my God that I speak through the gift of tongues more than all of you." Nevertheless he could not speak Greek in a manner worthy of the majesty and grandeur of our mysteries. Therefore Titus served as an interpreter, as Saint Mark used to serve Saint Peter, with whom he wrote his Gospel. Also we see that the two epistles attributed to Saint Peter have different styles and turn phrases differently, by which it is discerned that it

was sometimes necessary for him to use different interpreters.[16]

The conversion of Peter could also be of interest:

> Peter was brought to Christ by his brother Andrew. They were both fishermen, plying their trade on the sea of Galilee. Peter was a young man when he first met Christ, and certainly he was interested in the Messiah. When his brother Andrew announced that he had found the Messiah, Peter eagerly dropped his nets and went along to see for himself. Then he returned to his trade.
>
> It was sometime later that Jesus came to the shores of Galilee and there found Peter who had talked with Him before. There the invitation of Christ came, "Follow me and I will make you to become fishers of men" (Matthew 4:19). Peter and Andrew straightway left their nets and boats and followed Jesus. He was married and his mother-in-law apparently lived with him and his wife.[17]

Based on the above comments, one might imagine that Peter ("a young man") was perhaps in his twenties toward the beginning of Jesus's ministry. Peter is indeed known to have been married, as the three synoptic gospels all relate how Jesus healed Peter's mother-in-law (Matthew 8:14-17, Mark 1:29-31, Luke 4:38). It may also be reasonable to suppose that Peter had one or more children. He is known to have taken his wife with him on at least some of his journeys (1 Corinthians 9:5), and she was martyred in Rome just before he was:

> Peter's parting words to his wife as she was being led out to martyrdom are recorded by Clement of Alexandria in his Miscellanies and repeated by Eusebius in his Church History: "They say that when the blessed Peter

saw his own wife led out to die, he rejoiced because of her summons and her return home, and called to her very encouragingly and comfortingly, addressing her by name and saying. 'O thou, remember the Lord!' "[18]

These martyrdoms occurred in about 65 A.D. under Emperor Nero.

It is possible that Peter may also have visited Rome several years before the reign of Nero:

> Was Paul's the only mission to the West? The Acts tell us that in 43, after the death of James, Peter left Jerusalem "for another place" (Acts 12:17). He is lost from sight until 49, when we find him at the Council of Jerusalem. No canonical text has anything to say about his missionary activity during this time. But Eusebius writes that he came to Rome about 44, at the beginning of Claudius's reign (HE II, 14, 61). It seems certain that Rome was evangelised during the period from 43 to 49. Suetonius says that Claudius expelled the Jews in 50, because they were growing agitated "at the prompting of Chrestos." This shows that discussions between Jews and Judaeo-Christians were taking place, leading to conflicts which came to the ear of the emperor.[19]

Besides his wife, there is also the possibility that Peter may have been accompanied in his travels by his children. Thus, in his visit to Rome (mentioned above) in about 44 A.D., he is said to have traveled with his fair daughter Petronilla. There is an intriguing "legend" about Petronilla, and she is considered by the Catholic Church to be a saint:

> The Apostle Peter had a daughter born in lawful wedlock, who accompanied him in his journey from the

18. Before the Beginning

East. Being at Rome with him, she fell sick of a grievous infirmity which deprived her of the use of her limbs. And it happened that as the disciples were at meat with him in his house, one said to him, "Master, how is it that thou, who healest the infirmities of others, dost not heal thy daughter Petronilla?" And St. Peter answered, "It is good for her to remain sick:" but, that they might see the power that was in the word of God, he commanded her to get up and serve them at table, which she did; and having done so, she lay down again helpless as before; but many years afterwards, being perfected by her long suffering, and praying fervently, she was healed. Petronilla was wonderfully fair; and Valerius Flaccus, a young and noble Roman, who was a heathen, became enamored of her beauty, and sought her for his wife; and he being very powerful, she feared to refuse him; she therefore desired him to return in three days, and promised that he should then carry her home. But she prayed earnestly to be delivered from this peril; and when Flaccus returned in three days with great pomp to celebrate the marriage, he found her dead. The company of nobles who attended him carried her to the grave, in which they laid her, crowned with roses; and Flaccus lamented greatly.[20]

If Petronilla and Flaccus in this story were of a more modern era, one could imagine her replying to him in the words of a once popular song:

No! No! A thousand times no!
You cannot buy my caress!
No! No! A thousand times no!
I'd rather die than say yes.[21]

Flaccus was too intimidating for her to answer him in this manner, but when he returned she was dead.

One of the reasons for including the above brief discussion of Petronilla is her possible association with a seemingly unrelated subject in these considerations. This subject concerns again the adventures of the Argonauts in the more distant past. Chapters 15 to 17 involved the semi-mythological story of a young Greek warrior named Jason and his companions as they overcame even supernatural obstacles in their efforts to reclaim a golden fleece and thereby restore Jason to his rightful kingship. The principal text on which the story of their quest, as investigated here, was based is the *Argonautica* written by Apollonius Rhodius in about the middle of the third century B.C. As noted previously, there were other written versions of Argonaut-related stories, but only a few of those have survived in relatively complete form.

Besides the version of Apollonius Rhodius, there is another surviving Argonaut story that curiously may have a connection with the family of Peter. We were introduced above to Valerius Flaccus, who was described as a "young and noble" Roman who was also "very powerful." The connection between the Petronilla story and the Argonauts is that a writer named Valerius Flaccus composed his own substantial version of the *Argonautica*,[22] and that version survives today. The writer Flaccus is said to have flourished during the period 70-90 A.D. That period is not incompatible with the Flaccus of the Petronilla story. Like the would-be husband of Petronilla, the writer is said also to have been "probably not a poor man."[23]

Ironically, a significant part of the *Argonautica* story as developed by Flaccus relates to Medea's infatuation with Jason. It is tempting to imagine that in Flaccus's version, Medea's extreme desire for Jason reflects in some way what Flaccus may

18. Before the Beginning

have fantasized had been Petronilla's interest in himself. In any case, a single excerpt from this *Argonautica* may convey the general idea:

> Now doth late evening sunder thee, maiden [Medea], from the Thessalian stranger [Jason], and now do thy joys leave thee, while night comes on apace with balm for all save for the lover alone. So when, heart-sick, with feet that hesitated on the threshold's verge, she gained her chamber and in the darkness her imaginings took fire, long time she lay unsleeping, brooding on various plaints and ignorant of what plague was vexing her; at last at the height of her distress she dares avow the cause, and thus she speaks: "What mishap, what wilful deluding error holds me that so I lie ever sleepless? Not such for sure were my nights ere I had seen thy countenance, gallant youth. What madness makes me recall it again and yet again, though oceans lie between us? Why are my thoughts upon the stranger only? Nay, let him rather even now receive his kinsman Phrixus' fleece, his only quest and sole cause of all his toil. For when will he see this abode again? or when will my father visit Haemonia's cities? Happy they who braved the intervening seas, nor feared so long a voyage but straightway followed so valiant a hero to this land: for all that, valiant though he be, let him begone." Then as restlessly she tosses and tries now this, now that side of her couch, lo! she sees the doorway shimmering white as the daystar fades, nor less did the risen dawn refresh the love-sick girl than when a light shower lifts drooping ears of corn, or a welcome breeze descends upon weary oarsmen.[24]

18.9 Conclusion

This chapter has reviewed briefly some of the ways in which the date and mechanism of the creation of the earth (or world, or

Chronology In Context

universe) might be determined. Using a very literal interpretation of a particular version (Masoretic) of Genesis (the first book in the Bible), a specific date of 4004 B.C. was derived in the seventeenth century by Archbishop James Ussher. Using other versions and other interpretations of Genesis, other dates for creation are obtained.

Similarly, there is also a wide range of dates that might be inferred from biblical sources for the end of the earth. Using the biblical book Revelation (the last book in most Bibles), the prophecies seem to suggest that the earth should have ended "soon" after the writing of that book, thus perhaps in about 100 A.D. On the other hand, biblical sources besides Revelation predict a much later date for the end of the earth. In particular, it seems to be indicated in the Bible that one day in the distant future the earth will be burned with fire. The prophecies of Revelation will be considered in more detail in Chapter 19.

It may be noted that there is a curious symmetry in the comments just made on the timing of the creation and destruction of the earth. The first book in the Bible suggests that the earth was created much later than would seem to be indicated by other biblical and geological sources. On the other hand, the last book in the Bible suggests that the earth would come to an end much earlier than is indicated by other biblical and astrophysical sources.

As a final observation here, it is probably safe to say that many scientists do not consider the Bible to be a credible source, at least for questions concerning the origin or destruction of the earth. If those scientists thought about biblical chronology at all, they would probably disagree most particularly with the

18. Before the Beginning

chronological implications of Genesis and Revelation. The dates and mechanisms that scientists have recently been inferring for the creation and demise of the earth will be reviewed briefly in Chapter 20.

---

1. "Uniformitarianism," *Wikipedia*. Web. 20 June 2015.
2. "Catastrophism," *Wikipedia*. Web. 20 June 2015.
3. Ibid.
4. J. Finegan, *Handbook of Biblical Chronology: Principles of Time Reckoning in the Ancient World and Problems of Chronology in the Bible* (Princeton University Press, Princeton, New Jersey, 1964), p. 191.
5. J. J. O'Connor and E. F. Robertson, "Galileo Galilei," *JOC/EFR* (November 2002).
6. Pope John Paul II, "Faith can never conflict with reason," *L'Osservatore Romano*, N. 44, (1264) (4 November 1992).
7. S. Neuman, "1 in 4 Americans thinks the sun goes around the earth, survey says," *The Two-Way, Breaking News from NPR*, 14 February 2014, 5:55 PM ET.
8. "Giordano Bruno," *Wikipedia*. Web. 29 December 2017.
9. S. M. Miller and R. V. Huber, *The Bible: A History* (Good Books, Intercourse, PA, 2004), pp. 94-97.
10. "The Revelation to John," *The New Oxford Annotated Bible*, Third Edition, with the Apocryphal/Deuterocanonical Books, Edited by Michael D. Coogan (Oxford University Press, New York, 2001), p. 420 (New Testament).
11. D. McCasland, "Coming soon," *Our Daily Bread*, Saturday, 22 March 2014 (Our Daily Bread Ministries, Grand Rapids, Michigan).
12. S. K. Tee, "Waiting," *Our Daily Bread*, 3 December 2017 (Our Daily Bread Ministries, Grand Rapids, Michigan).

13. M. Stroud, "Walking on Water," *Our Daily Bread*, Wednesday, 24 June 2015 (Our Daily Bread Ministries, Grand Rapids, Michigan).
14. W. S. McBirnie, *The Search for the Twelve Apostles* (Tyndale House Publishers, Wheaton Illinois, 1977), p. 70.
15. A. Smith, *The Twelve Christ Chose* (Harper and Brothers, New York, 1958), pp. 21-22.
16. Saint Jerome, *Epistle 120 – To Hedibia*, Question #11, http://www.tertullian.org/fathers/jerome_hedibia_2_trans.htm. Web. 3 December 2016.
17. W. S. McBirnie, op. cit., p. 50.
18. E. J. Goodspeed, *The Twelve*, (The John C. Winston Company, Philadelphia, Pennsylvania, 1967), p. 157.
19. J. Danielou and H. Marrou, *The Christian Centuries* (Darton, Longman and Todd, London, 1964), p. 28.
20. A. Jameson, *Sacred and Legendary Art*, Third Edition (Houghton, Mifflin and Company, Boston and New York, 1957), Volume 1, p. 215.
21. A Sherman, A. Levis, and A. Silver, "No! No! A thousand times no!," (Fleischer Studios, New York, NY, 1935).
22. Valerius Flaccus, *Argonautica*, with an English Translation by J. H. Mozley, Loeb Classical Library (Harvard University Press, Cambridge, Massachusetts, 1998).
23. Ibid., Introduction, p. vii.
24. Ibid., Book VII, p. 361.

19. THE DISCIPLE AND THE PROPHET

"I am the Alpha and the Omega, the first and the last, the beginning and the end" (Revelation 22:13).

19.1 Introduction

The present chapter focuses first on some of the traits and accomplishments of the disciple John, and most of our information about him is to be found in the gospel of John. The later focus of the chapter is on writings and prophecies of John of Patmos, sometimes considered to be a different person from the disciple John. The plagues described in Revelation, as those in Exodus, seem to have several connections to aspects of volcanic eruptions, and the massive eruption of Mount Vesuvius in 79 A.D. may be of particular significance.

Several of the activities and achievements of the disciple John are summarized in Section 19.2. The question of whether this disciple would ever experience martyrdom like most other disciples is also raised, and a rumor that John would never die is considered in Section 19.3. In Revelation John of Patmos reports the prediction of seven plagues that would be visited upon earth by seven angels, and these plagues are reviewed in Section 19.4. The date and circumstances of the eruption of Mount Vesuvius are reviewed in Section 19.5. Similarities between the predicted plagues and events that occurred at the Vesuvius eruption and at the time of the exodus are considered in Section 19.6. The possibility that actual astrophysical occurrences are sometimes being described by John of Patmos is considered in Section 19.7.

19.2 Credentials of John

The disciple John had much in common with Peter, but there are also differences. While both had been early disciples of Jesus, it could be argued that Peter's accomplishments toward the establishment of the Christian religion may have been more substantial than those of John. Several aspects of Peter's work were considered previously in Chapter 18. John was a participant in many of the gospel stories, and he is the only apostle to have personally recorded those stories. He avoids use of his own name and sometimes refers to himself as the disciple whom Jesus loved. John had a unique status with respect to the high priest, and this gave him a special perspective on some of the events of holy week.

1. John and his brother James sought leadership positions with Jesus.

> And James and John, the sons of Zeb'edee, came forward to him, and said to him, "Teacher, we want you to do for us whatever we ask of you." And he said to them, "What do you want me to do for you?" And they said to him, "Grant us to sit, one at your right hand and one at your left, in your glory." But Jesus said to them, "You do not know what you are asking. Are you able to drink the cup that I drink, or to be baptized with the

> baptism with which I am baptized?" And they said to him, "We are able. And Jesus said to them, "The cup that I drink you will drink; and with the baptism with which I am baptized, you will be baptized; but to sit at my right hand or at my left is not mine to grant, but it is for those for whom it has been prepared." And when the ten heard it, they began to be indignant at James and John. And Jesus called them to him and said to them, "You know that those who are supposed to rule over the Gentiles lord it over them, and their great men exercise authority over them. But it shall not be so among you; but whoever would be great among you must be your servant, and whoever would be first among you must be slave of all. For the Son of man also came not to be served but to serve, and to give his life as a ransom for many" (Mark 10:35-45).

Thus, John and his brother James boldly asked Jesus to give them places of authority when he arrived at his leadership position. Jesus informed them that true greatness must be based upon service and sacrifice. That message isn't always understood by people seeking positions of authority. James is traditionally considered the first of the twelve apostles to be martyred, while John is often considered to be the only one of the apostles not to be martyred. Essentially the same story is quoted in Luke 20:20-28. However, in that version it is the mother of the sons of Zebedee that approached Jesus concerning her sons.

2. John served as an intermediary for the other disciples at the last supper.

> When Jesus had thus spoken, he was troubled in spirit, and testified, "Truly, truly, I say to you, one of you will betray me." The disciples looked at one another, uncertain

of whom he spoke. One of his disciples, whom Jesus loved [John], was lying close to the breast of Jesus; so Simon Peter beckoned to him and said, "Tell us who it is of whom he speaks." So lying thus, close to the breast of Jesus, he said to him, "Lord, who is it?" Jesus answered, "It is he to whom I shall give this morsel when I have dipped it." So when he had dipped the morsel, he gave it to Judas, the son of Simon Iscariot (John 13:21-26).

3. John was one of only two disciples that were present at the inquiry.

> Simon Peter followed Jesus, and so did another disciple [John]. As this other disciple was known to the high priest, he entered the court of the high priest along with Jesus, while Peter stood outside at the door. So the other disciple, who was known to the high priest, went out and spoke to the maid who kept the door, and brought Peter in (John 18:15-16).

In the quotation above John refers to himself as "another disciple" (John 18:15). As noted in Section 18.7, Peter was fearing for his life, while John was apparently on amicable terms with the family of the high priest Caiaphas. It was this family that was pressing for the crucifixion of Jesus. Thus, John knew the name Malchus of the slave of the high priest whose ear Peter had cut off (John 18:10-14). John even identified the kinsman of Malchus who had questioned Peter in the courtyard of the high priest (John 18:26).

John's status seems to have allowed him to come and go as he pleased and without fear, from the time of the events preceding the trial to the crucifixion itself. He tells us that he found a way for Peter to be brought into the courtyard while the inquiry was

19. The Disciple and the Prophet

taking place (John 18:15-16). In spite of this entry, Peter continued to be afraid for his life, and this led him to deny his identity as a disciple of Jesus.

4. John was present at the cross.

> But standing by the cross of Jesus were his mother, and his mother's sister, Mary the wife of Clopas, and Mary Mag'dalene. When Jesus saw his mother, and the disciple whom he loved [John] standing near, he said to his mother, "Woman, behold, your son!" Then he said to the disciple, "Behold, your mother!" And from that hour the disciple took her to his own home (John 19:25-27).

As implied in the quotation above, John may have been the only disciple who was safe and present in close proximity at the crucifixion. Other followers of Jesus were obliged to look on from a safe distance: "And all his acquaintances and the women who had followed him from Galilee stood at a distance and saw these things" (Luke 23:49). According to John, as quoted above, he was asked by Jesus from the cross to care for his mother. John's curious relationship with the family of the high priest might have helped to ensure Mary's safety.

5. John was the first disciple to reach the tomb.

> Now on the first day of the week Mary Mag'dalene came to the tomb early, while it was still dark, and saw that the stone had been taken away from the tomb. So she ran, and went to Simon Peter and the other disciple, the one whom Jesus loved [John], and said to them, "They have taken the Lord out of the tomb, and we do not know where they have laid him." Peter then came out with the

other disciple, and they went toward the tomb. They both ran, but the other disciple outran Peter and reached the tomb first; and stooping to look in, he saw the linen cloths lying there, but he did not go in. Then Simon Peter came, following him, and went into the tomb; he saw the linen cloths lying, and the napkin, which had been on his head, not lying with the linen cloths but rolled up in a place by itself. Then the other disciple, who reached the tomb first, also went in, and he saw and believed; for as yet they did not know the scripture, that he must rise from the dead. Then the disciples went back to their homes (John 20:1-10).

Thus, among other things, John emphasizes for us that he could run faster than Peter.

6. John seems not to have suffered martyrdom. An indication that John's special status continued long after the resurrection is the fact that he is often considered to be the only disciple, besides Judas Iscariot, not to have died a martyr's death: "Almost all the other Apostles met violent deaths, but John died peacefully in Ephesus at an advanced age, around the year 100 A.D."[1] Perhaps John was a more cautious advocate of the new Christian religion than were the other disciples, who are believed to have been killed because of their teachings.

19.3 Rumor that John wouldn't die

A more complex subject than the question of the possible martyrdom of John is the question of whether he was expected to die at all, or whether he would instead remain alive until the return of Jesus:

19. The Disciple and the Prophet

> Peter turned and saw following them the disciple whom Jesus loved [John], who had lain close to his breast at the supper and had said, "Lord, who is it that is going to betray you?" When Peter saw him, he said to Jesus, "Lord, what about this man?" Jesus said to him, "If it is my will that he remain until I come, what is that to you? Follow me!" The saying spread abroad among the brethren that this disciple was not to die; yet Jesus did not say to him that he was not to die, but, "If it is my will that he remain until I come, what is that to you?"
>
> This is the disciple who is bearing witness to these things, and who has written these things; and we know that his testimony is true (John 21:20-24).

As indicated in the above quotation, some people might have believed that the disciple John wouldn't die before the return of Jesus. This subject was presented as being raised by Jesus but not resolved by him. Clearly, John is not claiming in this text that he had been promised never to die. This helps support the distinction between John the disciple and John of Patmos (the author of Revelation mentioned previously in Section 18.5). John of Patmos seems to have been expecting and pleading for the hasty return of Jesus, which suggests perhaps that he had been hoping to avoid death.

By the time of the quoted conversation the disciples would already have known that Jesus would be leaving them soon:

> "A little while, and you will see me no more; again a little while, and you will see me." Some of his disciples said to one another, "What is this that he says to us, 'A little while, and you will not see me, and again a little while, and you will see me'; and, 'because I go to the Father'?" (John 16:16-17).

In this context the disciples could have been filled with questions for Jesus about their fates after his departure. In fact, in a later conversation, Jesus informed Peter of his coming martyrdom:

> "Truly, truly, I say to you, when you were young, you girded yourself and walked where you would; but when you are old, you will stretch out your hands, and another will gird you and carry you where you do not wish to go." (This he said to show by what death he was to glorify God.) And after this he said to him, "Follow me" (John 21:18-19).

That Peter could even think to ask Jesus his question about the disciple John in particular (John 21:20-24), and that John would quote Peter on this subject, suggests that there may already have been something different about John in comparison with the other disciples. This is indicated above (Section 19.2) in his acceptance into the company of the high priest and his presence at the cross where other disciples might have feared for their lives. A quotation given above indicates that John "is the disciple who is bearing witness to these things, and who has written these things; and we know that his testimony is true" (John 21:24). This statement might seem to suggest that John can verify the truth of his own testimony. However, the biblical standard for testimony is that when a person talks about himself his testimony cannot be trusted. Thus, on two occasions Jesus made statements (according to John) concerning testimony relating to himself:

> "I can do nothing on my own authority; as I hear, I judge; and my judgment is just, because I seek not my own will but the will of him who sent me. If I bear witness to myself, my testimony is not true; there is

19. The Disciple and the Prophet

another who bears witness to me, and I know that the testimony which he bears to me is true. You sent to John [the Baptist], and he has borne witness to the truth. Not that the testimony which I receive is from man; but I say this that you may be saved. He was a burning and shining lamp, and you were willing to rejoice for a while in his light. But the testimony which I have is greater than that of John; for the works which the Father has granted me to accomplish, these very works which I am doing, bear me witness that the Father has sent me. And the Father who sent me has himself borne witness to me. His voice you have never heard, his form you have never seen; and you do not have his word abiding in you, for you do not believe him whom he has sent. You search the scriptures, because you think that in them you have eternal life; and it is they that bear witness to me; yet you refuse to come to me that you may have life. I do not receive glory from men. But I know that you have not the love of God within you. I have come in my Father's name, and you do not receive me; if another comes in his own name, him you will receive. How can you believe, who receive glory from one another and do not seek the glory that comes from the only God? Do not think that I shall accuse you to the Father; it is Moses who accuses you, on whom you set your hope. If you believed Moses, you would believe me, for he wrote of me. But if you do not believe his writings, how will you believe my words?" (John 5:30-47).

Again Jesus spoke to them, saying, "I am the light of the world; he who follows me will not walk in darkness, but will have the light of life." The Pharisees then said to him, "You are bearing witness to yourself; your testimony is not true." Jesus answered, "Even if I do bear witness to myself, my testimony is true, for I know whence I have come and whither I am going, but you do not know whence I come or whither I am going. You judge according to the flesh, I judge no one. Yet even if I do judge, my judgment is true, for it is not I alone that judge, but I and he who sent me. In your law it is written that the testimony of two men is true; I bear witness to myself, and the Father who sent me bears witness to me." They said to him therefore, "Where is your Father?" Jesus answered, "You know neither me nor my Father; if you knew me, you would know my Father also." These words he spoke in the treasury, as he taught in the temple; but no one arrested him, because his hour had not yet come (John 8:12-20).

At this point it may be helpful to mention again a subject introduced above. The disciple John has quoted Jesus's statement on the reliability of testimony as follows: "If I bear witness to myself, my testimony is not true" (John 5:31). On the other hand, this assumed author of John has also written the following: "This is the disciple who is bearing witness to these things, and who has written these things; and we know that his testimony is true" (John 21:24). These statements are consistent if the "we" in John 21:24 includes later commentators on John's book.

19.4 Plagues of Revelation

A substantial portion of Revelation is devoted to plagues that are prophesied to inflict humanity prior to the second coming of Jesus. Many of these plagues are dramatic and horrific in nature, involving major geological, meteorological, and perhaps astronomical events. While questions have already been raised here concerning the accuracy of the chronology of the predictions by John of Patmos, his colorful and awesome descriptions are still of interest. One may wonder whether and how a Christian

19. The Disciple and the Prophet

writer of long imprisonment and doubtful identity could possibly have imagined events of the magnitude that he describes. To begin this consideration, it may be recalled that seven angels with seven trumpets in succession are said to have unleashed seven plagues upon the earth, and the descriptions of these plagues are included in Chapters 8 to 11 of Revelation. The following brief excerpts from these chapters, emphasizing geophysical aspects, will suggest the general concepts. The headings have been added here for clarity:

Introduction
Then I saw the seven angels who stand before God, and seven trumpets were given to them. And another angel came and stood at the altar with a golden censer. . . . Then the angel took the censer and filled it with fire from the altar and threw it on the earth; and there were peals of thunder, loud noises, flashes of lightning, and an earthquake.
Now the seven angels who had the seven trumpets made ready to blow them (Revelation 8:2-3,5-6).

First angel
The first angel blew his trumpet, and there followed hail and fire, mixed with blood, which fell on the earth; and a third of the earth was burnt up, and a third of the trees were burnt up, and all green grass was burnt up (Revelation 8:7).

Second angel
The second angel blew his trumpet, and something like a great mountain, burning with fire, was thrown into the sea; and a third of the sea became blood, a third of the living creatures in the sea died, and a third of the ships were destroyed (Revelation 8:8-9).

Third angel

The third angel blew his trumpet, and a great star fell from heaven, blazing like a torch, and it fell on a third of the rivers and on the fountains of water. The name of the star is Wormwood. A third of the waters became wormwood, and many men died of the water, because it was made bitter (Revelation 8:10-11).

Fourth angel

The fourth angel blew his trumpet, and a third of the sun was struck, and a third of the moon, and a third of the stars, so that a third of their light was darkened; a third of the day was kept from shining, and likewise a third of the night (Revelation 8:12).

Fifth angel

And the fifth angel blew his trumpet, and I saw a star fallen from heaven to earth, and he was given the key of the shaft of the bottomless pit; he opened the shaft of the bottomless pit, and from the shaft rose smoke like the smoke of a great furnace, and the sun and the air were darkened with the smoke from the shaft (Revelation 9:1-2).

Sixth angel

Then the sixth angel blew his trumpet. . . . [A]nd the heads of the horses were like lions' heads, and fire and smoke and sulphur issued from their mouths. By these three plagues a third of mankind was killed, by the fire and smoke and sulphur issuing from their mouths (Revelation 9:13,17-18).

Seventh angel

Then the seventh angel blew his trumpet. . . . [A]nd the ark of his covenant was seen within his temple; and there were flashes of lightning, loud noises, peals of

19. The Disciple and the Prophet

thunder, an earthquake, and heavy hail (Revelation 11:15,19).

The Revelation plagues are similar in some respects to the much earlier exodus plagues, described in Exodus 5-12. The exodus plagues have often been interpreted, at least in part, as consequences of volcanic eruptions at about the time of the exodus. Some comments on this possibility are included in Sections 8.5 "Pre-exodus volcanic events" and 8.6 "Post-exodus volcanic events".

19.5 Volcanic interpretation

The idea that the exodus plagues had some association with a volcanic eruption encourages the thought that the similar plagues described in Revelation may have been inspired in part by some much later volcanic eruption. Therefore it is of interest to see whether any significant eruption might have occurred in the world of the Roman Empire with which John of Patmos would have had some awareness. As it happens, there was a major eruption near the heart of the Empire at a time that may have been during or before the lifetime of the author of Revelation. The most deadly volcanic eruption known to have occurred in the ancient world was that of Mount Vesuvius in 79 A.D.:

> Mount Vesuvius is best known for its eruption in AD 79 that led to the burying and destruction of the Roman cities of Pompeii, Herculaneum and several other settlements. That eruption ejected a cloud of stones, ash and fumes to a height of 33 km (20.5 mi), spewing molten rock and pulverized pumice at the rate of 1.5 million tons per second, ultimately releasing a hundred thousand times the

thermal energy released by the Hiroshima bombing. An estimated 16,000 people died due to hydrothermal pyroclastic flows. The only surviving eyewitness account of the event consists of two letters by Pliny the Younger to the historian Tacitus.[2]

It is difficult to imagine anyone in the empire near the time of the eruption in 79 A.D. who would not have seen or heard about the occurrence and effects of this catastrophic event.

One of the ways to look for any relationship between the Revelation plagues and the eruption of Vesuvius would be to compare the plague descriptions with any records that might be available concerning the eruption. Eye-witness records of the eruption were left to us by Pliny the Younger, and his testimony is contained in two letters that he wrote. Letter 6.16 is the response by Pliny the Younger to a request from the historian Cornelius Tacitus for an account of the death of Pliny's uncle Pliny the Elder during the eruption of Vesuvius.[3] The letter reports observations of the beginning of the eruption as seen from Misenum. After these observations Pliny the Elder sailed with a fleet of warships on a mission of investigation and a rescue of victims from the eruption area. Reports from that mission provided the substance of the further observations that are recorded. Letter 6.20 reports the adventures of Pliny the Younger and his mother in Misenum during the eruption.

6.16. Pliny to Cornelius Tacitus

Thank you for asking me to send you a description of my uncle's death so that you can leave an accurate account of it for posterity;[4] I know that immortal fame awaits him if his death is recorded by you. It is true that he perished in a catastrophe which destroyed the loveliest regions of the earth, a fate shared by whole cities and their people, and one so memorable that is likely to make his name live for ever: and he himself wrote a number of books of lasting value: but you write for all time and can still do much to perpetuate his memory. The fortunate man, in my opinion, is he to whom the gods have granted the power either to do something which is worth recording or to write what is worth reading, and most fortunate of all is the man who can do both. Such a man was my uncle, as his own books and yours will prove. So you set me a task I would choose for myself, and I am more than willing to start on it.

My uncle was stationed at Misenum, in active command of the fleet.[5] On 24 August, in the early afternoon, my mother drew his attention to a cloud of unusual size and appearance. He had been out in the sun, had taken a cold bath, and lunched while lying down, and was then working at his books. He called for his shoes and climbed up to a place which would give him the best view of the phenomenon. It was not clear at that distance from which mountain the cloud was rising (it was afterwards known to be Vesuvius); its general appearance can be best expressed as being like an umbrella pine,[6] for it rose to a great height on a sort of trunk and then split off into branches, I imagine because it was thrust upwards by the first blast and then left unsupported as the pressure subsided, or else it was borne down by its own weight so that it spread out and gradually dispersed. Sometimes it looked white, sometimes blotched and dirty, according to the amount of soil and ashes it carried with it. My uncle's scholarly acumen[7] saw at once that it was important enough for a closer inspection, and he ordered a boat to be made ready, telling me I could come with him if I wished. I replied that I preferred to go on with my studies, and as it happened he had himself given me some writing to do.

As he was leaving the house, he was handed a message from Rectina, wife of Tascius whose house was at the foot of the mountain, so that escape was impossible except by boat. She was terrified by the danger threatening her and implored him to rescue her from her fate. He changed his plans, and what he had begun in a spirit of inquiry he completed as a hero. He gave orders for the warships[8] to be launched and went on board himself with the intention of bringing help to many more people besides Rectina, for this lovely stretch of coast was thickly populated. He hurried to the place which everyone else was hastily leaving, steering his course straight for the danger zone. He was entirely fearless, describing each new movement and phase of the portent to be noted down exactly as he observed them. Ashes were already falling, hotter and thicker as the ships drew near, followed by bits of pumice and blackened stones, charred and cracked by the flames: then suddenly they were in shallow water, and the shore was blocked by the debris from the mountain. For a moment my uncle wondered whether to turn back, but when the helmsman advised this he refused, telling him that Fortune stood by the courageous[9] and they must make for Pomponianus at Stabiae. He was cut off there by the breadth of the bay (for the shore gradually curves round a basin filled by the sea) so that he was not as yet in danger, though it was clear that this would come nearer as it spread. Pomponianus had therefore already put his belongings on board ship, intending to escape if the contrary wind fell. This wind was of course full in my uncle's favour, and he was able to bring his ship in.[10] He embraced his terrified friend, cheered and encouraged him, and thinking he could calm his fears by showing his own composure, gave orders that he was to be carried to the bathroom. After his bath he lay down and dined;[11] he was quite cheerful, or at any rate he pretended he was, which was no less courageous.

19. The Disciple and the Prophet

Meanwhile on Mount Vesuvius broad sheets of fire and leaping flames blazed at several points, their bright glare emphasized by the darkness of night. My uncle tried to allay the fears of his companions by repeatedly declaring that these were nothing but bonfires left by the peasants in their terror, or else empty houses on fire in the districts they had abandoned. Then he went to rest and certainly slept, for as he was a stout man his breathing was rather loud and heavy and could be heard by people coming and going outside his door. By this time the courtyard giving access to his room was full of ashes mixed with pumice-stones, so that its level had risen, and if he had stayed in the room any longer he would never have got out. He was wakened, came out and joined Pomponianus and the rest of the household who had sat up all night. They debated whether to stay indoors or take their chance in the open, for the buildings were now shaking with violent shocks, and seemed to be swaying to and fro, as if they were torn from their foundations. Outside on the other hand, there was the danger of falling pumice-stones, even though these were light and porous; however, after comparing the risks they chose the latter. In my uncle's case one reason outweighed the other, but for the others it was a choice of fears. As a protection against falling objects they put pillows on their heads tied down with cloths.

Elsewhere there was daylight by this time, but they were still in darkness, blacker and denser than any ordinary night, which they relieved by lighting torches and various kinds of lamp. My uncle decided to go down to the shore and investigate on the spot the possibility of any escape by sea, but he found the waves still wild and dangerous. A sheet was spread on the ground for him to lie down, and he repeatedly asked for cold water to drink. Then the flames and smell of sulphur which gave warning of the approaching fire drove the others to take flight and roused him to stand up. He stood leaning on two slaves and then suddenly collapsed, I imagine because the dense fumes choked his breathing by blocking his windpipe which was constitutionally weak and narrow and often inflamed. When daylight returned on the 26th – two days after the last day he had seen – his body was found intact and uninjured, still fully clothed and looking more like sleep than death.

Meanwhile my mother and I were at Misenum, but this is not of any historic interest, and you only wanted to hear about my uncle's death. I will say no more, except to add that I have described in detail every incident which I either witnessed myself or heard about immediately after the event, when reports were most likely to be accurate. It is for you to select what best suits your purpose, for there is a great difference between a letter to a friend and history written for all to read.

6.20. Pliny to Cornelius Tacitus

So the letter which you asked me to write on my uncle's death has made you eager to hear about the terrors and hazards I had to face when left at Misenum, for I broke off at the beginning of this part of my story. "Though my mind shrinks from remembering...I will begin."[12]

After my uncle's departure I spent the rest of the day with my books, as this was my reason for staying behind. Then I took a bath, dined, and then dozed fitfully for a while. For several days past there had been earth tremors which were not particularly alarming because they are frequent in Campania: but that night the shocks were so violent that everything felt as if it were not only shaken but overturned. My mother hurried into my room and found me already getting up to wake her if she were still asleep. We sat down in the forecourt of the house, between the buildings and the sea close by. I don't know

19. The Disciple and the Prophet

whether I should call this courage or folly on my part (I was only seventeen at the time) but I called for a volume of Livy and went on reading as if I had nothing else to do. I even went on with the extracts I had been making. Up came a friend of my uncle's who had just come from Spain to join him.[13] When he saw us sitting there and me actually reading, he scolded us both – me for my foolhardiness and my mother for allowing it. Nevertheless, I remained absorbed in my book.

Figure 19.1. Painting of Pliny the Younger working on a writing project while seated next to his mother in their home at Misenum. They are both being scolded by a friend of his uncle Pliny the Elder for their lack of concern for the imminent eruption of Mt. Vesuvius seen in the background.[14] (Reproduced with permission)

By now it was dawn, but the light was still dim and faint. The buildings round us were already tottering, and the open space we were in was too small for us not to be in real and imminent danger if the house collapsed. This finally decided us to leave the town. We were followed by a panic-stricken mob of people wanting to act on someone else's decision in preference to their own (a point in which fear looks like prudence), who hurried us on our way by pressing hard behind in a dense crowd. Once beyond the buildings we stopped, and there we had some extraordinary experiences which thoroughly alarmed us. The carriages we had ordered to be brought out began to run in different directions though the ground was quite level, and would not remain stationary even when wedged with stones. We also saw the sea sucked away and apparently forced back by the earthquake: at any rate it receded from the shore so that quantities of sea creatures were left stranded on dry sand. On the landward side a fearful black cloud was rent by forked and quivering bursts of flame, and parted to reveal great tongues of fire, like flashes of lightning magnified in size.

At this point my uncle's friend from Spain spoke up still more urgently: "If your brother, if your uncle is still alive, he will want you both to be saved; if he is dead, he would want you to survive him – why put off your escape?" We replied that we would not think of considering our own safety as long as we were uncertain of his. Without waiting any longer, our friend rushed off and hurried out of danger as fast as he could.

Soon afterwards the cloud sank down to earth and covered the sea; it had already blotted out Capri and hidden the promontory of Misenum from sight. Then my mother implored, entreated and commanded me to escape the best I could – a young man might escape, whereas she was old and slow and could die in peace as long as she had not been the cause of my death too. I refused to save

19. The Disciple and the Prophet

myself without her, and grasping her hand forced her to quicken her pace. She gave in reluctantly, blaming herself for delaying me. Ashes were already falling, not as yet very thickly. I looked round: a dense black cloud was coming up behind us, spreading over the earth like a flood. "Let us leave the road while we can still see," I said, "or we shall be knocked down and trampled underfoot in the dark by the crowd behind." We had scarcely sat down to rest when darkness fell, not the dark of a moonless or cloudy night, but as if the lamp had been put out in a closed room. You could hear the shrieks of women, the wailing of infants, and the shouting of men; some were calling their parents, others their children or their wives, trying to recognize them by their voices. People bewailed their own fate or that of their relatives, and there were some who prayed for death in their terror of dying. Many besought the aid of the gods, but still more imagined there were no gods left, and that the universe was plunged into eternal darkness for evermore. There were people, too, who added to the real perils by inventing fictitious dangers: some reported that part of Misenum had collapsed or another part was on fire, and though their tales were false they found others to believe them. A gleam of light returned, but we took this to be a warning of the approaching flames rather than daylight. However, the flames remained some distance off; then darkness came on once more and ashes began to fall again, this time in heavy showers. We rose from time to time and shook them off, otherwise we should have been buried and crushed beneath their weight. I could boast that not a groan or cry of fear escaped me in these perils, had I not derived some poor consolation in my mortal lot from the belief that the whole world was dying with me and I with it.

At last the darkness thinned and dispersed into smoke or cloud; then there was genuine daylight, and the sun actually shone out, but yellowish as it is during an eclipse. We were terrified to see everything changed, buried deep in ashes like snowdrifts. We returned to Misenum where we attended to our physical needs as best we could, and then spent an anxious night alternating between hope and fear. Fear predominated, for the earthquakes went on, and several hysterical individuals made their own and other people's calamities seem ludicrous in comparison with their frightful predictions. But even then, in spite of the dangers we had been through, and were still expecting, my mother and I had still no intention of leaving until we had news of my uncle.

Of course these details are not important enough for history, and you will read them without any idea of recording them; if they seem scarcely worth putting in a letter, you have only yourself to blame for asking them.

19.6 Comparison of plagues

There have been many comparisons between the Apostle John and the author of Revelation, sometimes referred to as John of Patmos, and it is often now accepted that these two persons named John might represent two distinct individuals. In the book Revelation, the person called John of Patmos is often understood to be predicting dramatic climate events that might occur at a time near the end of the earth. Several of these climate events or plagues are found to have close parallels in the eruption-related plagues that had been associated with the exodus.

With the information provided by Pliny the Younger, it is possible to undertake a comparison of the plagues described by John of Patmos and the events relating to the deadly eruption of Mount Vesuvius. The idea here is that John may have witnessed the actual eruption, or alternatively he may have heard eyewitness reports. The information gained from such interactions could have provided valuable insights to use in his writing of Revelation. For comparison a few statements of eruption-like events from Revelation are indicated here together with somewhat similar quotations from Pliny and also from Exodus:

A. Thunder and lightning

John of Patmos

[A]nd there were peals of thunder, loud noises, flashes of lightning, and an earthquake (Revelation 8:5).

Vesuvius, 79 A.D.

On the landward side a fearful black cloud was rent by forked and quivering bursts of flame, and parted to reveal great tongues of fire, like flashes of lightning magnified in size (Pliny 6:20).

Exodus

Then Moses stretched forth his rod toward heaven; and the Lord sent thunder and hail, and fire ran down to the earth. And the Lord rained hail upon the land of Egypt; there was hail, and fire flashing continually in the midst of the hail, very heavy hail, such as had never been in all the land of Egypt since it became a nation (Exodus 9:23-24).

B. Darkness

John of Patmos

[H]e opened the shaft of the bottomless pit, and from the shaft rose smoke like the smoke of a great furnace, and the sun and the air were darkened with the smoke from the shaft (Revelation 9:2).

Vesuvius, 79 A.D.

Elsewhere there was daylight by this time, but they were still in darkness, blacker and denser than any ordinary night, which they relieved by lighting torches and various kinds of lamp (Pliny 6.16).

We had scarcely sat down to rest when darkness fell, not the dark of a moonless or cloudy night, but as if the lamp had been put out in a closed room (Pliny 6.20).

Exodus

Then the Lord said to Moses, "Stretch out your hand toward heaven that there may be darkness over the land of Egypt, a darkness to be felt" (Exodus 10:21)

C. Hail or pumice falling

John of Patmos

Then the angel took the censer and filled it with fire from the altar and threw it on the earth (Revelation 8:5).

The first angel blew his trumpet, and there followed hail and fire, mixed with blood, which fell on the earth (Revelation 8:7).

[A]nd there were flashes of lightning, loud noises, peals of thunder, an earthquake, and heavy hail (Revelation 11:19).

Vesuvius, 79 A.D.

Outside on the other hand, there was the danger of falling pumice-stones, even though these were light and porous; however, after comparing the risks they chose the latter. In my uncle's case one reason outweighed the other, but for the others it was a choice of fears. As a protection against falling objects they put pillows on their heads tied down with cloths (Pliny 6.16).

Exodus

The hail struck down everything that was in the field throughout all the land of Egypt, both man and beast; and

19. The Disciple and the Prophet

the hail struck down every plant of the field, and shattered every tree of the field (Exodus 9:25).

D. Earthquakes

John of Patmos

[A]nd there were peals of thunder, loud noises, flashes of lightning, and an earthquake (Revelation 8:5).

Vesuvius, 79 A.D.

He was wakened, came out and joined Pomponianus and the rest of the household who had sat up all night. They debated whether to stay indoors or take their chance in the open, for the buildings were now shaking with violent shocks, and seemed to be swaying to and fro, as if they were torn from their foundations (Pliny 6:16).

For several days past there had been earth tremors which were not particularly alarming because they are frequent in Campania: but that night the shocks were so violent that everything felt as if it were not only shaken but overturned (Pliny 6.20).

The buildings round us were already tottering, and the open space we were in was too small for us not to be in real and imminent danger if the house collapsed (Pliny 6.20).

Once beyond the buildings we stopped, and there we had some extraordinary experiences which thoroughly alarmed us. The carriages we had ordered to be brought out began to run in different directions though the ground was quite level, and would not remain stationary even when wedged with stones. We also saw the sea sucked away and apparently forced back by the earthquake: at any rate it receded from the shore so that quantities of sea creatures were left stranded on dry sand (Pliny 6.20).

Fear predominated, for the earthquakes went on, and several hysterical individuals made their own and other people's calamities seem ludicrous in comparison with their frightful predictions (Pliny 6:20).

E. Loss (or jeopardy) of ships and poisoning of water

John of Patmos

[S]omething like a great mountain, burning with fire, was thrown into the sea; and a third of the sea became blood, a third of the living creatures in the sea died, and a third of the ships were destroyed (Revelation 8:8).

Vesuvius, 79 A.D.

Ashes were already falling, hotter and thicker as the ships drew near, followed by bits of pumice and blackened stones, charred and cracked by the flames: then suddenly they were in shallow water, and the shore was blocked by the debris from the mountain. For a moment my uncle wondered whether to turn back, but when the helmsman advised this he refused, telling him that Fortune stood by the courageous (Pliny 6:16).

Exodus [assuming that ash falls may have colored and poisoned the water]

Moses and Aaron did as the Lord commanded; in the sight of Pharaoh and in the sight of his servants, he lifted up the rod and struck the water that was in the Nile, and all the water that was in the Nile turned to blood. And the fish in the Nile died; and the Nile became foul, so that the Egyptians could not drink water from the Nile; and there was blood throughout all the land of Egypt (Exodus 7:20-21).

And all the Egyptians dug round about the Nile for water to drink, for they could not drink the water of the Nile (Exodus 7:24).

Among the many parallels indicated above are the weather events of thunder, lightning, and hail. Darkness is also said to have occurred at all of these locations, and suggests that the cause of this darkness (perhaps volcanic ash and dust) was near enough to the observers that it could be carried to them by air currents in the atmosphere. Parallels in the dramatic language of darkness employed by different sources are also striking. In the words of John, "We had scarcely sat down to rest when darkness fell, not the dark of a moonless or cloudy night, but as if the lamp had been put out in a closed room." In the exodus account we find the following: "Then the Lord said to Moses, 'Stretch out your hand toward heaven that there may be darkness over the land of Egypt, a darkness to be felt' " A similar darkness event was experienced by the Argonauts in Egypt at about the time of the exodus, as discussed in Section 15.6.

Earthquakes also seem to have occurred, as described in Section 19.6 above. The materials or objects darkening the atmosphere may also have been poisoning the water. In the words of John, "a third of the sea became blood, a third of the living creatures in the sea died." In the exodus account, "all the water that was in the Nile turned to blood. And the fish in the Nile died; and the Nile became foul, so that the Egyptians could not drink water from the Nile; and there was blood throughout all the land of Egypt."

19.7 An astrophysical possibility

Some descriptions of the Revelation plagues as reported by John of Patmos could have an astrophysical interpretation. It may be recalled that in the *Argonautica* Apollonius seems to be reporting the appearance of an asteroid as it burns and disintegrates while crossing the sky: "all the hosts of the stars appear shining through the gloom. . . . And even as a fiery star leaps from heaven, trailing a furrow of light, . . . "[15] John's description of the plagues, as quoted especially in Section 19.4, resembles a report of the burning remnants of an asteroid as they fell to earth:

> Then I saw the seven angels who stand before God, and seven trumpets were given to them. And another angel came and stood at the altar with a golden censer. . . . Then the angel took the censer and filled it with fire from the altar and threw it on the earth; and there were peals of thunder, loud noises, flashes of lightning, and an earthquake . . . (Revelation 8:2-3,5).

After that "there followed hail and fire, mixed with blood, which fell on the earth; and a third of the earth was burnt up, and a third of the trees were burnt up, and all green grass was burnt up" (Revelation 8:7). Then "a great star fell from heaven, blazing like a torch" (Revelation 8:10):

> I saw a star fallen from heaven to earth, and he was given the key of the shaft of the bottomless pit; he opened the shaft of the bottomless pit, and from the shaft rose smoke like the smoke of a great furnace, and the sun and the air were darkened with the smoke from the shaft (Revelation 9:1-2).

These quotes can be imagined to describe the impact of an asteroid with the surface of the earth.

This chapter has noted many parallels between the plagues described in Revelation and the volcanic eruption of Mount Vesuvius in 79 A.D. as well as an unidentified event at the much earlier time of the exodus. The preceding paragraph now suggests that the writings of Apollonius and John of Patmos could

19. The Disciple and the Prophet

alternatively be representing the impact of an asteroid on the surface of the earth. Related considerations are included in Section 9.2. The absence of similar indications in the detailed reports of Pliny from the vicinity of Mount Vesuvius may be taken to suggest that such interactions probably didn't happen near the time and place of that eruption. This leaves open the possibility that a significant impact might have occurred earlier in the eastern Mediterranean area, where observations could have been made and recorded. The resulting records may have found their way to the island of Rhodes (where Apollonius had studied and written) and to the island of Patmos (where John did his writing). Both Patmos and Rhodes are members of the Dodecanese island group in the southeast part of the Aegean Sea.

19.8 Conclusion

Dramatic events involving the climate experienced by the Israelites occur at various places in the Bible. Plagues and supernatural-seeming phenomena were reported for times around the exodus, for the times of interest in Revelation, and for several more-or-less clearly identified volcanic eruption events. Eruptions at about the time of the exodus seem to provide an interpretation for some of the exodus plagues. In principle, the well-documented eruption of Mount Vesuvius in 79 A.D. could explain some of the geophysical and climatological effects recorded in the Bible and elsewhere. Asteroid impacts are less familiar, but they could provide an alternative interpretation for some of the reported observations.

Chronology In Context

1. See for example, W. S. McBirnie, *The Search for the Twelve Apostles* (Tyndale House Publishers, Wheaton, Illinois, 1977), p. 109.
2. "Mount Vesuvius," *Wikipedia*. Web. 23 July 2015.
3. Pliny the Younger, *Letters* (Loeb Classical Library 55, Harvard University Press, Cambridge, Massachusetts, 1969), Translated by Betty Radice, Notes by Futrell, Volume I: Books 1-7.
4. "Tacitus' description of the eruption of Vesuvius and the death of Pliny the Elder would have been in the part of his Histories that does not survive; remember his narrative breaks off in 70 CE, some years before the volcanic disaster."
5. "Pliny the Elder was praefectus classi, praefect of the fleet, appointed by the emperor and holding imperium."
6. "A particular kind of pine tree known in the Mediterranean, shaped in outline like an umbrella (hence the name); we would call this a 'mushroom cloud'."
7. "As the author of the encyclopedic Natural History, Pliny would naturally have been interested in this unusual phenomenon."
8. "Pliny originally was going to head toward Vesuvius in a small light galley; after receiving the note, he ordered out the larger quadriremes, as they were much larger and better able to take numbers of people to safety."
9. "Variations on this idiomatic saying, 'Fortune favors the brave,' can be found in a number of Roman authors."
10. "So at first, apparently, wind is blowing toward the mountain, allowing Pliny's sails to push the boats toward the shore, but making it difficult to launch sailboats in flight from Pompeii and/or Herculaneum. The wind would later shift, according to the debris pattern preserved archaeologically."
11. "Romans reclined on couches to dine formally."
12. "Aeneid, 2:12-13."

13. "Pliny the Elder had been a procuratorial governor in Spain."
14. This painting is entitled "Pliny the Younger and his Mother at Misenum, 79 A.D., 1785; and it was painted by Angelica Kauffmann, British, born in Switzerland, 1741-1807. It is owned by and on display at Princeton University Art Museum. Other information includes the following: Oil on canvas, 103 x 127.5 cm (40 9/16 x 50 3/16 in.), frame 140.5 x 116 cm (55 5/16 x 45 11/16 in.), Museum purchase, gift of Franklin H. Kissner, 1969-89.
15. *Apollonius Rhodius: The Argonautica*, With an English translation by R. C. Seaton (Loeb Classical Library, Harvard University Press, Cambridge, Massachusetts, 1980), Book III, p. 287.

20. SCIENCE AND CHRONOLOGY

"[T]he heavens and earth that now exist have been stored up for fire, being kept until the day of judgment and destruction of ungodly men" (2 Peter 3:7).

20.1 Introduction

As indicated in Chapter 1, there are many different stories of the origin of the universe and its first inhabitants. These stories appear to have been composed at different times and in different places, and they also differ widely in their chronologies of the creation events. Of particular interest in this study has been, of course, the biblical version of creation. Perhaps not surprisingly, this version was suggested to have close parallels in cultures that developed in the geographical areas that at one time or another were associated with the ancient Hebrews. The Bible-related creation stories have been accepted as literal history by their advocates for several centuries. One of the more popular of the creation chronologies is associated with Archbishop James Ussher as mentioned previously in Chapter 18. Archbishop Ussher estimated that the starting date of creation was in 4004 B.C. Assuming that more typical modern life expectancies would also apply for the early personalities of the Bible, somewhat later

20. Science and Chronology

dates for the biblical flood and garden of Eden stories than those derived by Ussher were proposed in Tables 12.2 and 12.4. In addition to Bible-related versions, several other ancient cultures have provided us with creation accounts and chronologies.

The idea that the universe has not always existed has been the understanding of most Bible readers because of the statements to that effect in Genesis. There it is indicated explicitly that God created the heavens and the earth during the very short time span of about six days. Creation activities culminated in the creation of human beings, and thus the history of the universe and of humanity began abruptly at essentially a moment in time. While it is tempting and straightforward to establish the creation date of the universe according to Genesis, this is not the only approach to finding a date for creation. Thus, other Bible verses suggest very different results for the origin of the earth. As discussed in Section 18.4, Psalm 90 (ascribed to Moses) informs us that God doesn't measure time in the way that people do. Long ages before God created the earth he was said already to have been active in the universe: "For a thousand years in thy sight are but as yesterday when it is past, or as a watch in the night" (Psalm 90:4). Thus, the creation of the universe and its inhabitants may have been underway many thousands of years before the creation dates that could be inferred from Genesis.

During the past century the big bang theory became one of the key elements in many scientific discussions of the origin of the universe. A summary of this theory is included in Section 20.2. The explanation of the origin of life held by many scientists is mentioned in Section 20.3, and the ultimate fate of the universe is considered in Section 20.4. Of more particular interest might be the scientists' estimation of the ultimate fate of the earth, and this topic is addressed in Section 20.5. The fate of humanity is considered briefly in Section 20.6. The topics in this chapter are included for the possible interest of some readers, but the greater emphasis of these studies has been on the exploration of subjects and chronologies associated more directly with the Bible.

20.2 Big bang theory

The idea that the universe had no beginning was popular among many atheistic scientists and others until early in the twentieth century, but a different approach to the chronology of the universe has been advanced in more recent years. A striking feature of this newer approach is its claim that the universe had a beginning, but the beginning didn't require a creator. Thus, atheists have developed a chronology of the universe using experimental data obtained in the present, together with theoretical models developed by astronomers, cosmologists, geologists, biologists, etc., to extrapolate far into the past and the future. Though there isn't agreement on all details, there is now a fairly specific consensus on some aspects of such a model. According to these interpretations, all of the mass and energy in the universe emerged spontaneously long ago from a singularity (location in space-time where the gravitational field is infinite). This event is often called the big bang. By some estimates, the origin of the universe occurred about 13.7 billion years in the past. The model is intended to be useable in making estimates about the state of the universe at any location and at any time from its origin and on into the indefinite future. A concern is that one could be attempting to apply a model inferred from a limited

20. Science and Chronology

range of experiments to conditions that may become very different as new data are obtained.

The approach just outlined, accepted by many atheists, is of course radically different from the traditional beliefs of Bible readers. Recent generations of scientists seem mostly to believe that there is no God, or that if there is he was not an active participant in the creation of the universe. This section attempts to provide a brief summary of the big bang theory and also to consider whether that theory might have anything in common with the biblical creation story. The theory itself involves a substantial amount of esoteric mathematics, and only a qualitative sketch is attempted here.

Observations in astronomy during the twentieth century brought dramatic changes to the way in which many scientists tended to think about the history of the universe. As an example, a summary of those developments has been given by the noted physicist Stephen Hawking:

> When most people believed in an essentially static and unchanging universe, the question of whether or not it had a beginning was really one of metaphysics or theology. One could account for what was observed equally well on the theory that the universe had existed forever or on the theory that it was set in motion at some finite time in such a manner as to look as though it had existed forever. But in 1929, Edwin Hubble made the landmark observation that wherever you look, distant galaxies are moving rapidly away from us. In other words, the universe is expanding. This means that at earlier times objects would have been closer together. In fact, it seemed that there was a time, about ten or twenty thousand million years ago, when they were all at exactly the same place and when,

therefore, the density of the universe was infinite. This discovery finally brought the question of the beginning of the universe into the realm of science.

Hubble's observations suggested that there was a time, called the big bang, when the universe was infinitesimally small and infinitely dense. Under such conditions all the laws of science, and therefore all ability to predict the future, would break down. If there were events earlier than this time, then they could not affect what happens at the present time. Their existence can be ignored because it would have no observational consequences. One may say that time had a beginning at the big bang, in the sense that earlier times simply would not be defined. It should be emphasized that this beginning in time is very different from those that had been considered previously. In an unchanging universe a beginning in time is something that has to be imposed by some being outside the universe; there is no physical necessity for a beginning. One can imagine that God created the universe at literally any time in the past. On the other hand, if the universe is expanding, there may be physical reasons why there had to be a beginning. One could still imagine that God created the universe at the instant of the big bang, or even afterwards in just such a way as to make it look as though there had been a big bang, but it would be meaningless to suppose that it was created before the big bang. An expanding universe does not preclude a creator, but it does place limits on when he might have carried out his job![1]

Depending on the completeness and accuracy of current models of physics, it could be possible, in principle, to start from the current state of the universe and its observed rate of expansion and calculate the state of the universe for times into the distant past or distant future. Many discussions of the results

20. Science and Chronology

of such calculations have been published. Like the universe itself, scientists' understanding of the underlying physics is changing with time, and some of the implications of their physical models are also evolving with time. More information is readily available on the internet and elsewhere for those who might be interested.

There are very significant differences between the scientists' version of the origin of the universe and the biblical version. The most obvious difference is that for many scientists the universe was formed with no help from a creator. We are told that postulated conditions at the time of the big bang were bizarre and, of course, outside of our experience. Thus, we should not be surprised to also be told that our entire universe seems to have made itself out of nothing, but proof of this theory seems unlikely. Acceptance seems to depend in part on how determined one is to deny the existence of a creator, and the majority attitude today seems to be either to trust scientists or to not think about such questions at all.

Other differences between the big bang model and the biblical account are also inevitable, as the biblical authors and editors would not have had the experience or vocabulary to describe the objects, events, and time intervals that a scientist might employ today in postulating the early history of the universe. In the big bang model the time required for "creation" is of course vastly greater than the time period indicated in Genesis. However, the arrangement of creation events is somewhat similar between the two interpretations.

It should not be inferred from the above considerations that the big bang model is rejected by all Bible readers. In spite of discrepancies between the Bible and the latest models of science, many people who today would claim to be religious are willing to accept the tenets of the big bang model:

> Many people do not like the idea that time has a beginning, probably because it smacks of divine intervention. (The Catholic Church, on the other hand, seized on the big bang model and in 1951 officially pronounced it to be in accordance with the Bible.)[2]

In this position a religious person might insist that God exists, even though in the conventional big bang theory God has no verifiable involvement in the creation or day-to-day operation of the universe. Not all readers might consider this to be a reasonable point of view.

20.3 Origin of life

Many scientists have taken a special interest in the origins of plants, animals, and humans. They usually explain these origins by means of the theory of evolution. This theory was introduced by Charles Darwin and is often associated with the words "survival of the fittest." A modern one-paragraph version has been given by Hawking:

> The earth was initially very hot and without an atmosphere. In the course of time it cooled and acquired an atmosphere from the emission of gases from the rocks. This early atmosphere was not one in which we could have survived. It contained no oxygen but a lot of other gases that are poisonous to us, such as hydrogen sulfide (the gas that gives rotten eggs their smell). There are, however, other primitive forms of life that can flourish under such conditions. It is thought that they developed in the oceans, possibly as a result of chance combinations of

atoms into large structures, called macromolecules, which were capable of assembling other atoms in the ocean into similar structures. They would thus have reproduced themselves and multiplied. In some cases there would be errors in the reproduction. Mostly these errors would have been such that the new macromolecule could not reproduce itself and eventually would have been destroyed. However, a few of the errors would have produced new macromolecules that were even better at reproducing themselves. They would have therefore had an advantage and would have tended to replace the original macromolecules. In this way a process of evolution was started that led to the development of more and more complicated, self-reproducing organisms. The first primitive forms of life consumed various materials, including hydrogen sulfide, and released oxygen. This gradually changed the atmosphere to the composition that it has today and allowed the development of higher forms of life such as fish, reptiles, mammals, and ultimately the human race.[3]

A special aspect of some major scientific conjectures, such as the big bang theory or the theory of evolution, in contrast to most other scientific theories, is that acceptance or rejection can have an intimate connection with a person's religious beliefs. Many of the greatest scientists of modern times have been devout Christians who, as caretakers of God's "garden," seemingly did all they could to advance human understanding of the world about them. A survey conducted by the Pew Research Center for the People and the Press in 2009 revealed that 33 percent of scientists who are members of the American Association for the Advancement of Science (The World's Largest General Scientific Society) say that they believe in God.[4]

20.4 Fate of the universe

The preceding discussions of this chapter have dealt briefly with the most popular modern theories concerning the creation of the universe and its inhabitants, from galaxies to human beings. The continuing expansion of the universe has allowed scientists using theoretical models to extrapolate backward almost to the beginning of time. Some scientists have concluded that the universe originated about 13.7 billion years ago by means that are not necessarily understood.

In continuing, it may be of interest to consider more briefly what scientists originally believed might be the ultimate fate of the expanding universe. On the grandest scale this question formerly seemed to have three possible answers or Friedmann models:[5]

1. The gravitational forces arising from the mass distribution of the universe by well-known laws of physics may slow the expansion of the universe, but those forces will never be strong enough to stop that growth. Thus, the universe will continue expanding forever.

2. The gravitational forces are just sufficient to eventually bring the expansion of the universe to a complete stop at infinite size and after an infinite period of time.

3. The gravitational forces are so strong that they will reverse the expansion and cause the universe to collapse. No one knows what would happen after that collapse. One of the possibilities suggested is that another big bang would occur leading to a new universe.

The finding of the correct answer of these three was believed to be mostly a problem in general relativity. If one could determine from astronomy the present distribution of the mass and the velocity of the matter in the universe, one could predict its future distribution. More recently, however, it has been determined that the gravitational forces in the universe are not so simple as once believed. One difficulty is that the peculiar rotation properties of galaxies (rotation rate versus radius) require that the galaxies include gravitational mass that otherwise has never been seen. This postulated invisible mass is usually called "dark matter," and efforts are underway to confirm its existence, distribution, and consequences.

A second difficulty is that the expansion of the universe seems to be accelerating rather than slowing as one would expect from gravitational attraction. This suggests that the space between galaxies is filled with some form of energy that is pushing the galaxies apart. This is often now called "dark energy," and efforts are also underway to better understand this effect. In the meantime controversies remain concerning the nature of dark matter and dark energy.

A recently discovered difficulty with the big bang theory concerns what are called supermassive black holes. A black hole is an object (or region of spacetime) that contains so much mass that its gravitational field does not permit light or anything else to escape from its surface. Thus, a black hole would appear black to any observer. A gravitating object (such as a star or planet) tends to attract other objects to it. After a sufficient period of time some of these objects might gain enough mass to become black holes. Black holes are not particularly rare in the universe. For example, it is now believed that in our Milky Way Galaxy there may be as many as 20,000 black holes within one parsec (19 trillion miles) of the galaxy center.[6] For the most part, the processes leading to the creation of black holes had been considered to be well understood. As it happens, however, recent experimental results have shown that very fundamental mysteries about black hole formation still remain.

Black holes are thought to have begun their existence as more conventional stars, which then grew by accreting mass from their surrounding environment. Some of these growing stars eventually became what are known as supermassive black holes. However, there remain difficulties with this interpretation:

> A team of astronomers, including two from MIT, has detected the most distant supermassive black hole ever observed. The black hole sits in the center of an ultrabright quasar, the light of which was emitted just 690 million years after the Big Bang. That light has taken about 13 billion years to reach us – a span of time that is nearly equal to the age of the universe.
>
> The black hole is about 800 million times as massive as our sun – a Goliath by modern-day standards and a relative anomaly in the early universe. "This is the only object we have observed from this era," says physics professor Robert Simcoe of MIT's Kavli Institute for Astrophysics and Space Research. "It has an extremely high mass, and yet the universe is so young that this thing shouldn't exist. The universe was just not old enough to make a black hole that big. It's very puzzling."[7]

Yet another problem for the big bang theory concerns the value of the Hubble Constant. As noted in Section 20.2 above, the concept of the ongoing expansion of the universe was introduced in 1929 by Edwin Hubble, and the expansion rate is now characterized by what is called the Hubble constant. However, the calculated value of the Hubble constant seems to vary significantly depending on the procedures that one uses to determine that value. These discrepancies have not been resolved:

> Applying the standard model of cosmology – the Lambda Cold Dark Matter (λCDM) model – researchers used the [cosmic microwave background] CMB map to calculate the Hubble constant, a number that describes how quickly the universe is expanding. But that number disagreed with calculations based on telescope observations of supernovae and pulsating stars. . . .
>
> Unfortunately, no one knows where the discrepancies come from.[8]

Thus, in several regards the big bang model does not yet provide a complete and satisfactory explanation of the current observable behavior of the universe. One might reasonably suppose that any of the model's estimates of unobservable past and future behavior could have significant deficiencies as well.

20.5 Fate of the earth

The main interest of human beings is usually focused on the much smaller-scale "universe" of at most the dimensions of our solar system. The earth is thought by scientists to have formed about 4.5 billion years in the past.[9] The end of the universe as it pertains to humanity may reduce to the much narrower subject of the end of the earth. In commenting on the importance of paying attention to the "law and the prophets," Jesus seems to have indicated that the earth would not last forever: "Think not that I have come to abolish the law and the prophets; I have come not to abolish them but to fulfil them. For truly, I say to you, till heaven and earth pass away, not an iota, not a dot, will pass from the law until all is accomplished" (Matthew 5:17-18). Concerning those who have rejected the law and the prophets, it should be remembered that "it is appointed unto men once to die, but after this the judgment" (Hebrews 9:27 KJV). It may also be recalled that "there is no repentance after death."[10]

From the point of view of astronomy and cosmology, the end of the earth could happen at any time and without much warning. Scientists tell us that a collision between the earth and another planet-sized object could destroy the earth, and known individual past collisions between asteroids and the earth have killed much of the life on our planet. Such collisions are not common on the human time scale, and thus they are unlikely to occur in our own lifetimes. On the other hand, over longer periods of time, asteroid impacts could bring an end to humanity long before the end of the earth.

It has recently been suggested by some scientists that there may be a long-term periodicity to asteroid impacts that is correlated with the periodic crossing of the disk of our galaxy by our solar system. The probability of asteroid impacts with the earth is thought to be greatest when the solar system is crossing the plane of the galaxy, where the concentration of astrophysical debris is expected to be greatest. The inferred time between crossings of the plane is calculated to be about twenty-six million years:[11] "[T]here may be grand geological cycles driven by astronomical circumstances, as evidenced by the 26-million-year extinction cycle seen in the fossil record." Thus, some times for asteroid impacts would seem to be more likely than others.

However, if human beings could somehow manage to survive on earth for about five billion years and find nowhere else to go, then their fate would be sealed with the destruction of the earth by the sun. According to cosmologists, the outer layers of the sun will expand greatly, and the sun will become a so-called red giant star. Eventually the earth may be burned up or made uninhabitable by the growing sun. The sun (without the earth) will then shine for more billions of years before it consumes its remaining fuel and goes dark.

Astronomer A. Wolszczan is one of those who have been particularly interested in the fate of planets orbiting around expanding red giant stars. Two excerpts relating to his studies include the following:[12]

> More recently, Wolszczan has turned his attention to another class of doomed planets. He has begun finding and studying worlds around red giants, elderly stars that have nearly exhausted their nuclear fuel. In a last spasm of activity, they swell up, brighten tremendously and shed enormous clouds of gas. Any planets circling a red giant would get baked and buffeted in the process. That fate awaits Earth in about 5 billion years, when our sun will join the ranks of the red giants. For the distant planets Wolszczan is scrutinizing, the future is now.
>
> Only the most massive stars evolve into pulsars, and such stellar bigwigs are rare. Roughly 97 percent of the stars in the Milky Way are smaller ones that take another evolutionary path, leading eventually to the red giant stage and continuing beyond. A mainstream star like the sun grows gradually brighter and hotter as it ages (it's

happening right now) until the final spasm that inflates it into a red giant. In short order, astronomically speaking, the red giant blows off its outer layers and leaves behind a white dwarf – essentially the naked heart of the star – which slowly cools to eternal blackness.

If you want to understand what happens to most dying, dead and reborn planets, red giants are the place to start. Wolszczan has shifted much of his attention in this direction, and he enlisted a variety of collaborators to help. Since 2004, they've looked at a set of about a thousand stars, mostly red giants. "If we start detecting planets around stars like that, we can tell what happens to planetary systems when their stars begin to evolve and lose mass and swell up and do all those unpleasant things," he says.

Wolszczan and others have already found more than 40 planetary systems around red giants, allowing them to sketch out how the same process will play out here at home. As the sun grows more luminous, the solar system's habitable zone will shift outward. Earth will overheat in about a billion years, but Mars will become balmy. Then the moons of Jupiter and Saturn will melt, with Europa and Titan turning into temporary ocean worlds. At the sun's ruddy peak, it will radiate so much energy that even Pluto will reach comfy temperatures, according to Alan Stern, leader of the New Horizons mission that is heading there next year.

The red giant phase is make-or-break time for planetary survival. As the star sheds its outer layers, it becomes less massive, loosening its gravitational grip. In response, the planets migrate outward into new orbits, potentially turning the whole system chaotically unstable. "You may get really dramatic evolution in the system, including orbit crossings, planet collisions and all kinds of interesting things," Wolszczan says. All the while, the star also keeps expanding, threatening to consume its children.

20. Science and Chronology

> Wolszczan has seen these outcomes, planets with highly oval orbits, or others persisting only as phantom gas clouds in their star's outer layers. The data do not yet definitively show which way Earth will go. Probably it will survive as a ball of rock, but thoroughly sterilized. Mercury and Venus will almost surely be vaporized – about as ghostly as you can get.
>
> Wolszczan and colleagues recently caught both possible outcomes at work around a red giant star called BD+48 740. In a 2012 paper, the researchers report that the surface layers of BD+48 740 contain high levels of lithium, an element common in planets but almost never seen in stars. They interpret the lithium as the chemical remains of a cremated planet. At the same time, one major planet, slightly more massive than Jupiter, still orbits the star, but on a disturbed, elongated path."

The above scenario described by scientists is not actually new with them. In a general sort of way, the idea of the earth ending by being burned up has been described at several places in the Bible, as considered previously in Section 18.5. The following quotation was included in Section 18.6 in a discussion of the prophetic writings of the apostle Peter:

> This is now the second letter that I have written to you, beloved, and in both of them I have aroused your sincere mind by way of reminder; that you should remember the predictions of the holy prophets and the commandment of the Lord and Savior through your apostles. First of all you must understand this, that scoffers will come in the last days with scoffing, following their own passions and saying, "Where is the promise of his coming? For ever since the fathers fell asleep, all things have continued as they were from the beginning of creation." They deliberately ignore this fact,

that by the word of God heavens existed long ago, and an earth formed out of water and by means of water, through which the world that then existed was deluged with water and perished. But by the same word the heavens and earth that now exist have been stored up for fire, being kept until the day of judgment and destruction of ungodly men.

But do not ignore this one fact, beloved, that with the Lord one day is as a thousand years, and a thousand years as one day. The Lord is not slow about his promise as some count slowness, but is forbearing toward you, not wishing that any should perish, but that all should reach repentance. But the day of the Lord will come like a thief, and then the heavens will pass away with a loud noise, and the elements will be dissolved with fire, and the earth and the works that are upon it will be burned up (2 Peter 3:1-10).

As foretold by Peter in the quotation above (2 Peter 3:3-4), there certainly are many people today that scoff at the biblical prophecies concerning the return of Jesus. Peter's statement (2 Peter 3:5) concerning "an earth formed out of water and by means of water" brings to mind the scientists' statement about the formation of the earth quoted previously in Section 9.2:

Asteroids and the comets from the Jupiter-Saturn region were the first water deliverers, when the Earth was less than half its present mass. The bulk of the water presently on Earth was carried by a few planetary embryos, originally formed in the outer asteroid belt and accreted by the Earth at the final stage of its formation. Finally, a late veneer, accounting for at most 10% of the present water mass, occurred due to comets from the Uranus-Neptune region and from the Kuiper Belt.[13]

A following statement by Peter (2 Peter 3:6) may be a reference to the flood: "the world that then existed was deluged with water." The concluding statement by Peter in the above quotation (2 Peter 3:10) is "the earth and the works that are upon it will be burned up." This brief prophecy is compatible with the prediction of scientists, quoted earlier in this section, that the earth would either be "vaporized" or at least "thoroughly sterilized" during the anticipated expansion of the sun.

20.6 Fate of humanity

As this study comes to an end, it may be appropriate to repeat a few words about the possibly gloomy future of humanity as anticipated by many scientists. The preceding section has suggested that the earth may become too hot for human habitation in about a billion years, before it is completely sterilized or vaporized in about five billion years. However, it seems likely from past history that our civilization may end much sooner than that as a result of collisions with asteroids. More unfortunately still, humanity may not survive long enough to experience any of these disasters.

We are living in a time of great material abundance, together with unprecedented advances in science, medicine, energy production, and food resources. In spite of this wealth, we are rapidly consuming our irreplaceable energy resources, polluting our environment, neglecting the threat of asteroid encounters, and facing the prospect of disastrous future wars. Our ignoring of global warming, especially at present in the U.S., may hasten the occurrence of irreversible climate changes with resulting misery for our descendants. In the area of religion, many of the important beliefs and behaviors associated with our Bible-based religious institutions are being abandoned or corrupted.

The underlying arrogance and greed of our human natures are not newly discovered phenomena. Observations by Josephus from the first century A.D. remain valid today:

> [U]nderstand and consider the disposition of men, that while they are private persons, and in a low condition, because it is not in their power to indulge nature, nor to venture upon what they wish for, they are equitable and moderate, and pursue nothing but what is just, and bend their whole minds and labours that way; then it is that they have this belief about God, that he is present to all the actions of their lives, and that he does not only see the actions that are done, but clearly knows those their thoughts also, whence those actions do arise.
>
> But when once they are advanced into power and authority, then they put off all such notions, and, as if they were no other than actors upon a theatre, they lay aside their disguised parts and manners, and take up boldness, insolence, and a contempt of both human and Divine laws, and this at a time when they especially stand in need of piety and righteousness, because they are then most of all exposed to envy, and all they think, and all they say, are in the view of all men; then it is that they become so insolent in their actions, as though God saw them no longer, or were afraid of them because of their power: and whatsoever it is that they either are afraid of by the rumours they hear, or they hate by inclination, or they love without reason, these seem to them to be authentic, and firm, and true, and pleasing both to men and to God; but as to what will come hereafter, they have not the least regard to it.[14]

20.7 Conclusion

In the past century there has been great interest among scientists in what is now commonly known as the big bang theory. The discovery of the expansion of the universe has opened a way for estimating probable characteristics of the universe for times from far in the past to the distant future. Estimations of this sort could not have been contemplated as long as there was no evidence for such expansion.

The origin of life in the universe has also long been a subject of particular interest among scientists. Living plants and animals are extremely important to the survival of human beings, and it was probably inevitable that their origin and evolution should have attracted attention. Discoveries related to geology have suggested changes in the complexity of life forms on the earth, ranging from the simplest and mostly extinct plants and animals of the ancient past to modern human beings. In the discipline of biology these developments are associated with terms like evolution and survival of the fittest. In recent years genetically engineered life forms are also being developed.

Understanding the history of the universe is in some respects a simpler task than predicting its future. The past seems to be an unchanging fact, and a scientist may try to better discover that past by applying with ever greater precision the laws of science. On the other hand, determining the precise future of the universe may not be possible. With the best available information, scientists have not yet been able to obtain answers for some of the most basic questions that they might contemplate. Thus, several possibilities have been proposed for interpreting even the observed expansion of the universe.

On a much smaller scale, the earth itself is not expected to last forever. At this scale the predictions of science are more secure, because many observable stars and their planets are at stages in their history that are still in the future for our own sun and earth. According to calculations, the sun will eventually expand and consume the earth, in approximate agreement with several biblical prophecies. Many people today aren't inclined to prepare for future events (predicted either by science or by the Bible) which they hope will not affect them. This was not always the case, and in the early church the return of Jesus was eagerly anticipated. "Therefore you also must be ready; for the Son of man is coming at an hour you do not expect" (Matthew 24:44).

---

1. S. W. Hawking, *A Brief History of Time: From the Big Bang to Black Holes* (Bantam Books, New York, 1989), pp. 8-9.
2. Ibid., pp. 46-47.
3. Ibid., pp. 120-121.
4. "Scientists and Belief," http://www.pewforum.org/2009/11/05/scientists-and-belief. Web. 9 February 2018.
5. S. W. Hawking, op. cit., pp. 40-51.
6. C. J. Hailey, K. Mori, F. E. Bauer, M. E Berkowitz, J. Hong, and B. J. Hord, "A density cusp of quiescent X-ray binaries in the central parsec of the Galaxy," *Nature*, Volume 556, pp. 70-73 (05 April 2018).
7. J. Chu, "A supermassive black hole that 'shouldn't exist'," *MIT News*, p. 6 (March/April 2018).
8. S. Chen, "Hubble Trouble: A Crisis in Cosmology?," *APS News*, Volume 27, Number 5, p. 1 (May 2018).
9. "History of the Earth," Wikipedia. Web. 9 February 2018.

10. *The Secrets of Enoch* (2 Enoch), included in *The Forgotten Books of Eden*, Edited by Rutherford H. Platt, Jr. (Alpha House, Inc., 1927), 62:2, p. 102.
11. M. R. Rampino, "Reexamining Lyell's Laws," *American Scientist*, Volume 105, Number 4, pp. 224-231 (July-August 2017); p. 230.
12. Corey S. Powell, "Phantom worlds," *Discover Magazine*, pp. 38-43 (November 2014).
13. A. Morbidelli, J. Chambers, J. I. Lunine, J. M. Petit, F. Robert, G. B. Valsecchi, and K. E. Cyr, "Source regions and timescales for the delivery of water to the Earth," *Meteoritics and Planetary Science*, Volume 35, pp. 1309-1320 (2000); p. 1309.
14. *The Life and Works of Flavius Josephus*, Translated by William Whiston (Holt, Rinehart and Winston, New York), *The Antiquities of the Jews*, Book VI, 12:7, p. 194.

www.ingramcontent.com/pod-product-compliance
Lightning Source LLC
Chambersburg PA
CBHW030256080526
44584CB00012B/341